MODERN
POWER DEVICES

MODERN POWER DEVICES

B. JAYANT BALIGA

General Electric Company
Schenectady, New York

A Wiley-Interscience Publication

JOHN WILEY & SONS

New York • Chichester • Brisbane • Toronto • Singapore

Library of Congress Cataloging in Publication Data:

Baliga, B. Jayant, 1948-
  Modern power devices.

  "A Wiley-Interscience publication."
  Includes bibliographies and index.
  1. Power electronics. 2. Power semiconductors.
3. Power transistors. 4. Semiconductor rectifiers.
I. Title.
TK7881.15.B35 1987   621.38   86-15891
ISBN 0-471-81986-7

Printed in the United States of America

10 9 8 7 6 5 4 3 2 1

To

My wife

PREFACE

The development of electronic power systems based on semiconductor devices can be traced to the early 1950s. At that time, the first rectifiers and thyristors capable of operating at high current and voltage levels were introduced into the marketplace. During the ensuing 30 years, the technology for these bipolar devices reached a high degree of maturity. Devices capable of operating at up to 3000 volts and controlling 1000 amperes of current were in production by the early 1970s. The physics and technology of these bipolar power devices have been treated by Professor S. K. Ghandhi in a book titled *Semiconductor Power Devices*, published by John Wiley & Sons in 1977.

Since that time considerable progress has been made in discrete power semiconductor device technology. The most important development is the creation of a new family of power devices with high-input-impedance metal oxide–semiconductor (MOS) gates. These devices greatly reduce the size and complexity of the control circuitry. This has allowed a large reduction in system cost, making the power electronics attractive for many new applications such as control of home appliances and for automotive electronics. In addition, several other new device concepts, such as the field-controlled diode (FCD) and the high-voltage junction field-effect transistor (JFET), have been explored for applications in power circuits. Despite the growing interest in this field, no book has been available that treats the physics and technology of these modern power devices.

This book is written at the tutorial level to fill this need. It can be used for self-study by professional engineers practicing the art of power device design and fabrication. It can also serve as a textbook for a graduate level course on power devices when used together with Professor Ghandhi's book. In writing the book, it has been assumed that the reader is familiar with the fundamental concepts of current transport in semiconductors and has a basic knowledge of process technology.

Each chapter is organized to first introduce the basic device structure and its electrical output characteristics. A detailed discussion of the physics of device operation in the static blocking and conduction states is then provided. Following this, the dynamic switching behavior and frequency response are analyzed.

Throughout this discussion, emphasis is placed on deriving simple analytical expressions that describe the device characteristics, because they provide a good physical understanding of device performance. For a complete characterization of the electrical properties of devices, it is necessary to resort to numerical techniques using computer programs that solve the fundamental semiconductor transport equation in two dimensions (and sometimes in three dimensions), with time dependence included for the transient case. By use of the device analysis presented in each chapter, in conjunction with the silicon material properties provided in Chapter 2, such computer programs can be generated if desired.

No book on power devices would be complete without a detailed discussion of the physics, design, and technology for the realization of high breakdown voltage as discussed in Chapter 3. The device breakdown voltage is usually limited by the edge termination. The importance of this technology has stimulated the conception of a wide variety of termination techniques. The choice of the device termination technique is interrelated to device power rating, chip size, and the fabrication process. The study of this chapter is essential to the design of all the devices discussed in this book.

Chapters 4–7 are devoted to the three-terminal power devices that have evolved since 1975. In Chapter 4, the power JFET is treated. These devices are attractive for very-high-frequency applications. The power FCD is discussed in Chapter 5. These devices possess unusually good radiation tolerance and can be used in high-temperature applications. Chapter 6 provides a detailed analysis of the power MOSFET, whereas Chapter 7 is focused on the MOS–bipolar device concepts. These two developments are the most important aspects of the new device technology. They can be expected to play an increasing role in power circuits in the future.

Concurrent with these revolutionary advances in three-terminal power devices, remarkable strides have been made in improving the performance of power rectifiers. The new concepts that are responsible for the availability of lower forward drop and higher speed in power rectifiers are discussed in Chapter 8. Finally, in Chapter 9, guidelines for making a choice between these devices are provided.

Over the course of the many years that I have worked in this field, it has been my privilege to have the opportunity to discuss the operation of devices and their process development with my many colleagues at the General Electric Research and Development Center, Schenectady, New York. I would like to acknowledge the assistance received from them and the unselfish dedication of the process technicians who were often called upon to perform seemingly impossible tasks. My gratitude goes to Dr. T. S. P. Chow for his assistance with the preparation of the illustrations used in this book, and to Mrs. Y. Nakahigashi for her typing of the manuscript on a demanding schedule.

A special note of gratitude must be expressed to Professor S. K. Ghandhi. He had the foresight to encourage me to start my career in the field of power semiconductor devices at a time when it appeared to be stagnant. In addition to his encouragement in the preparation of this book, I am indebted to him for his detailed and constructive criticism of the manuscript.

The preparation of this book has been made possible by the generosity of the General Electric Company in conferring on me the Coolidge Fellowship Award. During the one-year sabbatical that is provided for in the award, I found the time to pursue the preparation of this book, which would have been impossible under the pressures of my normal working schedule.

In closing, special thanks are due to my wife, Pratima, and my son, Avinash, for tolerating my preoccupation with this project, which absorbed a large proportion of my time and detracted from my giving them the attention that they certainly deserve.

B. JAYANT BALIGA

Schenectady, New York
December 1985

CONTENTS

9. SYNOPSIS   451

INDEX   461

MODERN
POWER DEVICES

1 INTRODUCTION

Starting with the conception of the bipolar junction transistor (BJT) in 1947, the power handling capability of silicon devices has grown steadily. As the power handling capability and frequency response have improved, new applications for these devices have been created. Today, the market for silicon bipolar power devices exceeds $1 billion. The recent introduction of MOS-controlled power devices and power integrated circuits promises to lead to a further growth in this industry.

Power metal oxide semiconductor field-effect transistors (MOSFETs) were commercially introduced in the 1970s. Because of the high input impedance of the device, a significant reduction in the complexity and cost of the gate drive circuit in power control systems can be achieved by replacing the bipolar transistor with power MOSFETs. During the past 10 years there has been an increasing acceptance of the usage of power MOSFETs. However, the displacement of bipolar transistors in their applications by power MOSFETs has lagged far behind initial expectations because of the relatively high cost of the latter devices for the same power rating. As power MOSFET technology matures, the cost differential is continually diminishing making it more attractive. In 1985 the power MOSFET market was only 10% of the bipolar transistor market.

A new class of power devices has emerged in the 1980s. The operation of these devices is based on the fusion of the physics of the MOS gate structure and bipolar current conduction. The most commercially advanced device of this category is the insulated-gate transistor (IGT). The IGT exhibits the important features of a high input impedance and a very high power handling capability for a given chip size. In combination with power integrated circuits, these devices can be used in power systems to obtain cost reductions by factors of 10 or more over existing systems. This is expected to result in a very rapid growth in the overall market for power transistors in which the MOS-controlled devices will occupy an increasing segment.

The applications for power devices extend over a very large range of power levels and frequencies. Some of these applications are shown in the power–frequency spectrum in Fig. 1.1. From this figure, it can be seen that the power

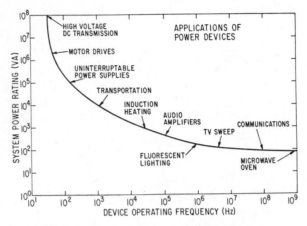

Fig. 1.1. Applications of power semiconductor devices.

levels demanded by systems are generally higher at lower frequencies. For very high power applications, such as high-voltage direct-current (HVDC) transmission systems, the power thyristor is being used very successfully. The recent ability to obtain excellent electrical characteristics combined with light triggering capability for device isolation has ensured the dominance of this technology. The physics of operation, design considerations, and fabrication technology for power thyristors have been adequately treated in other books [1, 2]. These references also provide a good analysis of the operation of bipolar transistors and gate turn-off thyristors that have been used for higher-frequency applications such as motor drives and lighting control. These devices, therefore, are not treated in this book.

During recent years considerable progress has occurred in the power device industry. New power devices, such as the static-induction transistor, FCD, and the IGT, have been conceived. These devices have matured to the point at which they are no longer laboratory curiosities. Their commercial availability creates the threat of displacing the bipolar transistor in many of its traditional applications. The purpose of this book is to elucidate the physics of the operation of these new devices, provide a guide to their design methodology, and describe the unique fabrication technology that has evolved for each device.

The advent of the new power devices discussed in this book is important not merely because these devices may replace the bipolar transistor, but also—and more significantly—because they are forcing a change in the way power circuits are being designed and applied. For example, the availability of MOS gate-controlled devices capable of operating at high switching speeds is changing the way in which motor control is performed. There is an increasing trend toward pulse-width-modulated, higher-frequency, adjustable-speed motor drives because of the significant improvement in efficiency and system performance. This shift in technology was not a viable option with the traditional bipolar transistor technology because of the high cost of implementing the power electronic circuits for adjustable-speed motor drives. Other examples of areas in which the new device technology is expected to play an important role are automotive electronics, switching power supplies, factory automation, new higher-efficiency lighting systems, and appliance control for the electronic home of the future.

REFERENCES

1. S. K. Ghandhi, *Semiconductor Power Devices*, Wiley, New York, 1977.
2. B. J. Baliga and D. Y. Chen, *Power Transistors: Device Design and Applications*, IEEE Press, New York, 1984.
3. A. Blitcher, *Field Effect and Bipolar Power Transistor Physics*, Academic, New York, 1981.
4. A. Blitcher, *Thyristor Physics*, Springer-Verlag, New York, 1976.

2 CARRIER TRANSPORT PHYSICS

At present all power devices used for high-voltage control applications are being fabricated by using silicon as the starting material. For high-voltage power FETs, it has been theoretically demonstrated [1] that other materials such as gallium arsenide and ternary compound semiconductors offer superior characteristics. However, the development of these devices is in its infancy because of the difficult technological problems associated with processing these compound semiconductor materials. The design of power devices and an understanding of their operating characteristics are, consequently, based on a sound knowledge of the fundamental properties of the silicon material used to fabricate these devices. This chapter reviews those properties of silicon that are relevant to the operation of the power devices discussed in this book. In addition to providing data on various properties of silicon such as intrinsic carrier concentration, and mobility, which can be used during numerical simulation of devices, simple expressions are provided wherever possible to facilitate device analysis.

The fundamental material characteristics of the semiconductor govern the operation of power devices; however, the processing techniques used to control these properties and the technological constraints imposed by the available processes are equally important in obtaining the desired device performance. For this reason, this chapter includes a discussion of current technologies such as neutron transmutation doping for controlling the resistivity and electron irradiation for controlling minority carrier lifetime. These technologies have been specifically developed for power devices and are used almost exclusively for their fabrication.

2.1. MOBILITY

By definition, mobility is a measure of the average velocity of free carriers in the presence of an impressed electric field. In the presence of an electric field in a semiconductor, the free electrons and holes are accelerated in opposite directions. Silicon is an indirect-gap semiconductor in which electron transport under an applied electric field occurs in six equivalent conduction band minima located along the $\langle 100 \rangle$ crystallographic directions, whereas the transport of holes occurs at two degenerate subbands located at the zero in k space. The inequality between the mobility of electrons and holes in silicon arises from differences between the shapes of the conduction and valence band minima.

As the free carriers are transported along the direction of the electric field, their velocity increases until they undergo scattering. In the bulk, the scattering can occur by either interaction with the lattice or at ionized donor and acceptor atoms. As a result of this, the mobility is dependent on the lattice temperature and the ionized impurity concentration. When the free-carrier transport occurs near the semiconductor surface, additional scattering is observed, which decreases the mobility to below the bulk value.

These scattering mechanisms are dominant when the concentrations of holes and electrons are never simultaneously large. In bipolar power devices, however, a high concentration of holes and electrons is simultaneously injected into the base region during forward current conduction. The probability for mutual scattering between electrons and holes is high under these conditions, resulting in the effective mobility of the free carriers decreasing with increasing injection level.

In analyzing the influence of the preceding parameters on the mobility, one assumes the electric field strength to be small. The mobility is then defined as the proportionality constant relating the average carrier velocity to the electric field. At high electric fields such as those commonly encountered in power devices the velocity is no longer found to increase in proportion to the electric field and in fact attains a saturation value. These effects have important implications to current flow in power devices.

This section discusses the dependence of the electrons and hole mobility on the preceding parameters. In addition to providing data that can be used during numerical simulation of power devices, analytical expressions have been derived whenever possible to simplify device analysis.

2.1.1. Temperature Dependence

At low doping concentrations, the scattering of free carriers in the bulk occurs predominantly by interaction with lattice vibrations. Lattice scattering can occur by means of either optical phonons or acoustical phonons. Optical phonon scattering is important at high temperatures, whereas acoustical phonon scattering is dominant at low temperatures. In addition, at around room temperature, intervalley scattering mechanisms become important. For high-purity silicon, it has been determined experimentally that at temperatures below 50 K acoustical phonon scattering is dominant and the mobility varies as $T^{-3/2}$, where T is the absolute temperature. At around room temperature, in the region of interest for device operation, intervalley scattering comes into effect and the mobility can be determined by using the following equations [2, 3]:

$$\mu_n = 1360 \left(\frac{T}{300} \right)^{-2.42} \tag{2.1}$$

$$\mu_p = 495 \left(\frac{T}{300} \right)^{-2.20} \tag{2.2}$$

where μ_n and μ_p are the electron and hole mobilities, respectively, in square centimeters per volt · second, and T is the absolute temperature in degrees Kelvin. The variation of the mobility of electrons and holes with temperature from 200 to 500 K for lightly doped silicon at low electric fields is shown by the uppermost curves in Fig. 2.1. Note the rapid reduction in both the electron and hole mobilities with increasing temperature. Since doping levels of below 10^{15} cm^3 are necessary to achieve high breakdown voltages, the reduction in carrier mobility with increasing temperature is an important characteristic that must be accounted for during the design and analysis of unipolar power devices.

2.1.2. Dopant Concentration Dependence

The presence of ionized donor or acceptor atoms in the silicon lattice results in a reduction in the mobility due to the addition coulombic scattering of the free carriers. The effect of ionized impurity scattering is dominant at low temperatures because the effect of lattice scattering becomes small. Some specific examples of the effect of ionized impurity scattering on the mobility of electrons and holes are provided in Fig. 2.1. The mobility as determined by ionized impurity scattering exhibits a positive temperature coefficient. Consequently, as the ionized impurity concentration increases, not only does the absolute mobility decrease, but the temperature coefficient of the mobility also decreases [4].

At room temperature, ionized impurity scattering effects are small for doping levels below 10^{16} atoms/cm^3 and the mobility is independent of doping level as described by Eqs. (2.1) and (2.2). At higher doping concentrations, the mobility of electrons and holes decreases with increasing dopant concentration until a doping level of 10^{19} atoms/cm^3 is reached. For dopant concentrations above 10^{19} atoms/cm^3, the mobility of electrons becomes independent of donor concentration at a value of about 90 cm^2/(V·sec), and that of holes becomes independent of acceptor concentration at a value of about 48 cm^2/(V·sec). The variation of electron and hole mobility with dopant concentration is provided in Fig. 2.2.

The following empirical relationships that relate the mobility to the doping concentrations can be derived from the measured data [5]:

$$\mu_n(N_D) = \frac{5.10 \times 10^{18} + 92N_D^{0.91}}{3.75 \times 10^{15} + N_D^{0.91}} \tag{2.3}$$

$$\mu_p(N_A) = \frac{2.90 \times 10^{15} + 47.7N_A^{0.76}}{5.86 \times 10^{12} + N_A^{0.76}} \tag{2.4}$$

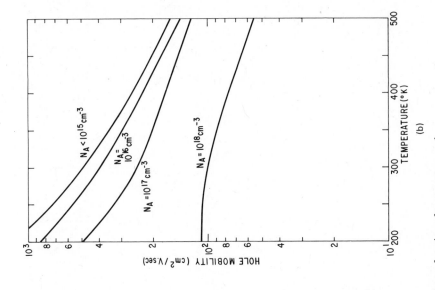

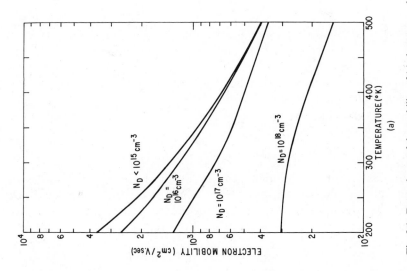

Fig. 2.1. Dependence of the mobility of (*a*) electrons and (*b*) holes as a function of temperature and doping concentration.

7

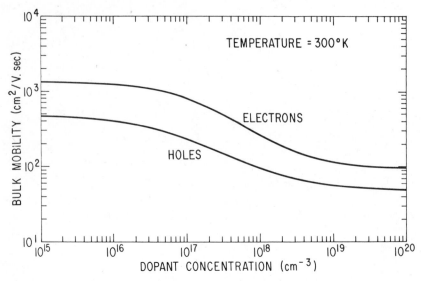

Fig. 2.2. Dependence of the mobility of electrons and holes on the dopant concentration (room-temperature data).

where N_D and N_A are the donor and acceptor concentrations, respectively, per cubic centimeter. These equations are useful for computing the mobility of N- and P-type silicon from the doping concentration and can be directly used for the analysis of device characteristics. They can also be used to derive an expression for the resistivity of N- and P-type silicon as a function of doping concentration.

2.1.3. Electric Field Dependence

In previous sections, the mobility was assumed to be independent of the magnitude of the applied electric field. This is found to be true only when the electric field strength is small. In the range of electric fields where the mobility is constant, the carrier velocity will increase linearly in proportion to the electric field. For silicon, at electric field strengths above 1×10^3 V/cm, it has been found that the velocity of both electrons and holes increases sublinearly with electric field. The variation of the velocity of electrons and holes as a function of electric field has been measured at various temperatures by using the time-of-flight technique [6]. Analytical expressions relating the drift velocity to the electric field and the ambient temperature have been derived by using these measured data.

For purposes of power device analysis, it is useful to define an average mobility as the ratio of the drift velocity to the applied electric field. At low doping levels, this average mobility (μ^{av}) can be related to the electric field $\mathscr{E}$ in volts per centimeter by the expressions

$$\mu_n^{av}(\mathscr{E}) = \frac{9.85 \times 10^6}{(1.04 \times 10^5 + \mathscr{E}^{1.3})^{0.77}} \tag{2.5}$$

$$\mu_p^{av}(\mathscr{E}) = \frac{8.91 \times 10^6}{(1.41 \times 10^5 + \mathscr{E}^{1.2})^{0.83}} \tag{2.6}$$

These expressions provide an analytical description of the decrease in the average mobility for electrons and holes with increasing electric field strength. The variation of the average mobility of electrons and holes with electric field is shown in Fig. 2.3. It can be seen that the average mobility is essentially equal to the low field mobility as long as the electric field remains below 10^3 V/cm but drops very rapidly at fields above 10^4 V/cm. In fact, at electric fields above 10^5 V/cm, the average mobility decreases inversely with the electric field strength, indicating that the drift velocity of the free carriers remains essentially constant. This phenomenon is called *drift velocity saturation*.

The saturated drift velocity is an important parameter that is required for the analysis of the characteristics of power devices operating in the presence of very high fields. At room temperature, the saturated drift velocity is 9.9×10^6 cm/sec for electrons in silicon and 8.4×10^6 cm/sec for holes in silicon. Just as the mobility of the electrons and holes changes with temperature, the saturated

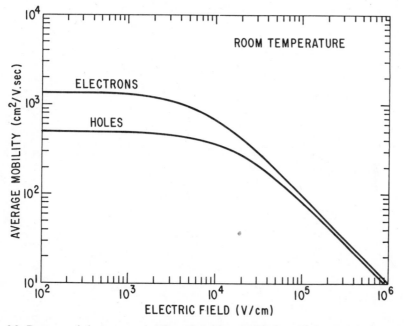

Fig. 2.3. Decrease of the average mobility of electrons and holes with increasing electric field strength at 300 K. Note that at very high electric fields, the average mobility decreases inversely with electric field, indicating a saturation of the velocity.

drift velocity is also a function of temperature. The drift velocity of both electrons and holes in silicon has been measured over a broad range of electric field strengths and ambient temperatures [6]. From these data, empirical expressions relating the drift velocity of both electrons (v_{dn}) and holes (v_{dp}) in centimeters per second to the electric field strength $\mathscr{E}$ in volts per centimeter and the absolute temperature T in degrees Kelvin can be derived:

$$v_{dn} = \frac{1.42 \times 10^9 \, T^{-2.42} \mathscr{E}}{[1 + (\mathscr{E}/1.01 \, T^{1.55})^{2.57 \times 10^{-2} T^{0.66}}]^{(2.57 \times 10^{-2} T^{0.66})^{-1}}} \tag{2.7}$$

$$v_{dp} = \frac{1.31 \times 10^8 \, T^{-2.2} \mathscr{E}}{[1 + (\mathscr{E}/1.24 \, T^{1.68})^{0.46 T^{0.17}}]^{(0.46 T^{0.17})^{-1}}} \tag{2.8}$$

From these equations, the saturated drift velocity for electrons ($v_{sat,n}$) and holes ($v_{sat,p}$) in centimeters per second in silicon can be derived as a function of the ambient temperature in degrees Kelvin:

$$v_{sat,n} = 1.434 \times 10^9 \, T^{-0.87} \tag{2.9}$$

$$v_{sat,p} = 1.624 \times 10^8 \, T^{-0.52} \tag{2.10}$$

Power devices are usually rated for operation between -25 and $150°C$. The variation of the saturated drift velocity of electrons and holes in silicon over a slightly broader range of temperatures is provided in Fig. 2.4. Interestingly,

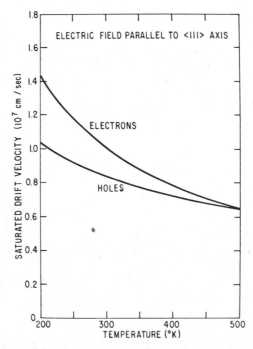

Fig. 2.4. Dependence of the saturated drift velocity of electrons and holes on temperature.

the saturated drift velocity of electrons is higher than that of holes at and below room temperature. However, this difference decreases with increasing temperature up to 500 K, where the saturated drift velocity of electrons becomes equal to that of holes. It should be noted that these results are valid only for lightly doped silicon. Monte Carlo calculations of the effect of ionized impurity atoms on high field transport in silicon at 300 K indicate that, in addition to the decrease in the mobility, the saturated drift velocity of electrons and holes also decreases when the doping concentration exceeds 10^{18} atoms per cubic centimeter. Since the active base regions of power devices, where the high electric fields are present, are doped well below this level, the use of the saturated drift velocity data for electrons and holes as provided in Fig. 2.4 is adequate for analysis.

These results apply to the transport of free carriers when the electric field is applied along the $\langle 111 \rangle$ crystallographic direction. Although the low field mobility of electrons and holes has been found to be independent of the orientation of the electric field with respect to the crystal, anisotropic carrier transport is observed at high electric fields [5]. This anisotropy is caused by the ellipsoidal shape of the six conduction band valleys and the warped nature of the valency subbands. The anisotropic transport of free carriers in silicon is seldom taken into account during the analysis of current flow in power devices.

2.1.4. Injection-Level Dependence

In general, current transport in semiconductors occurs by means of both drift of the free carriers, resulting from the presence of an electric field, and diffusion, resulting from the presence of a concentration gradient. In highly doped regions, majority carrier transport is dominant. In lightly doped regions, the transport of both electrons and holes (i.e., both majority and minority carriers) must be accounted for in computing the current flow. Minority carrier transport can be expressed in the form

$$\frac{\partial p'}{\partial t} = -\frac{p'}{\tau_a} + D_a \frac{\partial^2 p'}{\partial x^2} \tag{2.11}$$

for an N-type semiconductor and

$$\frac{\partial n'}{\partial t} = -\frac{n'}{\tau_a} + D_a \frac{\partial^2 n'}{\partial x^2} \tag{2.12}$$

for a P-type semiconductor. In these expressions, n' and p' are the excess electron and hole concentrations, and D_a and τ_a are defined as the ambipolar diffusion coefficient and the ambipolar lifetime, respectively. The ambipolar diffusion coefficient is a function of the free-carrier concentration and is given by

$$D_a = \frac{(n + p)D_n D_p}{(nD_n + pD_p)} \tag{2.13}$$

where n and p are the electron and hole concentrations and D_n and D_p are the electron and hole diffusion constants, which can be determined from the mobility by using the Einstein relationship, which is valid for nondegenerate semiconductors:

$$D = \frac{kT}{q}\mu \tag{2.14}$$

At low injection levels and for doping levels of several orders of magnitude above the intrinsic carrier concentration, the ambipolar diffusion coefficient is equal to the minority carrier diffusion constant. This is no longer true at high injection levels.

At low injection levels, where the density of the minority carriers is far less than that of the majority carriers, the transport is controlled by scattering by either phonons or ionized impurity centers. At high injection levels, which occur, for instance, in the lightly doped base regions of bipolar power devices during forward conduction, the densities of electrons and holes become approximately equal to satisfy charge neutrality. Now, the probability for the mutual coulombic interaction of the mobile carriers about a common center of mass becomes significant, resulting in a decrease in the mobility. With the inclusion of carrier–carrier scattering, the diffusion coefficient and the mobility decrease as the injected carrier density increases. By treating carrier–carrier scattering in a manner similar to ionized impurity scattering, the mobility can be shown to vary inversely with the injected carrier concentration when the injection level exceeds 10^{17} cm^3. The variation of the mobility with injected carrier density ($\Delta n = \Delta p$) is shown in Fig. 2.5 and takes the form

$$\frac{1}{\mu} = \frac{1}{\mu_0} + \frac{\Delta n \ln(1 + 4.54 \times 10^{11}\, \Delta n^{-0.667})}{1.428 \times 10^{20}} \tag{2.15}$$

where Δn is the excess carrier density and μ_0 (either μ_n or μ_p) is the majority carrier mobility. The decrease in the mobility at high injection levels has important implications in determining the forward voltage drop of bipolar power devices operating at high current densities and their surge current ratings.

2.1.5. Surface Scattering Effects

The transport of free carriers near the surface is of importance to power MOSFETs. The forward current conduction mode of the n-channel power MOSFET is induced by the inversion of a P-type layer in the device in such a manner as to create a n-channel region connecting the source to the drain. This inversion region is created by applying an electric field normal to the semiconductor surface by using a metal gate electrode separated from the semiconductor by a silicon dioxide layer. The conductivity of the inversion layer is dependent on the total number of free carriers in the inversion layer and their transport velocity along the surface under an applied transverse electric field.

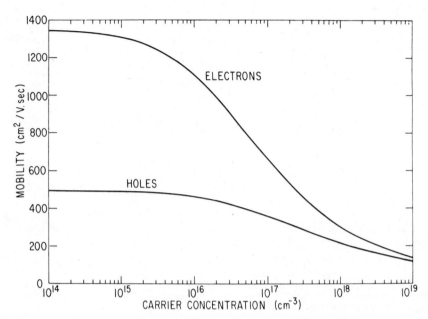

Fig. 2.5. Decrease of electron and hole mobility with increasing injection level due to carrier–carrier scattering.

Current transport in these devices also occurs by way of an accumulation layer induced by the gate bias. The computation of the number of free carriers in the inversion layer, available for current conduction, is discussed in the chapter on the MOS gated FET. In this section on mobility, the influence of the surface on the velocity–electric field relationship is discussed.

At the high electric fields applied across the oxide to create the inversion layer, the surface region within which current transport occurs is very thin. As a result, at high electric fields normal to the surface and at low temperatures, the motion of carriers within the inversion layer is quantized in a direction perpendicular to the surface. At room temperature, the kinetic energy of the electrons spreads their wave function among the quantum levels and electron transport can be treated by using classical statistics.

A pictorial representation of a semiconductor surface containing an inversion layer is shown in Fig. 2.6. The electrons in the inversion layer have a Gaussian distribution located very close to the surface within a region that is typically in the range of 100 Å in thickness. In addition to coulombic scattering due to ionized acceptors, the electrons in the inversion layer undergo several additional scattering processes: (1) surface phonon scattering due to lattice vibration, (2) additional coulombic scattering due to surface charge in interface states and the fixed oxide charge, and (3) surface roughness scattering due to deviation of the surface from a specular interface. As in the case of the bulk, phonon scattering is important at higher temperatures, where the thermal energy in the lattice is

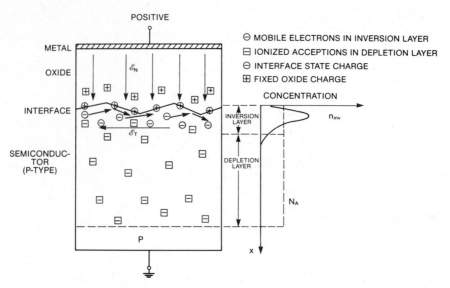

Fig. 2.6. Metal oxide–semiconductor surface with inversion layer formed by the application of a positive potential to the metal electrode.

large. The influence of coulombic scattering due to interface state and fixed charge is important for lightly inverted surfaces and when processing conditions produce high interface state or fixed charge densities. The surface roughness scattering is predominant under strong inversion conditions because the electric field normal to the surface must be increased for achieving higher inversion layer concentrations. This has the effect of increasing the velocity of the carriers toward the surface and bringing the inversion layer charge distribution closer to the surface. Both these factors enhance scattering at a rough interface such as that illustrated in Fig. 2.6. As expected, the surface roughness scattering is sensitive to processes that influence surface smoothness. The quantitative dependence of the carrier mobility in inversion and accumulation layers are of vital importance to the modeling and design of power MOS devices. The dependence of these mobilities on various parameters is discussed in the paragraphs that follow.

Before discussing these results it is useful to define an effective mobility for carriers in the inversion layer

$$\mu_e = \frac{\int_0^{x_i} \mu(x)n(x)\,dx}{\int_0^{x_i} n(x)\,dx} \tag{2.16}$$

where $\mu(x)$ and $n(x)$ are the local mobility and free-carrier concentrations, respectively, in the inversion layer and x_i is the inversion layer thickness. It must be emphasized that, defined in this manner, the effective mobility is a measure of the conductance of the inversion layer, so that this concept is eminently suitable for the calculation of the characteristics of MOS field-effect devices. In fact, determination of the effective mobility is performed by using MOSFET devices specially designed for these measurements. In these measurements, special care must be taken with regard to surface orientation, the direction of the current flow vector on the surface, and surface charge density, since these parameters have all been found to influence the effective mobility.

The surface mobility is primarily a function of the electric field $\mathscr{E}_N$ normal to the surface rather than the inversion layer concentration, which is dependent on the background doping level. A typical variation of the effective mobility with increasing surface field $\mathscr{E}_N$ is shown in Fig. 2.7. In general, it has been found that under weak inversion conditions, that is, at low electric fields normal to the surface, the effective mobility increases very rapidly with increasing inversion layer carrier concentration. As the inversion layer concentration increases, the effective mobility has been found to peak and then gradually decrease [7]. The increase in the effective mobility at low inversion layer concentrations has been related to coulombic scattering at charged surface states. In fact, if the surface state density is adequately lowered by suitable processing, the effective mobility remains independent of the inversion layer concentration in the weak inversion region [8]. When the electric field normal to the surface exceeds 10^4 V/cm, the effective mobility begins to decrease with increasing inversion layer concentration. The enhancement of surface scattering by the higher electric fields applied normal to the surface is responsible for this phenomenon. It has an important bearing on the transconductance of power MOS devices.

2.1.5.1. Substrate Doping Dependence.
The inversion layer mobility for electrons in silicon has been measured as a function of substrate doping [9]. The observed change in the maximum effective mobility with increasing substrate

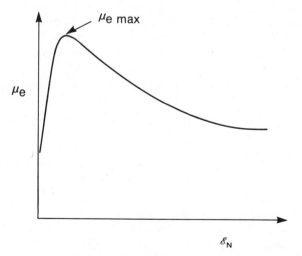

Fig. 2.7. Effect of electric field applied normal to the surface on effective mobility in an inversion layer.

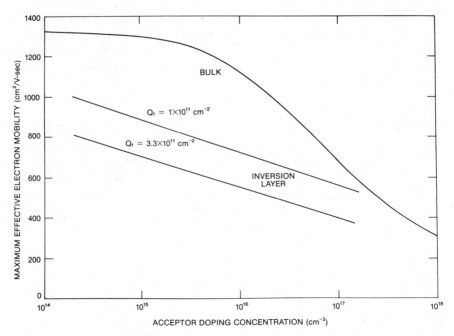

Fig. 2.8. Variation of the maximum effective mobility for electrons in an inversion layer with background acceptor doping level. The bulk mobility is shown for comparison. (After Ref. 9, reprinted with permission from the IEEE © 1980 IEEE.)

doping is shown in Fig. 2.8 for two cases of fixed charge density. It can be seen that the effective mobility decreases logarithmically with increasing substrate doping. This empirical observation will be incorporated in an analytical expression relating $\mu_{e,max}$ to the substrate doping N_A and the fixed charge Q_f.

The decrease in maximum effective mobility with increasing substrate doping is not due to enhanced ionized impurity scattering. Since the inversion layer thickness is on the order of 100 Å, the number of ionized impurities is less than $10^{10}/cm^2$ for doping levels up to $10^{16}/cm^3$. This is an order of magnitude smaller than the typical fixed charge density. This makes the influence of ionized impurity scattering negligible. Thus the observed reduction in the maximum effective mobility with increasing substrate doping arises primarily from the higher electric fields that must be applied normal to the surface to produce surface inversion.

The data in Fig. 2.8 also indicate that the maximum inversion layer mobility varies from 75 to 85% of the bulk value for substrate doping levels ranging from 10^{14} to $10^{17}/cm^3$ when low ($< 10^{11}/cm^2$) fixed charge density is achieved. If the fixed charge density is high, the maximum effective mobility will be substantially reduced.

2.1.5.2. Oxide Charge Dependence.   The inversion layer mobility for electrons in silicon is a strong function of the charge at the oxide interface [9] and has been experimentally found to decrease inversely with increasing surface charge. The hyperbolic variation of the maximum effective mobility with fixed charge Q_f is also theoretically predicted from calculation of coulombic scattering by charged centers [10]. On the basis of these findings, an empirical relationship between the maximum effective mobility $\mu_{e,max}$ and the fixed charge Q_f has been derived:

$$\mu_{e,max} = \frac{\mu_0}{1 + \alpha Q_f} \tag{2.17}$$

where the parameters μ_0 in square centimeters per volt·second and α are a function of the background doping level:

$$\mu_0 = 3490 - 164 \log(N_A) \tag{2.18}$$

and

$$\alpha = -0.104 + 0.0193 \log(N_A) \tag{2.19}$$

where N_A is the background acceptor concentration per cubic centimeter. The dependence of the maximum effective mobility on the background doping level described in Section 2.1.5.1 is incorporated in the last two equations. If the interface state charge is significant, it should be added to the fixed charge Q_f in Eq. (2.17) for calculation of the maximum effective mobility. However, it has been found that charges in the bulk of the oxide at distances of over 50 Å from the interface have negligible effect on the mobility.

2.1.5.3. Electric Field Dependence.

As illustrated in Fig. 2.6, the inversion layer charge is distributed near the surface with a Gaussian profile. The electric field experienced by the free carriers in the inversion layer varies from the peak electric field at the oxide–semiconductor interface to that at the depletion layer interface. For the same peak surface electric field, the average electric field in the inversion layer will vary with background doping level. To describe the effect of electric field strength normal to the surface on the effective mobility, it is appropriate to use the average electric field rather than the peak surface electric field. This effective field is given by the relationship

$$\mathscr{E}_{eff} = \frac{1}{\epsilon_s} \left(\frac{1}{2} Q_{inv} + Q_B \right) \tag{2.20}$$

where Q_{inv} is the inversion layer charge density per square centimeter, Q_B is the depletion layer charge per square centimeter, and ϵ_s is the dielectric constant in farads per centimeter. For the same peak surface electric field, the effective electric field will become smaller as the background doping level decreases.

As pointed out earlier with the aid of Fig. 2.7, the effective mobility in an inversion layer decreases with increasing electric field. For electrons in silicon inversion layers, the effective mobility can be described as a function of the effective electric field by the following expression:

$$\mu_e = \mu_{e,max}\left(\frac{\mathscr{E}_c}{\mathscr{E}_{eff}}\right)^{\beta} \qquad (2.21)$$

where $\mu_{e,max}$ is the maximum effective mobility for a given background doping level and fixed charge density as described by Eq. (2.17). The parameters $\mathscr{E}_c$ and β are empirical constants that are dependent on the fixed charge density and the background doping level. These constants have been measured for the commonly used oxidation techniques during silicon device fabrication employing steam and dry oxygen [9]. From these measurements it has been found that

$$\beta_{wet\,O_2} = 0.341 - 5.44 \times 10^{-3}\,\log(N_A) \qquad (2.22)$$

and

$$\beta_{dry\,O_2} = 0.313 - 6.05 \times 10^{-3}\,\log(N_A) \qquad (2.23)$$

where N_A is the background acceptor concentration per cubic centimeter.

The parameter β is a measure of the rate of degradation of the effective mobility with increasing electric field normal to the surface. It has a strong influence on the transconductance of power MOS devices. Under typical device processing conditions $\beta \approx 0.25$; in other words, the effective mobility decreases inversely as the fourth power of the effective electric field strength applied normal to the surface. The difference in β between wet and dry oxidation is believed to arise from the dissimilar surface roughness resulting from these processes. The rate of oxide growth during dry oxidation is lower than that for wet oxidation, resulting in a smoother interface. Since the surface roughness scattering is dominant at high surface fields, dry oxidation produces superior effective mobilities under strong inversion conditions.

The constant $\mathscr{E}_c$ is a function of the fixed charge density and background doping level. A general expression relating these parameters has been derived [9]:

$$\mathscr{E}_c = 2 \times 10^{-4}\,N_A^{0.25}\,A e^{BQ_f} \qquad (2.24)$$

where N_A is the background doping level in cubic centimeters and Q_f is the fixed charge density. The parameters A and B are process dependent. For wet oxidation $A = 2.79 \times 10^4$ V/cm and $B = 8.96 \times 10^{-2}$, whereas for dry oxidation $A = 2.61 \times 10^4$ V/cm and $B = 0.13$. In Eq. (2.24), Q_f is in units of $10^{11}/\text{cm}^2$. Thus the influence of substrate doping is small, whereas the fixed charge density has a strong effect on the effective mobility over a broad range of electric field strengths.

The mobility of holes in inversion layers has also been measured [7, 11], but in less detail compared with electrons. These early studies provide the data for the variation of the effective mobility for holes as a function of the inversion layer charge as shown in Fig. 2.9. As in the case of electrons, the effective mobility for holes goes through a peak. The variation of the effective mobility for

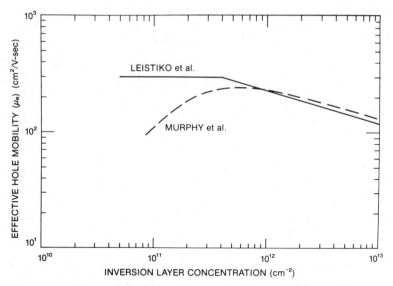

Fig. 2.9. Variation of effective mobility for holes in inversion layers with inversion layer charge.

holes in inversion layers as a function of the electric field strength normal to the surface can also be described by Eq. (2.21) for the region where the inversion layer concentration exceeds $5 \times 10^{11}/\text{cm}^2$. Through application of the data shown in Fig. 2.9, the parameters $\mathscr{E}_c$ and β in Eq. (2.21) can be determined to be $\mathscr{E}_c = 3 \times 10^4$ V/cm and $\beta = 0.28$. The dependence of these parameters on the substrate doping N_D and fixed charge Q_f has not been studied in detail.

2.1.5.4. Orientation Dependence.   The effective mobility in inversion layers has been found to be anisotropic. Its value depends not only on the orientation of the plane of the surface on which the inversion layer is formed, but also on the direction of the current flow vector along the surface being considered. Detailed measurements [12, 13] of the effective mobility on many surface orientations for both N- and P-type wafers have shown that

$$\mu_e(100) > \mu_e(111) > \mu_e(110) \tag{2.25}$$

for electrons in inversion layers and

$$\mu_p(110) > \mu_p(111) > \mu_p(100) \tag{2.26}$$

for holes in inversion layers. This dependence of the effective mobility on orientation has been correlated with the anisotropy of the conductivity effective mass in silicon.

Intraplanar anisotropy of the effective electron mobility in inversion layers has also been observed [9, 12]. For the same substrate doping and fixed charge density, the effective electron mobility tangential to the surface along the [001]

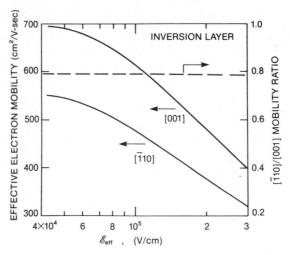

Fig. 2.10. Orientation dependence of effective mobility for electrons in inversion layers. (After Ref. 9, reprinted with permission from the IEEE © 1980 IEEE.)

direction is considerably higher than along the $[\bar{1}10]$ direction as shown in Fig. 2.10. It is worth noting that the ratio of the mobilities for the two surface orientations remains constant for the entire range of electric fields, indicating that the surface roughness is isotropic and that the difference in absolute magnitude between the mobilities arises from the anisotropy of the conductivity effective mass in silicon.

The results described in the preceding paragraphs apply to polished surfaces. Lower inversion layer mobilities have been observed on etched surfaces. One process developed for the fabrication of power MOSFET devices relies on the use of anisotropic etches to form V-shaped grooves on the top surface of (100)-oriented wafers. Because of the higher atomic density of (111) planes in silicon, the etch rate of the anisotropic etches is slowest for the (111) plane. Consequently, these etches expose (111) surfaces, creating the V-shaped grooves. The results of measurements of the effective mobility of electrons on the etched (111) surface are compared with those obtained on the unetched (i.e., polished) surface in Fig. 2.11. In this figure, two methods of anisotropic etching, based on a solution containing either potassium hydroxide (KOH) or ethylenediamine (ED), have been examined. For all three cases, the maximum effective mobility was measured to be about 600 cm²/V·cm. The lower effective mobilities for the etched surfaces are believed to arise from greater surface roughness scattering because the anisotropically etched silicon surface is rougher than a polished surface. These measurements indicate that planar double-diffused MOS (DMOS) devices fabricated on polished (100) wafers would have a 30% higher inversion layer mobility for electrons at high gate bias voltages when compared with V-groove MOS (VMOS) devices, which require anisotropic etching.

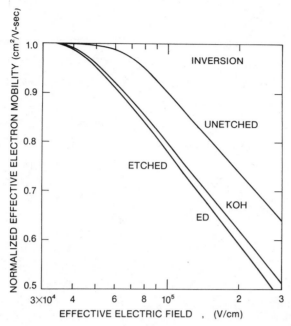

Fig. 2.11. Comparison of the effective mobility for electrons in inversion layers formed on (111) polished and anisotropically etched surfaces. (After Ref. 9, reprinted with permission from the IEEE © 1980 IEEE.)

2.1.5.5. Accumulation versus Inversion.
When the direction of the electric field normal to the surface is such as to attract the majority carriers to the surface an accumulation layer forms. The carriers in accumulation layers are distributed further from the surface than in the case of inversion layers. Because of this, the effective mobility in an accumulation layer can be expected to be higher than in an inversion layer because surface roughness scattering is not as severe. However, measurements indicate a difference of less than 5%. The measured effective mobility for electrons in silicon accumulation layers [9] is about 80% of the bulk as shown in Fig. 2.12. The variation of the effective electron mobility in an accumulation layer with increasing electric field normal to the surface is similar to that in an inversion layer.

The effective mobility for electrons in accumulation layers has also been found to exhibit the anisotropy observed in inversion layers [9]. At high electric fields normal to the surface, the mobility ratio for the [001] and [$\bar{1}$10] surface directions approaches that observed in inversion layers. This anisotropy of the effective mobility in accumulation layers is ascribed to the anisotropy in the effective mass for electrons in silicon.

Accumulation layers exist between device cells during current conduction in power MOSFETs. Current flow within the device is distributed by conduction through the accumulation layer. The conductivity of the accumulation layer

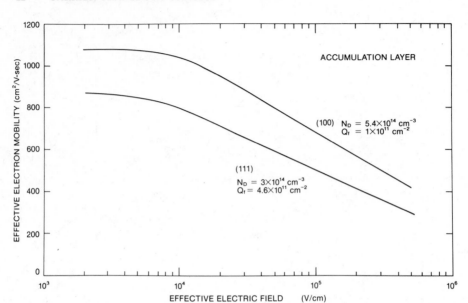

Fig. 2.12. Effective mobility for electrons in accumulation layers as a function of the electric field applied normal to the surface. (After Ref. 9, reprinted with permission from the IEEE © 1980 IEEE.)

plays an important part in determining the spreading resistance of these devices. The effective mobility data provided here is required for performing an accurate analysis of this parameter.

2.1.5.6. Velocity Saturation.

The preceding results apply at low electric field strengths applied parallel to the surface where the proportionality between the velocity of the free carriers and the electric field is maintained. At high field strengths, carrier velocity saturation is observed similar to that discussed for bulk silicon. The saturated drift velocity in inversion layers is a function of the surface orientation and is lower than that observed in bulk silicon. In the case of holes, the saturated drift velocity is the highest for the (100) orientation [14]. In the case of electron transport, the saturated drift velocity has been measured as a function of temperature for the (111), (110), and (100) surfaces [15]. An approximately linear decrease in the saturated drift velocity with increasing temperature is observed:

$$v^i_{sat,n} = v_0 - \delta T \tag{2.27}$$

where v_0 is the saturated drift velocity at 0 K. According to the measured data, $v_0(100) = 8.9 \times 10^6$ cm/sec, $v_0(111) = 7.4 \times 10^6$ cm/sec, and $v_0(110) = 6.2 \times 10^6$ cm/sec. The constant $\delta = 7 \times 10^3$ cm/sec·K forms a good fit to the measured data for all three of these orientations.

2.2. RESISTIVITY

The resistivity of the starting material used for the fabrication of power devices is an important parameter because it controls the maximum achievable break-down voltage. In the case of unipolar devices it also affects the maximum current handling capability. The resistivity in silicon is controlled by the transport of holes and electrons under the applied field. Silicon is an indirect-gap semi-conductor in which electron transport under applied electric fields occurs in six equivalent conduction band minima located along the $\langle 100 \rangle$ crystallographic directions, whereas the transport of holes occurs at two degenerate subbands located at the zero in k space. The resistivity is controlled by the concentration of free electrons in the conduction band and the concentration of holes in the valence band as well as the mobility of these carriers, which relates their average velocity to the applied electric field. The mobility of free carriers was treated in the previous section.

In high-purity material, the free-carrier density is determined by the generation of hole–electron pairs created by the thermal excitation of electrons from the valence band into the conduction band. These carriers determine the intrinsic resistivity. The resistivity can be controlled by the addition of dopants into the silicon lattice that contribute either electrons or holes to the conduction process when ionized. The dopants commonly used for power device fabrication are boron, gallium, and aluminum for P-type regions and phosphorus, arsenic, and antimony for the N-type regions. Since most of the resistance that controls the current flow in power devices during current conduction arises in that portion of the device that supports the high voltages during the blocking state of operation, these devices are generally fabricated from n-type starting material because the mobility of electrons is higher than that of holes.

2.2.1. Intrinsic Resistivity

In the absence of dopants, the resistivity is controlled by the creation of electrons and holes in the conduction and valence bands as a result of the thermal generation process, which allows the transfer of electrons from the valence band into the conduction band. This process produces both free electrons and free holes, which can take part in current conduction. The density of these intrinsically created carriers is dependent on the density of states in the conduction band (N_c) and valence band (N_v) and on the energy gap E_g:

$$n_i = \sqrt{np} = \sqrt{N_c N_v}\, e^{-(E_g/2kT)} \tag{2.28}$$

where k is Boltzmann's constant and T is the absolute temperature in degrees Kelvin. For silicon

$$n_i = 3.87 \times 10^{16}\, T^{3/2}\, e^{-(7.02 \times 10^3)/T} \tag{2.29}$$

per cubic centimeter as long as the temperature remains below 700 K [4, 16].

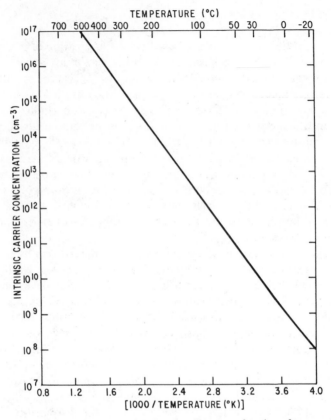

Fig. 2.13. Intrinsic carrier concentration for silicon as a function of temperature.

The variation of the intrinsic concentration with temperature is shown in Fig. 2.13. The intrinsic carrier concentration is an important parameter that determines the leakage current in power devices at elevated temperatures. It has also been found that the formation of current filaments (mesoplasmas) can be related to thermal runaway when the intrinsic concentration becomes comparable to the background concentration.

The intrinsic resistivity in the absence of dopants is related to the intrinsic carrier concentration and the mobility for holes and electrons:

$$\rho_i = \frac{1}{qn_i(\mu_n + \mu_p)} \tag{2.30}$$

Using Eq. (2.29) for the intrinsic concentration and the mobility values at low doping levels, it can be shown that

$$\rho_i = 1.75 \times 10^{-7} \, T^{0.8} \, e^{(7.02 \times 10^3)/T} \tag{2.31}$$

At room temperature, the intrinsic resistivity of silicon is $2.5 \times 10^5 \ \Omega \cdot$ cm. This resistivity decreases rapidly with increasing temperature.

The intrinsic concentration shown in Fig. 2.13 is purely a function of temperature as long as the doping level is below $1 \times 10^{17}/cm^3$. Above this value, the interaction between the dopant atoms becomes significant and the energy gap is no longer independent of the doping level. This phenomenon is called *band gap narrowing*. In the presence of band gap narrowing, an effective intrinsic concentration can be defined which is a function of doping level.

2.2.2. Band Gap Narrowing

At low doping levels, the energy-band diagram takes the form shown in Fig. 2.14*a*, where the density of states varies as the square root of energy. The donor and acceptor levels have discrete positions in the band gap that are separated from the conduction and valence bands. The well-defined separation between the conduction and valence band edges is called the *energy gap* (E_{g0}).

At high doping levels, three effects cause an alteration of the band structure. First, as the impurity density increases, the spacing between individual impurity atoms becomes small. The interaction between adjacent impurity atoms leads to a splitting of the impurity levels into an impurity band as shown in Fig. 2.14*b*. Second, the conduction and valence band edges no longer exhibit a parabolic shape. The statistical distribution of the dopant atoms introduces point-by-point differences in local doping and lattice potential leading to disorder. This results in the formation of band tails as illustrated in Fig. 2.14 as a result of the presence of disorder. Third, the interaction between the free carriers and more than one impurity atom leads to a modification of the density of states at the band edges. This is called *rigid band gap narrowing*.

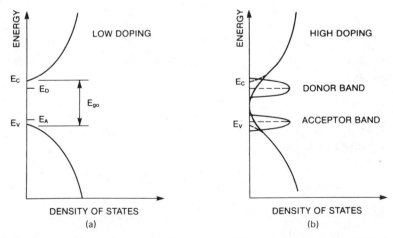

Fig. 2.14. Energy band diagram for a semiconductor at (*a*) low doping level and (*b*) high doping level.

The formation of the impurity band, the band tails, and the rigid band gap narrowing are majority carrier effects related to the reduced dopant impurity atom spacing. In addition to this, the electrostatic interaction of the minority carriers and the high concentration of majority carriers leads to a reduction in the thermal energy required to create an electron–hole pair. This is also a rigid band gap reduction that does not distort the energy dependence of the density of states. The rigid band gap reduction arising from the screening of the minority carriers by the high concentration of majority carriers can be derived by using Poisson's equation [17], written in spherical coordinates:

$$\frac{1}{r^2}\frac{d}{dr}\left(r^2\frac{dV}{dr}\right) = -\frac{q}{\epsilon_s}\Delta n \tag{2.32}$$

where Δn is the excess electron concentration given by

$$\Delta n = n_0(e^{qV/kT} - 1) \tag{2.33}$$

For low excess electron concentrations

$$\Delta n = \frac{qn_0 V}{kT} \tag{2.34}$$

Solution of Eq. (2.32) with the use of Eq. (2.34) gives

$$V(r) = \frac{q}{4\pi\epsilon_s}\left(\frac{e^{-(r/r_s)}}{r}\right) \tag{2.35}$$

where the screening radius r_s is given by

$$r_s = \sqrt{\frac{\epsilon_s kT}{q^2 n_0}} \tag{2.36}$$

The field distribution for the screened coulombic potential is given by

$$\mathcal{E}(r) = \frac{q}{4\pi\epsilon_s r}\left(\frac{1}{r} - \frac{1}{r_s}\right)e^{-(r/r_s)} \tag{2.37}$$

In comparison, at low doping concentrations, the unscreened field distribution is given by

$$\mathcal{E}_0(r) = \frac{q}{4\pi\epsilon_s r^2} \tag{2.38}$$

The band gap reduction is the difference in electrostatic energy between the screened and unscreened cases:

$$\Delta E_g = \frac{\epsilon_s}{2}\int[\mathcal{E}_0^2(r) - \mathcal{E}^2(r)]\,dr \tag{2.39}$$

Solving this equation, we obtain

$$\Delta E_g = \left(\frac{3q^2}{16\pi\epsilon_s}\right)\sqrt{\frac{q^2 n_0}{\epsilon_s kT}} \tag{2.40}$$

For low injection levels and assuming that all the dopant atoms are ionized

$$n_0 = N_D^+ = N_D \tag{2.41}$$

and

$$\Delta E_g = \left(\frac{3q^2}{16\pi\epsilon_s}\right)\sqrt{\frac{q^2 N_D}{\epsilon_s kT}} \tag{2.42}$$

This expression, valid for small changes in band gap, indicates that the band gap narrowing will be proportional to the square root of the doping level.

The variation of the energy band gap according to the preceding theory is shown in Fig. 2.15 for room temperature. In this case

Due to screening

$$\Delta E_g = 22.5 \times 10^{-3}\sqrt{\frac{N_I}{10^{18}}} \tag{2.43}$$

where N_I is the impurity (donor or acceptor) concentration per cubic centimeter In terms of modeling the electrical characteristics of devices, the influence of band gap narrowing on the product of the equilibrium electron and hole concentrations is required. The p–n product can be represented by an effective intrinsic concentration as in the case of lightly doped semiconductors:

$$n_{ie}^2 = np \tag{2.44}$$

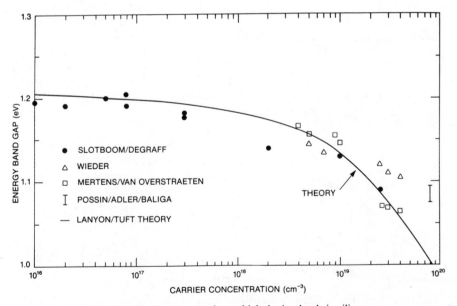

Fig. 2.15. Band gap narrowing at high doping levels in silicon.

The increase in the $p–n$ product arising from band gap narrowing is then given by

$$n_{ie}^2 = n_i^2 \exp \frac{q\,\Delta E_g}{kT} \tag{2.45}$$

where ΔE_g is the band gap narrowing due to the combined effects of impurity band formation, band tailing, and screening.

The results of measurements of band gap narrowing conducted using electrical determination of the $p–n$ product are provided in Fig. 2.15 [18–21]. It can be seen that the theoretical curve derived by using screening of the minority carriers by the majority carriers provided a good fit with the measured data up to a doping level of $10^{19}/\text{cm}^3$, indicating that this is the dominant effect. The spread in the data above this doping level needs to be experimentally resolved. An expression for the effective intrinsic carrier concentration at room temperature can be derived from Eq. (2.43):

$$n_{ie} = 1.4 \times 10^{10} \exp\left(0.433 \sqrt{\frac{N_1}{10^{18}}}\right) \tag{2.46}$$

where N_1 is the doping concentration per cubic centimeter. The increase in the intrinsic carrier concentration with doping is shown in Fig. 2.16. It is worth pointing out that there is an order of magnitude increase in the effective intrinsic carrier concentration when the doping level increases from 2×10^{19} to $10^{20}/\text{cm}^3$. These high doping levels are commonly encountered in the diffused regions of power devices. Since current conduction in bipolar power devices occurs in the presence of minority carrier injection into the diffused regions, the performance of these devices is affected by the band gap narrowing phenomena.

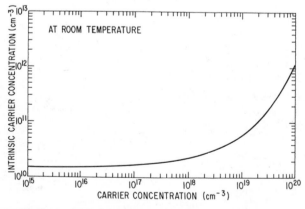

Fig. 2.16. Increase in effective intrinsic carrier concentration ($p–n$ product) at high doping levels in silicon.

2.2.3. Extrinsic Resistivity

The resistivity of a semiconductor region can be altered by the addition of controlled amounts of impurities. Elements from the fifth column of the periodic table produce N-type material, whereas elements from the third column produce P-type material. For device fabrication, the most commonly used N-type dopant is phosphorus. In addition, arsenic and antimony are used whenever their slower diffusion coefficient in silicon can be taken into advantage during device processing.

For creation of P-type regions, boron is the most commonly used impurity. For devices of very high voltage and large area, gallium and aluminum are also utilized. These elements have a much faster diffusion rate in silicon compared with boron. The diffusion coefficients of aluminum and gallium can be compared with those of boron with the aid of Fig. 2.17. It has been found experimentally that the diffusion coefficient of aluminum is a factor of 5 times larger than that of boron at typical diffusion temperatures. Thus a deep diffusion depth of 100 μm can be achieved with aluminum diffusions at 1200°C in 98 hr (4 days), whereas the same depth would require 556 hr (23 days) of diffusion time in the case of boron. It is obvious that a considerable savings in process time can be achieved by performing the diffusions with aluminum instead of

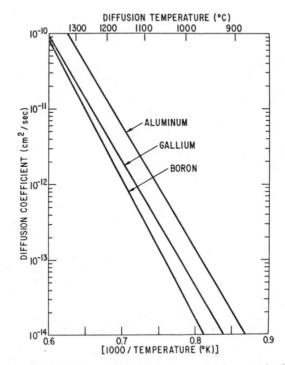

Fig. 2.17. Diffusion coefficients of P-type impurities—boron, aluminum, and gallium—in silicon.

boron. Other advantages of using aluminum instead of boron are that its ionic radius is comparable to that of silicon, thus minimizing misfit strain in the lattice. Furthermore, it does not form intermetallic compounds with silicon, and superior junction breakdown characteristics are observed with aluminum diffusions because of the ability to obtain low surface concentrations with the use of sealed-tube diffusion techniques.

The relationship between the phosphorus concentration and the resulting resistivity of silicon has been recently measured over a very wide range of doping densities [22]. The results of these measurements are provided in Fig. 2.18 for reference. Similar measurements of the resistivity of boron doped, P-type silicon have been made [23]. These data are also provided in Fig. 2.18. The difference in the resistivity between N- and P-type silicon arises from a difference in the mobility for electrons and holes. Using the relationship between mobility and doping concentration provided by equations (2.3) and (2.4), the resistivity ρ_n in ohm·centimeters can be related to the doping concentration N_D per cubic centimeter:

$$\rho_n = \frac{3.75 \times 10^{15} + N_D^{0.91}}{1.47 \times 10^{-17} \, N_D^{1.91} + 8.15 \times 10^{-1} \, N_D} \tag{2.47}$$

Similarly, for the case of P-type silicon, the resistivity ρ_p in ohm · centimeters is

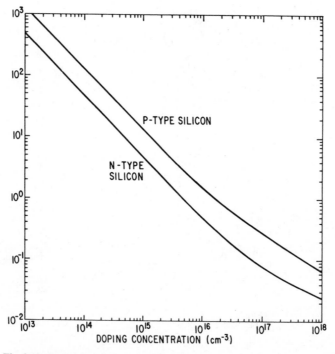

Fig. 2.18. Resistivity of N- and P-type silicon as a function of doping level.

related to the doping concentration N_A per cubic centimeter by

$$\rho_p = \frac{5.86 \times 10^{12} + N_A^{0.76}}{7.63 \times 10^{-18} N_A^{1.76} + 4.64 \times 10^{-4} N_A} \tag{2.48}$$

It is worth pointing out that, although these equations indicate a rather complex relationship between the resistivity and the doping concentration, the resistivity of starting material used for the fabrication of power devices with breakdown voltages above 100 V exceeds 1 $\Omega\cdot$cm in N-type silicon and 3 $\Omega\cdot$cm in P-type silicon. At these higher resistivities, the mobility becomes independent of the doping concentration and the resistivity can be obtained from the following expressions:

$$\rho_n = 4.596 \times 10^{15} N_D^{-1} \tag{2.49}$$

and

$$\rho_p = 1.263 \times 10^{16} N_A^{-1} \tag{2.50}$$

The starting material for power device fabrication is preferably bulk material with low defect density. In most very high voltage power devices, N-type starting material is used. Since a low concentration of oxygen is required to avoid the formation of precipitates that adversely affect the breakdown characteristics, float-zone (FZ) silicon is preferred over Czochralski (CZ) silicon.

Power devices are generally required to carry large currents. The area of a single device is much larger than that of equivalent devices in integrated circuits. In fact, for very-high-power devices, a single device is fabricated from a 3–4 in.-diameter wafer of silicon today. Consequently, variations in the resistivity across the diameter of a wafer as well as its thickness can have a strong influence on the device characteristics. For example, in devices containing back-to-back $P-N$ junctions, the breakdown voltage is limited to that of an open-base transistor. As a result, if a large variation in the resistivity occurs in the wafer about its nominal value, the device breakdown voltage may be limited by either punch-through effects at the higher resistivity locations or by premature avalanche breakdown at the lower resistivity locations as discussed in Chapter 3. Since an allowance must be made for both extremes in the resistivity distribution, a nonoptimum device design is unavoidable.

The inhomogeneous resistivity in silicon wafers grown by the FZ process occurs in the form of striations that have been attributed to local fluctuations in the growth rate caused by temperature variations. These changes in the growth rate cause a variation in the local dopant incorporation in the growing crystal because the effective segregation coefficient is determined by the growth rate. The fluctuations in the resistivity can be decreased by improving the homogeneity of the temperature distribution. Advanced crystal growth methods have been successful in reducing resistivity variations from the $\pm15\%$ observed by the conventional techniques to about $\pm7\%$. Although this has resulted in a significant improvement in device characteristics, even superior devices can be obtained by using extremely homogeneously doped N-type silicon produced by the neutron transmutation doping process.

2.2.4. Neutron Transmutation Doping

This technique for producing N-type silicon is based on the conversion of the ^{30}Si isotope into phosphorus atoms by the absorption of thermal neutrons [24]. In the last few years, the control over the transmutation process has been improved to the point where it is extensively used for the production of high-resistivity silicon for large-area power devices. In addition, the annealing kinetics of the silicon after the neutron transmutation process are now well understood, allowing its widespread usage in the power device industry.

The neutron transmutation doping process occurs by the absorption of thermal neutrons. The thermal neutron capture by all three of the naturally occurring stable silicon isotopes takes place in proportion to their natural abundance and the capture cross sections of the respective atoms. The abundance of the three stable isotopes has been found to be 92.27% for the ^{28}Si isotope, 4.68% for the ^{29}Si isotope, and 3.05% for the ^{30}Si isotope. The absorption of a neutron by these isotopes leads to the following reactions:

$$^{28}\text{Si } (n,\gamma) \; ^{29}\text{Si} \tag{2.51}$$

$$^{29}\text{Si } (n,\gamma) \; ^{30}\text{Si} \tag{2.52}$$

and

$$^{30}\text{Sin } (n,\gamma) \; ^{31}\text{Si} \rightarrow \; ^{31}\text{P} + \beta \tag{2.53}$$

The thermal neutron absorption cross sections for these reactions are 0.08, 0.28, and 0.13 barn, respectively (1 barn equals 10^{-24} cm^2). The first two reactions merely alter the isotope ratios. The third reaction is the one that can be utilized to create phosphorus atoms. Since the naturally occurring ^{30}Si isotope can be expected to be homogeneously distributed in the silicon wafer, a homogeneous distribution of phosphorus can also be created if a uniform neutron flux can be ensured within all points of the wafer being irradiated. In addition to the neutron flux distribution inside the reactor during irradiation, two other factors can influence the uniformity of the resistivity: (1) the inhomogeneous neutron flux distribution created by the absorption of the neutrons in the silicon itself and (2) the presence of a highly inhomogeneous distribution of dopants in the starting material.

The influence of the absorption of the thermal neutrons has been treated for the case of a silicon ingot surrounded by a medium with the same absorption properties as silicon [25]. In this case, if a uniform neutron flux is assumed to emanate from the reactor core, the neutron flux in the silicon takes the form

$$\phi = \phi_0 e^{(-x/b)} \tag{2.54}$$

where b is the decay length for neutron absorption in silicon. For an ingot of diameter d, a phosphorus concentration proportional to ϕ_0 is created on one side of the ingot, whereas a phosphorus concentration proportional to $\phi_0 \exp(-d/b)$ is created on the other side as illustrated in Fig. 2.19. The resulting inhomogeneous distribution of the neutron flux can be shown to be equivalent to a resistivity variation of $\pm \sinh(d/2b)$.

The decay length b of neutrons in silicon has been calculated to be 19 cm. With the use of this value, the effect of increasing the ingot diameter on the resistivity variation is shown in Fig. 2.20. It can be seen that neutron absorption in the silicon can lead to a significant resistivity variation when the ingot diameter exceeds 40 mm. To overcome this problem, the silicon ingots are rotated in the neutron flux during irradiation. The maximum resistivity now occurs at the axis of the ingot. The ratio of the neutron dose at the periphery to that at the axis of a cylindrical ingot is given by [25]

$$\frac{\phi(d)}{\phi(0)} = 1 + \frac{1}{16}\left(\frac{d}{b}\right)^2 \tag{2.55}$$

The resistivity variation within the crystal will then be $\pm\{[32(b/d)^2 - 1]^{-1}\}$. This variation is also plotted in Fig. 2.20 to contrast it with the variation in the resistivity without rotation. It can be seen that a significant improvement in the doping homogeneity results from ingot rotation and that a resistivity variation of less than 2% can be achieved for ingot diameters of up to 100 mm by this method. In actual practice, the ingot is surrounded by heavy water or graphite that tends to increase the decay length and results in even superior homogeneity than predicted by the preceding expressions.

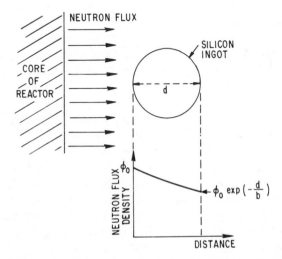

Fig. 2.19. Schematic illustration of silicon ingot irradiation during neutron transmutation doping.

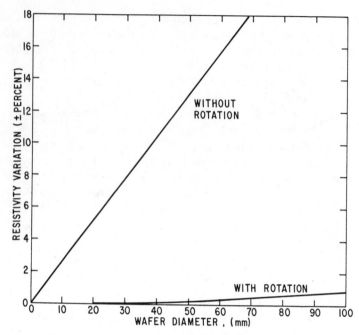

Fig. 2.20. Impact of ingot diameter on resistivity variation produced during neutron transmutation doping. Note the significant improvement produced by ingot rotation during irradiation.

The second important consideration in obtaining a homogeneous resistivity is the influence of the distribution of dopants in the starting material prior to irradiation. Because of the high segregation coefficient of boron in silicon, more homogeneous doping can be achieved in P-type starting material than in N-type material of the same resistivity. However, the lower mobility for holes results in a greater background doping level in the P-type starting material as compared with N-type starting material. Thus both P- and N-type starting materials have been used in the preparation of neutron transmutation doped silicon. In both cases, the concentration of phosphorus created by neutron transmutation must be significantly larger than the dopant concentration in the starting material.

If the starting material has a resistivity variation of $\pm\alpha$ percent, it can be shown that after irradiation, the percentage variation in the resistivity will be $\pm\alpha(\rho_f/\rho_i)$ for N-type starting material and $\pm\alpha(\mu_n/\mu_p)(\rho_f/\rho_i)$ for P-type starting material, where ρ_f is the postradiation resistivity, ρ_i is the preirradiation resistivity, μ_n is the mobility of electrons, and μ_p is the mobility of holes. The influence of resistivity fluctuation in the starting material on the resistivity variation after irradiation is shown in Fig. 2.21 for various post- to preirradiation resistivity ratios. Since the preirradiation resistivity variation typically lies in the cross-hatched region of this figure, the minimum initial resistivity required for achieving the desired uniformity in the final resistivity can be determined from Fig. 2.21.

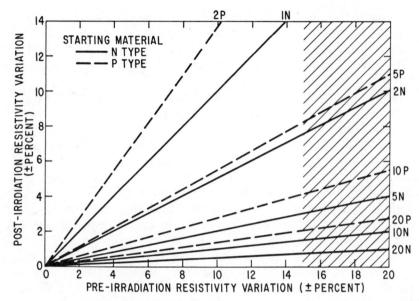

Fig. 2.21. Impact of starting material resistivity on the resistivity variation after neutron transmutation doping.

As an example, if the starting material has a resistivity fluctuation of $\pm 15\%$ and the resistivity variation after irradiation must be less than $\pm 3\%$, N-type starting material must have a resistivity of greater than 5 times the postirradiation value, and P-type starting material must have a resistivity of greater than 15 times the postirradiation value. Since starting material can be grown today with P-type resistivities of over 5000 $\Omega \cdot$ cm, homogeneous neutron doped N-type silicon can be readily produced with resistivities of up to 200 $\Omega \cdot$ cm.

In addition to these uniformity considerations, the occurrence of other nuclear reactions during irradiation and the creation of lattice damage must also be considered. The absorption of thermal neutrons by the phosphorus atoms, created during the transmutation process, can produce sulfur. Further, the irradiation invariably takes place in the presence of fast (high-energy) neutrons. Fast neutron absorption by the silicon leads to the production of aluminum that would compensate the donors created by the transmutation reaction [Eq. (2.53)], or to the creation of magnesium that is a recombination center in silicon. Recent studies [26] have shown that neither of these reactions has a significant effect on the transmutation process. Consequently, the net donor concentration in the irradiated material can be computed solely on the basis of phosphorus atoms created by the transmutation reaction described in Eq. (2.53) and is given by

$$C_{\text{donor}} = 2.06 \times 10^{-4} \, \phi t \tag{2.56}$$

where ϕ is the neutron flux density and t is the irradiation time.

The neutron transmutation process is accompanied by severe lattice damage. The lattice damage arises from the displacement of the silicon atoms due to (1) gamma recoil during the transmutation reaction, (2) crystal irradiation from the high-energy β particles emitted during the decay of ^{31}Si to ^{31}P, and (3) collisions of fast neutrons with silicon atoms. Although the effect of fast neutron damage can be minimized by performing the irradiation at locations in the reactor where the thermal:fast neutron ratio is high, it has been shown that the displacement effects due to fast neutrons are a thousand times greater than in the other processes [26]. Furthermore, the damage due to the fast neutrons has been found to be more difficult to anneal out because it occurs in the form of clusters with diameters of less than 100 °A. Because of these damage mechanisms, the resistivity of the silicon is over 10^5 $\Omega \cdot$cm after irradiation and the minority carrier lifetime is low. To recover the desired resistivity due to the phosphorus doping and to obtain a high minority carrier lifetime, it is necessary to anneal the silicon after irradiation. The effect of annealing on the resistivity and the minority carrier lifetime has been correlated with the dissociation of crystal defects in the temperature range of 400 to 900°C [26]. These studies have shown that, although the resistivity can be recovered by annealing at 600°C for 1 h, the minority carrier lifetime is still low and requires an annealing temperature above 750°C to reach its equilibrium value. Therefore, annealing the silicon at temperatures between 750 and 800°C is commonly used to obtain material from which devices can be fabricated.

In summary, neutron transmutation doping can be used to achieve superior homogeneiety in resistivity as compared with conventionally doped silicon. The advantages of neutron transmutation doped silicon for power devices can be briefly stated as (1) a greater precision in the control of the avalanche breakdown voltage, (2) a more uniform avalanche breakdown distribution across the wafer that leads to a greater capacity to withstand voltage surges, (3) a narrower base region width for achieving the desired breakdown voltage that lowers the forward voltage drop during current conduction, and (4) a more uniform current flow during forward conduction that improves the surge current handling capability of the device.

2.3 LIFETIME

Under thermal equilibrium conditions, a continuous balance between the generation and recombination of electron–hole pairs occurs in semiconductors. Any creation of excess carriers by an external stimulus disturbs this equilibrium. On removal of the external excitation, the excess carrier density decays and the carrier concentration returns to the equilibrium value. The lifetime is a measure of the duration of this recovery. The recovery to equilibrium conditions can occur by recombination occurring as a result of (1) an electron dropping directly from the conduction band into the valence band, (2) an electron dropping from the conduction band and a hole dropping from the valence band into a recombination center, and (3) electrons from the conduction band and holes from the valence band dropping into surface traps. During these recombination processes, the energy of the carriers must be dissipated by one of several mechanisms: (1) the emission of a photon (radiative recombination), (2) the dissipation of energy in the lattice in the form of phonons (multiphonon recombination), and (3) the transmission of the energy to a third particle that can be either an electron or a hole (Auger recombination). The transitions that occur during these recombination processes are schematically illustrated in Fig. 2.22. All these processes simultaneously assist in the recovery of the carrier density to its equilibrium value.

Since silicon is a semiconductor with an indirect band gap structure, the probability of direct transitions from the conduction band to the valence band is small. Consequently, the direct radiative recombination lifetime for silicon is on the order of 1 sec. In comparison to this, the density of recombination centers, even in the high-purity silicon used to fabricate power devices, is sufficiently high so as to reduce the lifetime associated with recombination through deep centers in the energy gap to less than 100 μsec. This recombination process, therefore, proceeds much more rapidly and is the predominant one under most device operating conditions.

2.3.1. Shockley–Read–Hall Recombination

The statistics of the recombination of electrons and holes in semiconductors through recombination centers was treated for the first time by Shockley, Read, and Hall [27, 28]. Their theory shows that the rate of recombination (U) in the

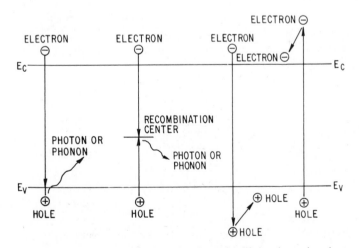

Fig. 2.22. Recombination mechanisms that determine lifetime in semiconductors.

steady state through a single-level recombination center is given by

$$U = \frac{\delta n\, p_0 + \delta p\, n_0 + \delta n\, \delta p}{\tau_{p0}(n_0 + \delta n + n_1) + \tau_{n0}(p_0 + \delta p + p_1)} \qquad (2.57)$$

where, respectively, δn and δp are the excess electron and hole concentrations, n_0 and p_0 are the equilibrium concentrations of electrons and holes, τ_{n0} and τ_{p0} are the electron and hole minority carrier lifetimes in heavily doped P- and N-type silicon, and n_1 and p_1 are the equilibrium electron and hole densities corresponding to the Fermi-level position coincident with the recombination level position in the band gap. These concentrations are given by the expressions

$$n_1 = N_c \exp \frac{E_r - E_c}{kT} \qquad (2.58)$$

and

$$p_1 = N_v \exp \frac{E_v - E_r}{kT} \qquad (2.59)$$

where N_c and N_v are the density of states in the conduction and valence bands, respectively; E_c, E_r, and E_v are the conduction band, recombination level, and valence band locations, respectively; k is Boltzmann's constant; and T is the absolute temperature. Under conditions of space charge neutrality, the excess electron (δn) and hole (δp) concentrations are equal, and the lifetime can be defined as

$$\tau = \frac{\delta n}{U} = \tau_{p0} \left(\frac{n_0 + n_1 + \delta n}{n_0 + p_0 + \delta n} \right) + \tau_{n0} \left(\frac{p_0 + p_1 + \delta n}{n_0 + p_0 + \delta n} \right) \qquad (2.60)$$

Since silicon power devices are generally made from N-type material because of the higher mobility of electrons, the remaining treatment in this section is confined to N-type silicon. Analogous equations can be derived for P-type silicon.

In the case of N-type silicon, the electron density n_0 is much larger than the hole density p_0. Combining this factor with Eqs. (2.58)–(2.60), and defining a normalized injection level as $h = (\delta n/n_0)$, it can be shown [31] that

$$\frac{\tau}{\tau_{p0}} = \left[1 + \frac{1}{(1+h)} \exp\frac{E_r - E_F}{kT} \right] + \zeta \left[\frac{h}{(1+h)} + \frac{1}{(1+h)} \exp\frac{2E_i - E_r - E_F}{kT} \right] \qquad (2.61)$$

where E_i and E_F are the positions of the intrinsic and Fermi levels, respectively, and ζ is the ratio of the minority carrier lifetimes in heavily doped P- and N-type silicon, respectively:

$$\zeta = \frac{\tau_{n0}}{\tau_{p0}} \qquad (2.62)$$

In deriving Eq. (2.61) it has been assumed that the density of states in the conduction and valence band are equal.

The minority carrier lifetime in heavily doped N- and P-type material used in the preceding equations is dependent on the capture rate for holes (C_p) and electrons (C_n) at the recombination center:

$$\tau_{n0} = \frac{1}{C_n N_r} = \frac{1}{v_{Tn}\sigma_{cn}N_r} \tag{2.63}$$

and

$$\tau_{p0} = \frac{1}{C_p N_r} = \frac{1}{v_{Tp}\sigma_{cp}N_r} \tag{2.64}$$

where v_{Tn} and v_{Tp} are the thermal velocities of electrons and holes, σ_{cn} and σ_{cp} are the capture cross sections for electrons and holes at the recombination center, and N_r is the density of the recombination center. Using these expressions:

$$\zeta = \frac{\tau_{n0}}{\tau_{p0}} = \frac{v_{Tp}\sigma_{cp}}{v_{Tn}\sigma_{cn}} = 0.827 \frac{\sigma_{cp}}{\sigma_{cn}} \tag{2.65}$$

Thus ζ is independent of the recombination center density and temperature but depends on the capture cross sections of the center.

Since well-developed techniques exist for the measurement of the capture cross sections and the position of the recombination center in the energy gap, the lifetime can be computed from these data. As an example, the variation in the lifetime with injection level is shown in Fig. 2.23 for N-type silicon for a doping level of 1×10^{14} atoms/cm^3. In Fig. 2.23, a range of values of the parameter ζ and the recombination center position E_r are considered. Several important observations can be made from these plots. First, at very low injection levels ($h \ll 1$) the lifetime is observed to be independent of the injected carrier

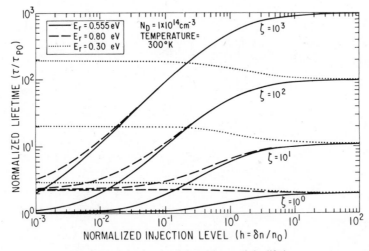

Fig. 2.23. Injection level dependence of the lifetime.

density. This low-level lifetime τ_{LL} is given by the expression

$$\tau_{LL} = \tau_{p0}\left[1 + \exp\left(\frac{E_r - E_F}{kT}\right)\right] + \tau_{n0}\exp\left(\frac{2E_i - E_r - E_F}{kT}\right) \qquad (2.66)$$

It can be seen from Fig. 2.23 that this low-level lifetime is dependent on the recombination-level position in the energy gap and the capture cross-section ratio ζ.

The variation of the low-level lifetime with recombination center position E_r and capture cross section for holes and electrons can be seen more clearly in the plots provided in Fig. 2.24 for N-type silicon with a doping level of 1×10^{14} cm^3. The low-level lifetime is observed to have its smallest value when the recombination center lies in a broad region centered around midgap. As the center is shifted toward either the conduction or the valence band edges, the

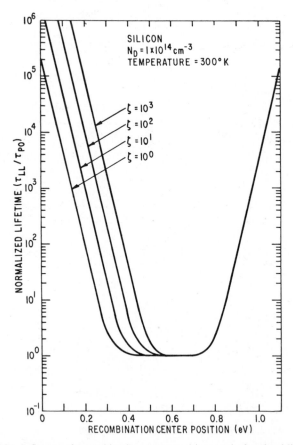

Fig. 2.24. Influence of recombination center position on the low-level lifetime.

low-level lifetime increases as a result of the decreasing probability of capturing either holes or electrons, respectively.

Referring to Fig. 2.23, it can be seen that when the injection level becomes very large ($h \gg 1$), the lifetime asymptotically approaches a constant value. This high-level lifetime is given by the expression

$$\tau_{HL} = \tau_{p0} + \tau_{n0} = \tau_{p0}(1 + \zeta) \tag{2.67}$$

The high-level lifetime is not dependent on the position of the recombination center and the temperature. However, it is directly dependent upon the capture cross-section ratio ζ.

The variation in the lifetime with injection level is important to bipolar devices operating over a very large range of injection levels. The plots in Fig. 2.23 demonstrate that the lifetime can either increase or decrease with injection level depending on the position of the recombination center in the energy gap. The magnitude of the change in lifetime with injection level is controlled by the relative capture cross sections at the recombination center, that is, the capture cross-section parameter ζ. In most practical cases, the lifetime is observed to increase with injection level by approximately a factor of 10.

2.3.2. Space-Charge Generation

When a power device is blocking current flow, the voltage is supported across the depletion layer of a *P–N* junction. Under these conditions, an ideal device should exhibit no current flow. In actual reverse biased *P–N* junctions, a finite current flow is always observed. This "leakage" current arises from a diffusion of free carriers that are generated within a minority carrier diffusion length from the edge of the depletion layer (diffusion current) and from the drift of carriers generated within the depletion layer itself (space-charge generation current). In the case of silicon, which has a large energy gap, the generation current is much larger than the diffusion current at around room temperature. At higher temperatures, the diffusion current becomes comparable to the generation current.

The contribution to the leakage current from space-charge generation can be derived under the assumption of a uniform generation rate within the depletion region [29]. From this analysis, the leakage current is given by

$$I_{SC} = \frac{qAWn_i}{\tau_{SC}} \tag{2.68}$$

where q is the electron charge, A is the junction area, W is the width of the depletion region, n_i is the intrinsic carrier concentration, and τ_{SC} is the space-charge generation lifetime. This lifetime can be derived from Eq. (2.57) for a single-level recombination center:

$$\tau_{SC} = \tau_{p0} \exp\left(\frac{E_r - E_i}{kT}\right) + \tau_{n0} \exp\left(\frac{E_i - E_r}{kT}\right) \tag{2.69}$$

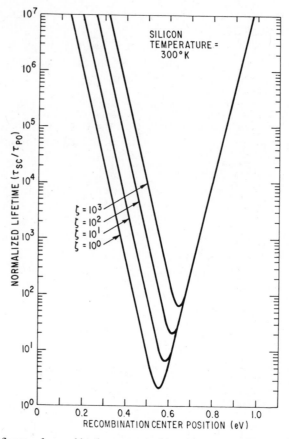

Fig. 2.25. Influence of recombination center position on space-charge generation lifetime.

The variation of the space-charge generation lifetime with the recombination center position is shown in Fig. 2.25. The lowest space-charge generation lifetimes, and consequently the highest leakage currents, occur when the recombination center is located close to midgap. Thus midgap recombination centers are not only most effective in reducing the minority carrier (low-level) lifetime as shown by Fig. 2.24, but they also control the rate of generation of leakage current in power devices.

The preceding analysis predicts that the leakage current is proportional to the width of depletion layer; consequently, the product of the leakage current and the junction capacitance should be independent of the junction reverse bias. Experimental measurements on silicon junctions indicate that the uniform generation of carriers occurs over only a fraction of the depletion region width [44]. The variation of this generation layer width W_i is a function of the applied junction bias and the substrate resistivity. Figures 2.26a and 2.26b show the variation in this generation region width, normalized to the total depletion width W,

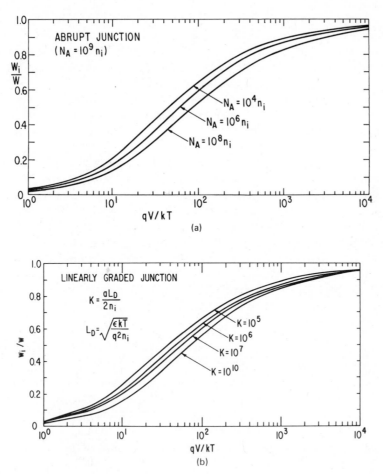

Fig. 2.26. Space-charge generation layer width as a function of the reverse bias voltage applied to (*a*) an abrupt junction and (*b*) a linearly graded junction. (After Ref. 44, reprinted with permission from *Solid State Electronics.*)

as a function of the reverse junction bias voltage for abrupt and linearly graded junctions. It is important to account for this variation in generation layer width with bias to adequately calculate the leakage current in power devices.

2.3.3. Recombination-Level Optimization

In the previous sections, it was shown that the recombination and generation lifetimes are a strong function of the deep-level position within the band gap and its capture cross section for electrons and holes. For obtaining the best device performance, it is necessary to develop criteria for the selection of the optimum location of the deep level and its capture cross sections.

In the case of unipolar devices operating in the absence of minority carrier injection, the only criterion is that the space-charge generation lifetime must be large to minimize the leakage current. This can be achieved by maintaining a clean process that prevents the introduction of deep-level impurities into the active area of the device.

In the case of devices where some low-level minority carrier injection occurs, it becomes necessary to introduce deep-level recombination centers to speed up or control the device characteristics. A small low-level lifetime is highly desirable here for achieving fast switching speed. As indicated by the curves in Fig. 2.24, the recombination level should be located close to midgap to achieve the smallest low-level lifetime. It is also desirable to have a large ratio ζ for the capture cross sections of holes and electrons. At the same time, it is important to minimize the leakage current in the device during reverse blocking. For low leakage current, it is necessary to locate the recombination center away from midgap as indicated by the curves in Fig. 2.25. These two requirements impose conflicting demands on the recombination center characteristics. To resolve this conflict and find an optimum location for the deep level, it is necessary to consider the ratio of the space-charge generation lifetime to the low-level lifetime [30]. Using equations (2.66) and (2.69) for the case of N-type silicon:

$$\frac{\tau_{SC}}{\tau_{LL}} = \frac{e^{(E_r - E_i)/kT} + \zeta\, e^{(E_i - E_r)/kT}}{1 + e^{(E_r - E_F)/kT} + \zeta\, e^{(2E_i - E_F - E_r)/kT}} \tag{2.70}$$

To achieve the best performance, it is necessary to maximize the ratio (τ_{SC}/τ_{LL}). Consider a typical example of silicon doped at a carrier concentration of $5 \times 10^{15}/cm^3$. The variation of the ratio (τ_{SC}/τ_{LL}) with recombination-level position relative to the valence band is shown in Fig. 2.27 for 300 and 400 K. It can be seen that the ratio has its highest value when the deep level is located near either the conduction band or valence band edges. The ratio can also be seen to rapidly decrease with increasing temperature. The best trade-off between low-level lifetime reduction and minimizing leakage current due to space-charge generation is achieved when the deep level is located near the band edges and the device operating temperature is low.

The ratio (τ_{SC}/τ_{LL}) remains high over a range of recombination-level positions close to the band edges. This range is a function of the carrier concentration and the capture cross-section ratio. If the range is defined to extend to 10% below the maximum ratio as indicated by the arrows in Fig. 2.27, the influence of doping on the optimum position of the deep level can be quantified. For the case of N-type silicon, the preferred locations of recombination levels within the band gap are indicated by the shaded areas in Fig. 2.28. This figure covers a spectrum of doping levels from 10^{13} to $10^{17}/cm^3$ and capture cross-section ratios from 0.01 to 100. The heavily shaded area located above midgap applies to all capture cross-section ratios. Recombination levels with larger cross-section ratios ζ are also preferable because they can be located over a

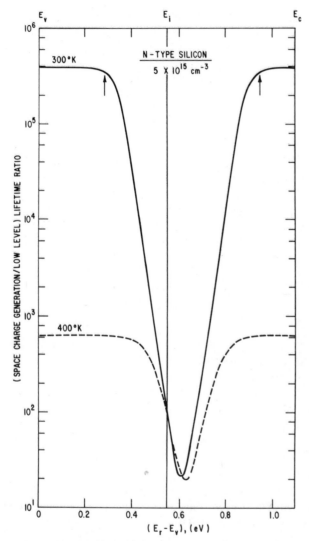

Fig. 2.27. Impact of recombination level position on the ratio of the space-charge generation and low-level lifetimes.

wider range within the band gap. For example, at a doping level of $1 \times 10^{14}/\text{cm}^3$, the recombination level must be located within 0.2 eV of the valence band for $\zeta = 0.01$ but can be located up to 0.4 eV from the valence band if $\zeta = 100$. Furthermore, note that a broader range of recombination levels is available for lightly doped material, thus facilitating the selection of the lifetime controlling impurity.

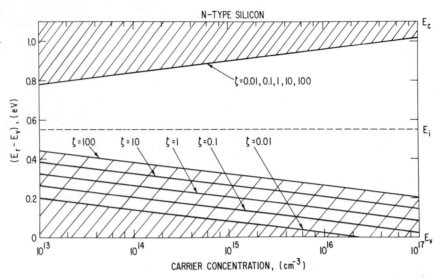

Fig. 2.28. Preferred locations (shaded areas) within the silicon band gap for recombination centers in the case of bipolar devices operating under low-level injection.

The maximum value of the ratio (τ_{SC}/τ_{LL}) occurs when the recombination level is located close to the band edges. The maximum value is a function of the doping level but independent of the capture cross section. It can be derived from Eq. (2.70):

$$\left(\frac{\tau_{SC}}{\tau_{LL}}\right)_{max} = \exp\left(\frac{E_F - E_i}{kT}\right) = \frac{n_0}{n_i} \qquad (2.71)$$

The variation in the maximum achievable ratio (τ_{SC}/τ_{LL}) with carrier concentration is shown in Fig. 2.29 for 300 and 400 K. It can be seen that the ratio (τ_{SC}/τ_{LL}) decreases with decreasing doping level and with increasing temperature. Consequently, it is more difficult to achieve a reduction of the low-level lifetime in high-resistivity silicon without substantial increase in leakage current when compared with low-resistivity silicon. The problem becomes aggravated as temperature increases.

The above discussion is relevant to devices operating with low-level injection during current conduction. In power devices, it is highly desirable to operate at high-level injection during current conduction. High-level injection increases the conductivity of the lightly doped regions and greatly enhances the current handling capability. In these devices, it is desirable to have a large high-level lifetime to maximize the conductivity modulation. At the same time, it is necessary to try to minimize the low-level lifetime to hasten the switching process by rapid recombination of the minority carriers during device turn-off. The optimization of the location of the deep level for these devices can be performed by considering the ratio of the high-level lifetime to the low-level lifetime [31].

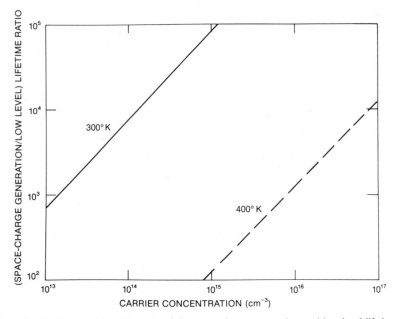

Fig. 2.29. Maximum achieveable ratio of the space-charge generation and low-level lifetimes.

Using Eqs. (2.66) and (2.67):

$$\frac{\tau_{HL}}{\tau_{LL}} = \frac{1 + \zeta}{\left[1 + e^{(E_r - E_F)/kT} + \zeta\, e^{(2E_i - E_F - E_r)/kT}\right]} \tag{2.72}$$

The ratio (τ_{HL}/τ_{LL}) can be maximized with respect to recombination-level position E_r by setting

$$\frac{d}{dE_r}\left(\frac{\tau_{HL}}{\tau_{LL}}\right) = 0 \tag{2.73}$$

and solving for E_r. This produces the first optimization criterion:

$$E_r = E_i + \frac{kT}{2}\ln(\zeta) \tag{2.74}$$

The optimum deep-level location is found to be independent of doping level and is purely a function of the capture cross-section ratio ζ and temperature.

To achieve the highest degree of conductivity modulation while maintaining a fast switching speed, it is necessary to achieve a large absolute value for τ_{HL}/τ_{LL}. For N-type silicon, this will occur when the capture cross-section ratio ζ is large. It should be noted that achieving large values for the ratio (τ_{HL}/τ_{LL}) becomes more difficult at lower doping levels and at higher temperatures. This can be illustrated by plotting the ratio (τ_{HL}/τ_{LL}) as a function of the doping

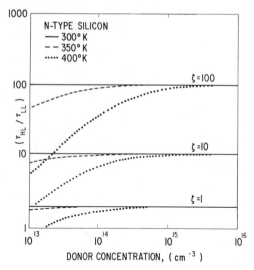

Fig. 2.30. High:low-level lifetime ratio as a function of background doping level.

concentration. In Fig. 2.30 calculated curves are provided for several values of ζ at three operating temperatures. Since the high-level lifetime is independent of temperature, the temperature dependence of the ratio (τ_{HL}/τ_{LL}) in Fig. 2.30 arises from the variation of the low-level lifetime with temperature.

In devices operating with high-level injection, it is desirable to achieve a rapid transition from low- to high-level lifetime with increasing injection level. This will maximize the region of injection levels over which the high-level lifetime is large, thus providing good current conduction characteristics, while simultaneously providing a small lifetime at low injection levels. The variation of the lifetime with injection level is described by Eq. (2.61). The transition from low- to high-level injection can be maximized by choosing a recombination-level position such that

$$\frac{d}{dE_r}\left(\frac{d\tau}{dh}\right) = 0 \tag{2.75}$$

The solution to this equation is also given by Eq. (2.74). The optimum recombination-level positions obtained by using this equation are found to lie close to the center of the band gap. It was shown in Section 2.3.2 that when the recombination center lies close to midgap, the space-charge leakage current will be high. The optimum level location defined for achieving a large (τ_{HL}/τ_{LL}) ratio conflicts with the need to maintain low leakage currents in power devices.

To resolve this conflict, it is necessary to examine the variation in the (τ_{HL}/τ_{LL}) ratio with recombination-level location. Consider two examples of background donor concentrations of $7 \times 10^{15}/\text{cm}^3$ and $2 \times 10^{13}/\text{cm}^3$. The variation of the ratio (τ_{HL}/τ_{LL}) with recombination-level position is shown in Fig. 2.31 for three

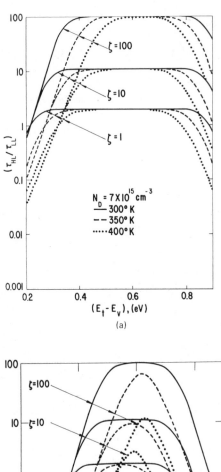

Fig. 2.31. Variation of the high:low-level lifetime ratio with recombination-level position.

values of ζ at different ambient temperatures. At the higher doping level (Fig. 2.31a), the ratio (τ_{HL}/τ_{LL}) exhibits a very broad maximum for all cases. Thus the recombination level can be located away from the absolute maximum point near midgap without suffering a significant drop in the ratio (τ_{HL}/τ_{LL}). The leakage current can then be significantly reduced without compromising the forward voltage drop. A similar behavior for the variation of the (τ_{HL}/τ_{LL}) ratio with recombination center position is observed for the case of the low doping level (Fig. 2.31b) at low temperatures, but the maximum is not as broad as in the case of the higher doping concentration. In fact, at higher temperatures the (τ_{HL}/τ_{LL}) ratio exhibits a sharp peak at a value less than $(1 + \zeta)$. An optimum recombination-level position that will provide a large (τ_{HL}/τ_{LL}) ratio while minimizing the leakage current can be chosen by selecting a location when the (τ_{HL}/τ_{LL}) is 10% below its maximum value for cases where the broad maximum is exhibited. Alternatively, at low doping levels and higher temperatures where the ratio (τ_{HL}/τ_{LL}) remains below 90% of the maximum achievable value of $(1 + \zeta)$ the optimum location will occur when the (τ_{HL}/τ_{LL}) ratio is at its peak value. The optimum recombination center location obtained in this manner is shown in Fig. 2.32 as a function of the donor concentration. It should be noted that although two optimum recombination-level locations can be defined on either side of the maximum in Fig. 2.31, only the position above the intrinsic level must be used because it produces the desired increase in lifetime with injection level as shown in Fig. 2.23. The optimum recombination-level location obtained from Fig. 2.32 can be seen to be closer to midgap when the doping

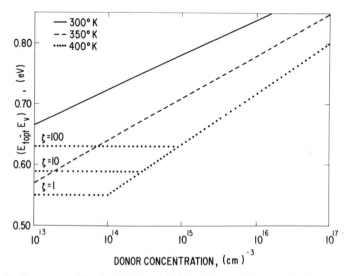

Fig. 2.32. Optimum recombination level position for bipolar devices operating under high-level injection.

level is reduced. Thus it becomes more difficult to achieve a low leakage current in devices designed for operation at higher voltages.

In addition to the preceding optimization criteria, other constraints can be defined for the deep-level impurity [32]. The introduction of a deep-level impurity causes compensation effects, resulting in changes in the background resistivity. When the deep-level impurity concentration approaches the background donor concentration, electrons are transferred from the donor into the deep level instead of the conduction band. The decrease in net free electron concentration in the conduction band causes an increase in the resistivity. Changes in resistivity have serious effects on both breakdown voltage and current conduction. The compensation can be minimized by achieving the lowest possible deep-level concentration for any desired lifetime. The lifetime is related to the deep-level concentration by Eqs. (2.63) and (2.64). From these expressions, it can be concluded that the least compensation will be achieved when the capture cross sections for electrons and holes are large.

Another constraint is that the same deep-level impurity should be useful for a broad range of resistivities. This allows a single lifetime control process to be used for a variety of devices. To use the same lifetime control process over a broad range of resistivities, the ratio (τ_{HL}/τ_{LL}) should remain high over a wide range of doping levels. In Fig. 2.33, the variation of the (τ_{HL}/τ_{LL}) ratio with background resistivity is shown for three recombination-level positions. It can

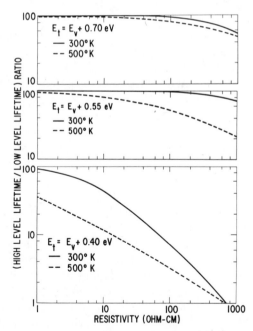

Fig. 2.33. Variation of the high:low-level lifetime ratio with resistivity for several recombination center positions.

be seen that the desired characteristics are exhibited when the deep level is located at 0.7 eV above the valence band. This technologically desirable feature is consistent with the optimum recombination center location obtained by using Fig. 2.32 over most of the range of doping levels.

2.3.4. Lifetime Control

Two fundamental processing methods have been developed to control lifetime in power devices: (1) a method involving the thermal diffusion of an impurity that exhibits deep levels in the energy gap of silicon and (2) a method based on the creation of lattice damage in the form of vacancies and interstitial atoms by bombardment of the silicon wafers with high-energy particles.

Historically, the introduction of recombination centers in silicon by the diffusion of impurities was adapted to power device processing well before the use of the high-energy radiation technique. A very large number of impurities have been found to exhibit deep levels in silicon. A catalog of these impurities has been compiled [31, 33]. Although many of these impurities could be used in the fabrication of power devices, only gold and platinum have been extensively used to control the characteristics of commercial devices. For this reason, the properties of gold and platinum in silicon are treated here.

The diffusion of gold and platinum into silicon occurs much more rapidly than does that of dopant atoms such as phosphorus or boron. Consequently, the diffusion of these impurities is always performed after the fabrication of the device structure but prior to metallization. To obtain a homogeneous distribution of the recombination centers throughout the wafer, it is necessary to perform the diffusion at temperatures between 800 and 900°C. The diffusion temperature determines the solid solubility of the impurity atoms in silicon and can be used to control the impurity density. A higher gold or platinum diffusion temperature should be used to obtain a lower lifetime. It must be kept in mind that because these diffusion temperatures exceed the eutectic point between silicon and the aluminum metallization used for power devices, the lifetime control must precede device metallization. Consequently, the devices cannot be tested immediately prior to or after impurity diffusion. Furthermore, small changes in the diffusion temperature are observed to cause a large variation in the device characteristics. These problems can be overcome by radiation processes through application of high-energy particles.

The bombardment of silicon with high-energy particles has been demonstrated to create lattice damage. This lattice damage consists of silicon atoms displaced from their lattice sites into interstitial positions, leaving behind vacancies. It has been found that these vacancies are highly mobile in silicon even at low temperatures. Consequently, the defects that remain after irradiation are composed of complexes of vacancies with impurity atoms such as phosphorus or oxygen and of two adjacent vacancy sites in the lattice (called *divacancies*). In heavily phosphorus doped silicon, the phosphorus-vacancy defect become predominant. However, in the high-resistivity material used to fabricate power devices, the divacancy defect has been shown to be the dominant one in both P- and N-type silicon [34, 35].

For controlling lifetime in power devices, both electron irradiation and gamma irradiation have been shown to be effective processing techniques [36, 37]. The advantages of using these techniques for controlling the lifetime in power devices are (1) the irradiation can be performed at room temperature following complete device fabrication and initial testing of the characteristics; (2) the lifetime is controlled by the radiation dose, which can be accurately metered by monitoring the electron current during irradiation; (3) a much tighter distribution in the device characteristics can be achieved because of the improved control over the recombination center density; (4) the irradiation process allows trimming of the device characteristics to the desired value by using several irradiation steps because the radiation is performed after complete device fabrication, thus allowing device testing between consecutive irradiations; (5) the irradiation damage can be annealed out by heating the devices above 400°C, thus allowing the recovery of devices that may have had an overdose during irradiation; and (6) the irradiation process is a cleaner and simpler process than impurity diffusion and avoids any possible contamination between the processing of devices requiring high lifetime with those requiring low lifetime. These advantages have made the electron irradiation process very attractive for manufacturing power devices. However, the application of gold diffusion is still in use today for two reasons: (1) the nature of the recombination centers introduced by gold diffusion are different from those produced by electron irradiation, and superior conduction characteristics are observed in gold doped devices; and (2) the annealing of electron irradiation induced defects at a few hundred degrees above room temperature has been of concern with regard to the long-term stability of the power devices. Annealing studies of the defect levels produced by electron irradiation in silicon have, however, demonstrated that, although the divacancy does anneal out at 300°C, the lifetime does not change appreciably because of the creation of a new recombination center during the annealing [34]. Thus electron irradiation does not result in long-term instability under typical device operating conditions.

Since gold–platinum diffusion and electron–gamma irradiation are being used today to control the characteristics of power devices, the recombination centers introduced by these techniques in silicon are provided in Fig. 2.34. It has been found that gold diffusion into silicon introduces an acceptor and a donor deep level in the energy gap [38]. The center controlling the minority carrier lifetime (i.e., the dominant center) in N-type silicon is the one at 0.54 eV below the conduction band. In the case of platinum diffusion into silicon, four deep levels have been found [39]. Of these levels, the one lying at 0.42 eV above the valency band edge has been found to control the minority carrier lifetime in N-type silicon. In the case of electron and gamma irradiation, three deep levels have been observed. The divacancy induced deep level at 0.40 eV below the conduction band has been shown to control the minority carrier lifetime in

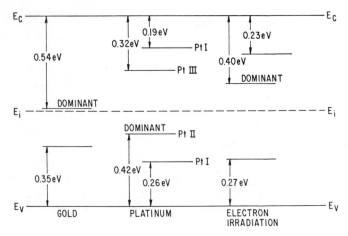

Fig. 2.34. Recombination levels produced by gold doping, platinum doping, and electron irradiation.

N-type silicon [34]. For calculation of the lifetime variation with resistivity and temperature, it is necessary to know the capture cross section for holes and electrons at these dominant recombination centers. These are provided in Table 2.1, together with the capture cross-section ratios ζ.

With the data provided in Fig. 2.34 and Table 2.1, the ratio (τ_{HL}/τ_{LL}) can be calculated for each of the preceding lifetime control techniques by assuming that only the dominant recombination center is present. This is a reasonable approximation for these multiple deep-level centers because in all three cases the level closest to midgap is found to control the recombination and generation rates. The calculated variation of the ratio (τ_{HL}/τ_{LL}) with resistivity is shown in Fig. 2.35 for several temperatures. Note that the ratio (τ_{HL}/τ_{LL}) is the lowest for electron irradiation and the highest for gold doping. From this calculation it can be concluded that gold doping will provide a superior trade-off between

TABLE 2.1. Dominant Levels of Gold, Platinum, and Electron Irradiation-Induced Deep Levels in Silicon and Capture Cross Sections for Holes and Electrons for N-type Silicon

Impurity	Energy-Level Position (eV)	Capture Cross Section		Ratio ζ
		Holes (cm^2)	Electrons (cm^2)	
Gold	$E_v + 0.56$	6.08×10^{-15}	7.21×10^{-17}	69.70
Platinum	$E_v + 0.42$	2.70×10^{-12}	3.20×10^{-14}	69.80
Electron irradiation	$E_v + 0.71$	8.66×10^{-16}	1.62×10^{-16}	4.42

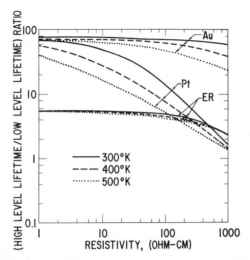

Fig. 2.35. Calculated high:low-level lifetime ratio as a function of resistivity for gold, platinum, and electron irradiation.

current conduction and switching speed when compared with electron irradiation. Experimental measurements performed using power rectifiers confirm the results of these calculations [36].

To evaluate the impact of these lifetime control methods, it is useful to calculate the ratio of the space-charge generation lifetime to the high-level lifetime. The calculated ratio (τ_{SC}/τ_{HL}) for each of the three cases is shown as a function of temperature in Fig. 2.36. The calculated ratio (τ_{SC}/τ_{HL}) for electron irradiation and for platinum doping are many orders of magnitude larger than for gold doping, especially at lower temperatures. For equal forward conduction characteristics as controlled by the high-level lifetime, the leakage current of electron-irradiated and platinum-doped devices will, therefore, be much lower than for gold-doped devices. Experimental measurements performed on rectifiers have confirmed the results of these calculations [36]. On the basis of the theoretical calculations and the experimental data, it can be concluded that the advantages of electron irradiation, in terms of lower leakage currents, superior uniformity, and processing simplicity make it a preferred technique for lifetime control.

2.3.5. Auger Recombination

The Auger recombination process occurs by the transfer of the energy and momentum released by the recombination of an electron–hole pair to a third particle that can be either an electron or a hole. This process becomes significant in heavily doped P- and N-type silicon such as the diffused end regions of power devices. It is also an important effect in determining recombination rates in the lightly doped base regions of power devices operating at high injection

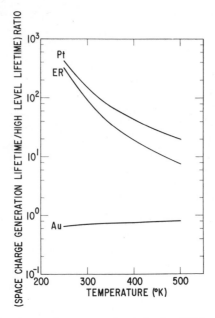

Fig. 2.36. Calculated ratio of space-charge generation to low-level lifetime ratio for gold, platinum, and electron irradiation.

levels during forward conduction because of the simultaneous presence of a high concentration of holes and electrons injected into this region. Theoretical investigations of Auger recombination [40, 41] in indirect-gap semiconductors indicate that the process can occur with or without phonon assistance, and may occur directly from band to band or through traps.

Measurements of Auger recombination have been performed by observing the decay of excess minority carriers at low injection levels in heavily doped N- and P-type silicon and the decay of excess free carriers at high injection levels in high-resistivity silicon [42, 43]. The measured Auger coefficients for these cases are not equal because of the difference in the structure of the conduction and the valence bands. In the case of heavily doped N-type silicon, the Auger process transfers energy and momentum to an electron in the conduction band and the Auger lifetime is given by

$$\tau_A^N = \frac{1}{2.8 \times 10^{-31} \, n^2} \tag{2.76}$$

where n is the majority carrier (electron) concentration. In the case of P-type silicon, the Auger process occurs by the transfer of energy and momentum to a hole in the valence band and the measured Auger lifetime is given by

$$\tau_A^P = \frac{1}{1 \times 10^{-31} \, p^2} \tag{2.77}$$

where p is the majority carrier (hole) concentration. For Auger recombination occurring at high injection levels where the density of electrons and holes is simultaneously large, the Auger lifetime, as measured in heavily excited silicon by laser radiation, is given by

$$\tau_A^{\Delta n} = \frac{1}{3.4 \times 10^{-31}(\Delta n)^2} \tag{2.78}$$

where Δn is the excess carrier concentration. The variation of the Auger lifetime with carrier concentration is shown in Fig. 2.37 for all three cases. These plots are extended to doping levels of up to 2×10^{20} for the heavily doped cases since such high concentrations are prevalent in the diffused regions of power devices. The extremely low Auger lifetimes at the high doping concentrations have a strong influence on the injection efficiency and recombination in the end regions of power devices. The Auger lifetime for the case of high injected carrier densities in lightly doped regions is even lower than in the case of heavily doped

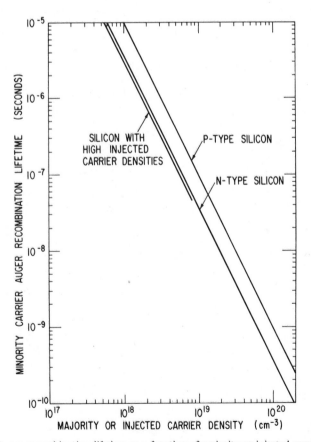

Fig. 2.37. Auger recombination lifetime as a function of majority or injected carrier density.

regions because the Auger processes requiring two electrons plus a hole and two holes plus an electron can occur simultaneously. At surge current conditions, the injected carrier concentration near the junctions of power devices can exceed $10^{18}/cm^3$. Since these devices are usually processed to achieve a high Shockley–Read–Hall lifetime (usually in excess of 10 μsec), it can be seen from Fig. 2.37 that the Auger recombination process can play a significant role in affecting current conduction and switching speed at high injection levels. It should be noted that the Auger recombination process occurs simultaneously with the Shockley–Read–Hall recombination process. The effective lifetime is then given by

$$\frac{1}{\tau_{\text{eff}}} = \frac{1}{\tau_{\text{SRH}}} + \frac{1}{\tau_{\text{A}}} \tag{2.79}$$

where τ_{SRH} is the Shockley–Read–Hall recombination lifetime discussed in Section 2.3.1.

REFERENCES

1. B. J. Baliga, "Semiconductors for high voltage vertical channel, field effect transistors," *J. Appl. Phys.*, **53**, 1759–1764 (1982).
2. C. Canali, C. Jacoboni, F. Nava, G. Ottaviani, and A. A. Quaranta, "Electron drift velocity in silicon," *Phys. Rev.*, **B12**, 2265–2284 (1975).
3. G. Ottaviani, L. Reggiani, C. Canali, F. Nava, and A. A. Quaranta, "Hole drift velocity in silicon," *Phys. Rev.*, **B12**, 3318–3329 (1975).
4. F. J. Morin and J. P. Maita, "Electrical properties of silicon containing arsenic and boron," *Phys. Rev.*, **96**, 28–35 (1954).
5. C. Jacoboni, C. Canali, G. Ottaviani, and A. A. Quaranta, "A review of some charge transport properties of silicon," *Solid State Electron.*, **20**, 77–89 (1977).
6. C. Canali, G. Majni, R. Minder, and G. Ottaviani, "Electron and hole drift velocity measurements in silicon," *IEEE Trans. Electron Devices*, **ED-22**, 1045–1047 (1975).
7. N. J. Murphy, F. Berz, and I. Flinn, "Carrier mobility in silicon MOST's," *Solid State Electron.*, **12**, 775–786 (1969).
8. A. A. Guzev, G. L. Kurishev, and S. P. Sinitsa, "Scattering mechanisms in inversion channels of MIS structures on silicon," *Phys. Status Solidi*, **A14**, 41–50 (1972).
9. S. C. Sun and J. D. Plummer, "Electron mobility in inversion and accumulation layers on thermally oxidized silicon surfaces," *IEEE Trans. Electron Devices*, **ED-27**, 1497–1508 (1980).
10. Y. C. Cheng and E. A. Sullivan, "Relative importance of phonon scattering to carrier mobility in silicon surface layer at room temperature," *J. Appl. Phys.*, **44**, 3619–3625 (1973).
11. O. Leistiko, A. S. Grove, and C. T. Sah, "Electron and hole mobilities in inversion layers on thermally oxidized silicon surfaces," *IEEE Trans. Electron Devices*, **ED-12**, 248–254 (1965).
12. T. Sato, Y. Takeishi, H. Hara, and Y. Okamoto, "Mobility anisotropy of electrons in inversion layers on oxidized, silicon surfaces," *Phys. Rev.*, **B-4**, 1950–1960 (1971).

13. A. Ohwada, H. Maeda, and T. Tanaka, "Effect of the crystal orientation upon electron mobility at the Si/SiO$_2$ interface," *Jpn. J. Appl. Phys.*, **8**, 629–630 (1969).

14. T. Sato, Y. Takeishi, H. Tango, H. Ohnuma, and Y. Okamoto, "Drift-velocity saturation of holes in silicon inversion layers," *J. Phys. Soc. Jpn.*, **31**, 1846 (1971).

15. F. F. Fang and A. B. Fowler, "Hot electron effects and saturation velocities in silicon inversion layers," *J. Appl. Phys.*, **41**, 1825–1831 (1970).

16. G. G. McFarlane, J. P. McLean, J. E. Quarrington, and V. Roberts, "Fine structure in the absorption edge spectrum of silicon," *Phys. Rev.*, **111**, 1245–1254 (1958).

17. H. P. D. Lanyon and R. A. Tuft, "Bandgap narrowing in heavily doped silicon," *IEEE Int. Electron Devices Meeting Digest*, Abstract 13.3, pp. 316–319 (1978).

18. J. W. Slotboom and H. C. DeGraf, "Measurements of bandgap narrowing in silicon bipolar transistors," *Solid State Electron.*, **19**, 857–862 (1976).

19. A. W. Wieder, "Arsenic emitter effects," *IEEE International Electron Devices Meeting Digest*, Abstract 18.7, pp. 460–462 (1978).

20. R. Mertens and R. J. Van Overstraeten, "Measurement of the minority carrier transport parameters in heavily doped silicon," *IEEE International Electron Devices Meeting Digest*, Abstract 13.4, pp. 320–323 (1978).

21. G. E. Possin, M. S. Adler, and B. J. Baliga, "Measurements of the *pn* product in heavily doping epitaxial emitters," *IEEE Trans. Electron Devices*, **ED-31**, 3–17 (1984).

22. F. Mousty, P. Ostoja, and L. Passari, "Relationship between resistivity and phosphorus concentration in silicon," *J. Appl. Phys.*, **45**, 4576–4580 (1974).

23. D. M. Caughey and R. F. Thomas, "Carrier mobilities in silicon empirically related to doping and field," *Proc. IEEE*, **55**, 2192–2193 (1967).

24. M. Tanenbaum and A. D. Mills, "Preparation of uniform resistivity *n*-type silicon by nuclear transmutation," *J. Electrochem. Soc.*, **108**, 171–176 (1961).

25. H. M. Janus and O. Malmos, "Application of thermal neutron irradiation for large scale production of homogeneous phosphorus doping of floatzone silicon," *IEEE Trans. Electron Devices*, **ED-23**, 797–802 (1976).

26. J. M. Meese, Ed., *Neutron Transmutation Doping in Semiconductors*, Plenum, New York, 1979.

27. R. N. Hall, "Electron–hole recombination in germanium," *Phys. Rev.*, **87**, 387 (1952).

28. W. Shockley and W. T. Read, "Statistics of the recombination of holes and electrons," *Phys. Rev.*, **87**, 835–842 (1952).

29. C. T. Sah, R. N. Noyce, and W. Shockley, "Carrier generation and recombination in *P–N* junctions and *P–N* junction characteristics," *Proc. IRE*, **45**, 1228–1243 (1957).

30. B. J. Baliga, "Recombination level selection criteria for lifetime reduction in integrated circuits," *Solid State Electron.*, **21**, 1033–1038 (1978).

31. B. J. Baliga and S. Krishna, "Optimization of recombination levels and their capture cross-section in power rectifiers and thyristors," *Solid State Electron.*, **20**, 225–232 (1977).

32. B. J. Baliga, "Technological constraints upon the properties of deep levels used for lifetime control in the fabrication of power rectifiers and thysistors," *Solid State Electron.*, **20**, 1029–1032 (1977).

33. A. G. Milnes, *Deep Impurities in Semiconductors*, Wiley, New York, 1973.

34. A. O. Evwaraye and B. J. Baliga, "The dominant recombination centers in electron-irradiated semiconductor devices," *J. Electrochem. Soc.*, **124**, 913–916 (1977).

35. B. J. Baliga and A. O. Evwaraye, "Correlation of lifetime with recombination centers in electron irradiated *P*-type silicon," *J. Electrochem. Soc.*, **130**, 1916–1918 (1983).

36. B. J. Baliga and E. Sun, "Comparison of gold, platinum, and electron irradiation for controlling lifetime in power rectifiers," *IEEE Trans. Electron Devices*, **ED-24**, 685–688 (1977).

37. R. O. Carlson, Y. S. Sun, and H. B. Assalit, "Lifetime control in silicon power devices by electron or gamma irradiation," *IEEE Trans. Electron Devices*, **ED-24**, 1103–1108 (1977).

38. J. M. Fairfield and B. V. Gokhale, "Gold as a recombination center in silicon," *Solid State Electron.*, **8**, 685–691 (1965).

39. K. P. Lisiak and A. G. Milnes, "Platinum as a lifetime control deep impurity in silicon," *J. Appl. Phys.*, **46**, 5229–5235 (1975).

40. L. Huldt, "Band-to-band Auger recombination in indirect gap semiconductors," *Phys. Status Solidi*, **A8**, 173–187 (1971).

41. A. Haug, "Carrier density dependence of Auger recombination," *Solid State Electron.*, **21**, 1281–1284 (1978).

42. J. Dziewior and W. Schmid, "Auger coefficients for highly doped and highly excited silicon," *Appl. Phys. Lett.* **31**, 346–348 (1977).

43. K. G. Svantesson and N. G. Nilson, "Measurement of Auger recombination in silicon by laser excitation," *Solid State Electron.*, **21**, 1603–1608 (1978).

44. P. U. Calzolari and S. Graffi, "A theoretical investigation of the generation current in silicon *pn* junctions under reverse bias," *Solid State Electron.*, **15**, 1003–1011 (1972).

PROBLEMS

2.1. Determine the ratio of the mobility at low doping concentrations to that at very high doping concentrations for electrons and holes in silicon.

2.2. Create a plot of the ratio of the mobility for electrons to that for holes as a function of doping concentration.

2.3. Consider a *p*-type semiconductor region with a doping concentration of $10^{17}/cm^3$. A gate oxide of 1000 Å is grown on its surface by dry oxidation. A metal electrode is deposited on the oxide. What is the maximum effective mobility in an inversion layer at the surface assuming a fixed charge of $3 \times 10^{11}/cm^2$? What is the effective surface mobility in the inversion layer when the surface electric field is 2×10^5 V/cm? This is representative of the situation in a typical power MOS device.

2.4. Calculate the intrinsic carrier concentration at 300 K under low doping concentrations.

2.5. Determine the intrinsic resistivity of silicon at 300 and 400 K.

2.6. Calculate the minority carrier density at 300 K for homogeneously doped silicon with doping concentrations of 1×10^{15}, 1×10^{17}, 1×10^{18}, 1×10^{19} and $1 \times 10^{20}/cm^3$ with and without the influence of band gap narrowing. What is the increase in minority carrier density due to this phenomenon?

2.7. What is the radiation time required to obtain 50 $\Omega \cdot cm$, n-type silicon in a reactor with thermal neutron flux density of $10^{14}/cm^2 \cdot sec$ by using the neutron transmutation doping process?

2.8. Compare the resistivity variation in neutron transmutation doped silicon with a resistivity of 50 $\Omega \cdot cm$ produced by using p- and n-type starting material with a resistivity of 1000 $\Omega \cdot cm$ with a resistivity variation of $\pm 15\%$.

2.9. Consider an n-type silicon region with a doping level of $10^{15}/cm^3$ containing 10^{13} atoms of gold/cm^3. Calculate the minority carrier low-level lifetime, high-level lifetime, and the space-charge generation lifetime at 300 K, assuming that a single (dominant) lifetime control center exists. What is the increase in lifetime with injection level?

2.10. Repeat Problem 2.9 for the case of platinum.

2.11. Repeat Problem 2.9 for the case of electron irradiation-induced defect center concentration of $10^{13}/cm^3$.

2.12. Consider an n-type silicon region with a doping concentration of $10^{19}/cm^3$ containing 10^{13} atoms of gold/cm^3. Calculate the minority carrier lifetime, including the effect of Auger recombination.

2.13. Consider an n-type silicon region with a doping concentration of $10^{14}/cm^3$ containing 10^{13} atoms of gold/cm^3. Calculate the minority carrier lifetime when the injected carrier density is 10^{16}, 10^{17}, and $10^{18}/cm^3$.

3 BREAKDOWN VOLTAGE

The breakdown voltage of a power semiconductor device is one of its most important characteristics. Together with its maximum current handling capability, this parameter determines the power rating of the device. Depending on the application of the power device, its breakdown voltage can range from as low as 25 V for high-speed output rectifiers used in switching power supplies for integrated circuits to over 6000 V for thyristors used in high-voltage DC transmission networks. In these devices, the voltage is supported by a depletion layer formed across either a $P–N$ junction, a metal–semiconductor (Schottky barrier) interface, or a metal oxide–semiconductor (MOS) interface. The high electric field that exists across the depletion layer is responsible for sweeping out any holes or electrons that enter this region, by either the process of space-charge generation or diffusion from the neighboring quasi-neutral regions. As more voltage is applied across the depletion layer, the electric field increases and the mobile carriers are accelerated to higher velocities. When the electric field exceeds 1×10^5 V/cm, the mobile carriers attain a saturated drift velocity of about 1×10^7 cm/sec. At higher electric fields, these carriers have sufficient energy that their collisions with the atoms in the lattice can excite valence band electrons into the conduction band. This process for the generation of electron–hole pairs is called *impact ionization*. Since the electron–hole pairs created in the depletion layer by the impact ionization process undergo acceleration by the existing electric field, they also participate in the creation of further electron–hole pairs. Consequently, impact ionization is a multiplicative phenomenon that leads to a cascade of mobile carriers transported through the depletion layer. The device is considered to undergo avalanche breakdown when the rate of impact ionization approaches infinity because it cannot support an increase in applied voltage. Avalanche breakdown represents a fundamental limitation to the maximum operating voltage of power devices.

The physics of the avalanche breakdown process is analyzed in this chapter. Its impact on the maximum voltage supported by abrupt and graded junctions is then discussed. Most of the power devices described in this book are based on shallow junction technology borrowed from integrated-circuit processing experience. For many of these devices, an abrupt junction approximation to the diffused junction is adequate for design and analysis. Using a low surface concentration diffusion to achieve a graded junction is also important because it can lead to as much as a 25% increase in breakdown voltage for the same background doping level.

The treatment of abrupt or graded junctions requires the assumption that these junctions are semi-infinite. In practical cases, these junctions must be terminated at the edges of the power devices. The art of junction termination has recently evolved into a science with the use of two-dimensional solution of Poisson's equation to characterize the electric field distribution at the termination. A large variety of junction terminations are reviewed in this chapter with consideration for their technological complexity. In addition, a comparison of the different approaches is provided to aid the application of these techniques to various devices.

3.1. AVALANCHE BREAKDOWN

Avalanche breakdown is defined as the condition under which the impact ionization process attains an infinite rate. Impact ionization results in the generation of electron–hole pairs during the transport of mobile carriers through the depletion layer. To characterize this process, it is useful to define ionization coefficients. The impact ionization coefficient for holes (α_p) is defined as the number of electron–hole pairs created by a hole traversing 1 cm through the depletion layer along the direction of the electric field. The impact ionization coefficient for electrons (α_n) is similarly defined as the number of electron–hole pairs created by an electron traversing 1 cm through the depletion layer along the direction of the electric field. Extensive measurements of these ionization coefficients for silicon have been conducted [1, 2]. These measurements indicate that

$$\alpha_n = a_n \, e^{-b_n/\mathscr{E}} \tag{3.1}$$

and

$$\alpha_p = a_p \, e^{-b_p/\mathscr{E}} \tag{3.2}$$

where $a_n = 7 \times 10^5/\text{cm}$, $b_n = 1.23 \times 10^6$ V/cm for electrons, and $a_p = 1.6 \times 10^6/\text{cm}$, $b_p = 2 \times 10^6$ V/cm for holes. These expressions apply for electric fields ranging from 1.75×10^5 to 6×10^5 V/cm. Plots of α_n and α_p as a function of the electric field $\mathscr{E}$ are shown in Fig. 3.1. Note the extremely rapid increase in the impact ionization coefficient with increasing electric field. This is a very important factor during the analysis of breakdown in power devices.

In many cases, an approximation to the ionization coefficient [3]

$$\alpha = 1.8 \times 10^{-35} \, \mathscr{E}^7 \tag{3.3}$$

is found to be useful. In a later section of this chapter, this approximation is

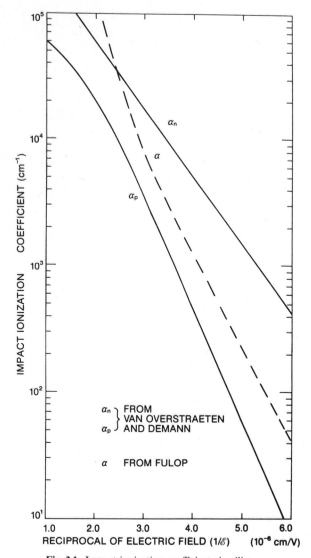

Fig. 3.1. Impact ionization coefficients in silicon.

shown to allow closed-form solutions to the avalanche breakdown voltage of abrupt junctions as well as for the case of cylindrical and spherical junction terminations. The dependence of the ionization coefficients for holes and electrons on the electric field according to this equation is also shown in Fig. 3.1 for comparison with the more accurate expressions.

Because of the strong dependence of the ionization coefficients on the electric field, when the avalanche breakdown of devices is analyzed by application of numerical integration techniques, the avalanche multiplication calculations are performed by selecting paths through points in the device structure containing the highest electric fields. This approach has been found to be effective in accurately predicting the avalanche breakdown of junctions for a large variety of terminations while avoiding the complexity of analyzing all possible avalanche multiplication paths.

To compute the avalanche breakdown voltage, it is necessary to determine the condition under which the impact ionization achieves an infinite rate. Consider an N^+-P junction with a reverse-bias voltage applied across it as illustrated in Fig. 3.2. Since the doping level on the N^+ side is significantly higher than on the P side, the depletion layer extends primarily in the lightly doped P material. To perform the avalanche breakdown analysis, assume that an electron–hole pair is generated at a distance x from the junction. In the presence of the electric field $\mathscr{E}$, the hole will be swept toward the P region and the electron will be swept toward the N^+ region. When traveling a distance dx, the hole will create $(\alpha_p\, dx)$ electron–hole pairs and the electron will create $(\alpha_n\, dx)$ electron–hole pairs. The total number of electron–hole pairs created in the depletion layer due to the single electron–hole pair initially generated at a distance x from the junction is then given by

$$M(x) = 1 + \int_0^x \alpha_n M(x)\, dx + \int_x^W \alpha_p M(x)\, dx \qquad (3.4)$$

where W is the depletion-layer width. A solution of this differential equation is

$$M(x) = M(0) \exp\left[\int_0^x (\alpha_n - \alpha_p)\, dx \right] \qquad (3.5)$$

where $M(0)$ is the total number of electron–hole pairs at the edge of the depletion

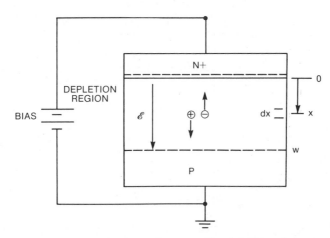

Fig. 3.2. Depletion layer of a reverse-biased P–N junction.

layer. Substituting this expression into Eq. (3.4) for the case of $x = 0$ and solution for $M(0)$ gives

$$M(0) = \left\{ 1 - \int_0^W \alpha_p \exp\left[\int_0^x (\alpha_n - \alpha_p)\, dx \right] dx \right\}^{-1} \tag{3.6}$$

Use of this expression in Eq. (3.5) gives

$$M(x) = \frac{\exp\left[\int_0^x (\alpha_n - \alpha_p)\, dx \right]}{1 - \int_0^W \alpha_p \exp\left[\int_0^x (\alpha_n - \alpha_p)\, dx \right] dx} \tag{3.7}$$

This equation can be used to compute the total number of electron–hole pairs created as a result of the generation of a single electron–hole pair at a distance x from the junction if the electric field distribution along the impact ionization path is known. Avalanche breakdown is defined to occur when the total number of electron–hole pairs $M(x)$, also commonly known as the *multiplication coefficient*, tends to infinity. From Eq. (3.7) it can be concluded that this will occur when

$$\int_0^W \alpha_p \exp\left[\int_0^x (\alpha_n - \alpha_p)\, dx \right] dx = 1 \tag{3.8}$$

The expression on the left-hand side of Eq. (3.8) is known as the *ionization integral*. With use of the approximation for the ionization coefficient given by Eq. (3.3), the ionization integral simplifies to

$$\int_0^W \alpha\, dx = 1 \tag{3.9}$$

Calculation of the breakdown voltage of devices is generally performed by evaluation of this integral either in closed form or by numerical techniques. However, some power device structures contain inherent parasitic transistor structures. For analysis of the breakdown voltage of these structures, it is necessary to include the current amplification characteristics of the transistor structure. In these cases, the evaluation of the ionization integral alone cannot be used. The multiplication coefficient must instead be calculated.

It has been demonstrated that the reverse current of a P^+–N diode approaches infinity at a steeper rate with increasing reverse-bias voltage when compared with a N^+–P diode [3]. The variation of the multiplication coefficient as a function of the reverse-bias voltage is given by

$$M_n = \frac{1}{[1 - (V/V_A)^4]} \tag{3.10}$$

for P^+–N diodes and

$$M_p = \frac{1}{[1 - (V/V_A)^6]} \tag{3.11}$$

for N^+–P diodes. Here V_A is the avalanche breakdown voltage and V is the applied reverse-bias voltage. These expressions are useful for obtaining closed-form analytical solutions to the breakdown voltage of devices containing parasitic bipolar transistors. For purposes of numerical analysis it is more accurate to use Eq. (3.7) with the appropriate electric field distribution in the device structure.

3.2. ABRUPT JUNCTION DIODE

Consider an abrupt junction diode in which the doping concentration on one side of the junction is very high compared to the homogeneously doped other side. Devices fabricated by use of shallow junctions can be represented by the abrupt junction case, especially when the doping level of the substrate is low. In these cases the depletion layer extends primarily into the lightly doped side.

For the case of a parallel-plane, abrupt N^+-P junction, with a reverse-bias voltage V_a applied, since the depletion layer extends only into the p side as a result of the very high doping level on the N^+ side, Poisson's equation needs to be solved only for the P side:

$$\frac{d^2V}{dx^2} = -\frac{d\mathscr{E}}{dx} = -\frac{Q(x)}{\epsilon_s} = \frac{qN_A}{\epsilon_s} \qquad (3.12)$$

where $Q(x)$ is the charge in the depletion layer due to ionized acceptors, ϵ_s is the dielectric constant of the semiconductor, q is the electronic charge, and N_A is the homogeneous acceptor doping density. Integration of Eq. (3.12) and use of the boundary condition that $\mathscr{E}(W) = 0$ gives the electric field distribution:

$$\mathscr{E}(x) = \frac{qN_A}{\epsilon_s}(W - x) \qquad (3.13)$$

The electric field varies linearly with distance as shown in Fig. 3.3. Integration of Eq. (3.13) results in the voltage distribution:

$$V(x) = \frac{qN_A}{\epsilon_s}\left(Wx - \frac{x^2}{2}\right) \qquad (3.14)$$

The potential varies quadratically with distance as illustrated in Fig. 3.3. From the boundary condition that the potential at $x = W$ must equal the applied reverse-bias voltage V_a, it can be shown that

$$W = \sqrt{\frac{2\epsilon_s V_a}{qN_A}} \qquad (3.15)$$

The maximum electric field in the abrupt junction case occurs at $x = 0$. Using Eq. (3.15) in Eq. (3.13), the maximum electric field is found to be given by

$$\mathscr{E}_m = \sqrt{\frac{2qN_A}{\epsilon_s}V_a} \qquad (3.16)$$

To evaluate the breakdown voltage of the abrupt junction diode, it is necessary to solve the ionization integral given by Eq. (3.8) using the electric field distribution given by Eq. (3.13). Numerical calculations of the breakdown voltages

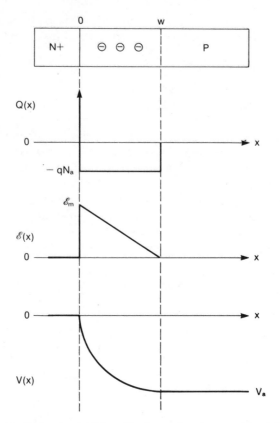

Fig. 3.3. Electric field and potential profiles for an abrupt reverse-biased *P–N* junction.

of abrupt junctions have been performed by using the measured ionization co-efficients [1]. Good agreement with the breakdown voltage of experimental diodes has been observed [4].

A closed-form analytical expression for the breakdown voltage of abrupt junction diodes can be derived by using Eq. (3.3) for the ionization coefficient. The ionization integral must be evaluated by using the electric field distribution given in Eq. (3.13):

$$\int_0^W 1.8 \times 10^{-35} \left[\frac{qN_A}{\epsilon_s} (W - x) \right]^7 dx = 1 \qquad (3.17)$$

From this equation, the depletion layer width at breakdown for the parallel-plane junction can be obtained:

$$W_{c,PP} = 2.67 \times 10^{10} \, N_A^{-7/8} \qquad (3.18)$$

Using this depletion width in Eq. (3.14) for the case of $x = W$, the avalanche breakdown voltage of the abrupt junction diode is found to be given by

$$BV_{PP} = 5.34 \times 10^{13} \, N_A^{-3/4} \qquad (3.19)$$

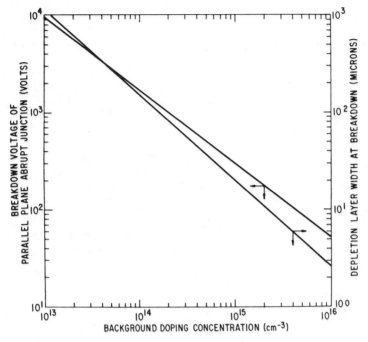

Fig. 3.4. Avalanche breakdown voltage and corresponding maximum depletion layer width for abrupt junction diodes.

where the letters PP are used to signify the parallel-plane or semi-infinite junction. The breakdown voltage and depletion layer width calculated from these analytical expressions are shown in Fig. 3.4. Another useful expression is that for the maximum electric field at breakdown, which occurs at $x = 0$:

$$\mathscr{E}_{c,PP} = 4010 \, N_A^{1/8} \tag{3.20}$$

This critical electric field is a useful parameter for determination of how close a device may be approaching avalanche breakdown. All the above expressions for the abrupt junction case will be shown to be useful normalization parameters for analysis of various junction terminations. These equations and the plots in Fig. 3.4 allow the determination of the background doping level and the minimum base width required to achieve the breakdown voltage being designed for any power device.

3.3. PUNCH-THROUGH DIODE

In the case of the abrupt junction diode treated in the previous section, it was assumed that the lightly doped side of the junction extends beyond the edge of the depletion layer under avalanche breakdown conditions. In bipolar devices, where the lightly doped side of the junction is flooded with minority carriers during current conduction, it is often preferable to achieve a desired breakdown voltage by using a punch-through structure. A comparison of the punch-through structure with the normal case designed with minimum depletion layer width can be performed from Fig. 3.5. In the punch-through case, the doping level of the lightly doped region is lower than that for the normal case. This causes a much slower rate of decrease in the electric field with distance as illustrated by the dashed lines in Fig. 3.5. In this figure, the variation in the critical electric field for breakdown with background doping level has been neglected. Since the breakdown voltage is given by the area under the electric field curves in Fig. 3.5, it is clear that the same avalanche breakdown voltage can be obtained for both cases by matching the shaded areas. This demonstrates that a narrower base region is needed in a punch-through diode to achieve a desired breakdown voltage.

The breakdown voltage of a punch-through diode can be derived by using the analysis presented for the normal diode to evaluate the electric field distribution. In the punch-through diode, the breakdown voltage is given by

$$V_{PT} = \tfrac{1}{2}(\mathscr{E}_c + \mathscr{E}_1)W_P \tag{3.21}$$

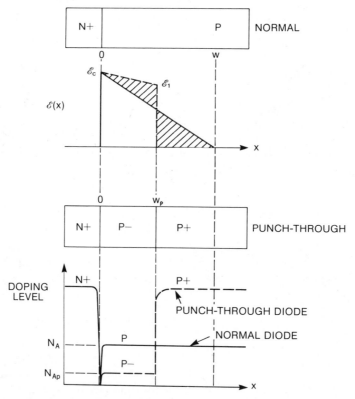

Fig. 3.5. Comparison of punch-through diode structure with ideal non-punch-through case.

where $\mathscr{E}_1$ is the electric field at distance W_P from the P–N junction. Using Eq. (3.13) and substituting for $\mathscr{E}_1$ in Eq. (3.21), it can be shown that

$$V_{PT} = \mathscr{E}_c W_P - \frac{1}{2}\frac{qN_{Ap}W_P^2}{\epsilon_s} \tag{3.22}$$

The breakdown voltage of punch-through diodes can be calculated by using this expression together with Eq. (3.20) for the critical electric field. The results obtained for several base region widths are provided in Fig. 3.6. Note that the breakdown voltage goes through a maxima when the doping level changes. This is due to a decrease in the critical electric field for breakdown when the doping level decreases. It can be seen that diodes with much narrower base regions can achieve a desired breakdown voltage in the punch-through case. For example, if a breakdown voltage of 1000 V is being designed, the normal diode would require a base width of 90 μm. In comparison, a punch-through diode with a base doping of $10^{14}/\text{cm}^3$ would require a base width of about 50 μm; that is, punch-through diodes can typically be half the width of normal diodes. This has

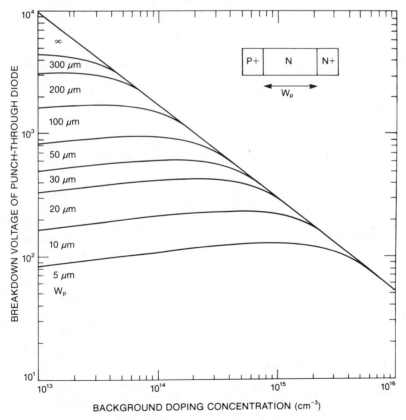

Fig. 3.6. Avalanche breakdown voltage of punch-through diodes.

important advantages in bipolar devices in terms of both current conduction and processing convenience.

3.4. LINEARLY GRADED JUNCTION

In the abrupt junction, the doping level on one side of the junction was much greater than on the other side, which was assumed to be homogeneously doped. In a linearly graded junction, the doping concentration is assumed to change linearly from N to P type throughout the extent of the depletion layer. For this type of junction it is useful to define a grade constant G that is the rate of change of the doping concentration with distance. The charge in the depletion layer is then given by

$$Q(x) = \pm qGx \tag{3.23}$$

where the sign of the charge is positive in the presence of ionized donors on the N side and negative in the presence of ionized acceptors on the P side of the junction. Poisson's equation for the linearly graded junction can be written as

$$\frac{d^2V}{dx^2} = -\frac{d\mathscr{E}}{dx} = -\frac{Q(x)}{\epsilon_s} = \frac{qGx}{\epsilon_s} \tag{3.24}$$

Integration of this equation once, using the boundary condition that the $\mathscr{E}(W) = 0$, gives the electric field distribution:

$$\mathscr{E}(x) = \frac{qG}{2\epsilon_s}(W^2 - x^2) \tag{3.25}$$

The electric field varies parabolically with distance for the linearly graded junction with the maximum electric field occurring at $x = 0$. The maximum electric field is given by

$$\mathscr{E}_m = \frac{qG}{2\epsilon_s}W^2 \tag{3.26}$$

The potential distribution can be obtained by integrating the electric field distribution and using the boundary condition that $V(-W) = 0$:

$$V(x) = \frac{qG}{\epsilon_s}\left(\frac{x^3}{6} - \frac{W^2x}{2} - \frac{W^3}{3}\right) \tag{3.27}$$

The voltage distribution in the linearly graded junction is a cubic function and has the form illustrated in Fig. 3.7. With use of the boundary condition that the voltage at $x = W$ must equal the applied bias V_a, Eq. (3.27) can be used to relate the depletion layer width to the applied voltage:

$$W = \left(\frac{3\epsilon_s V_a}{qG}\right)^{1/3} \tag{3.28}$$

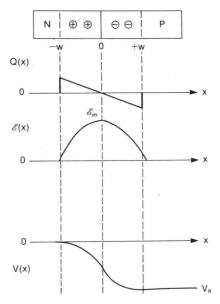

Fig. 3.7. Electric field and potential profiles for a linearly graded junction.

This depletion layer extends on both sides of the linearly graded junction to an equal amount. This must be allowed for in designing the contacts to both sides of the junction.

To solve for the breakdown voltage of the linearly graded junction, it is necessary to solve the ionization integral using the electric field distribution given by Eq. (3.25). By using the approximation given by Eq. (3.3) for the impact ionization coefficient, it is possible to obtain a closed-form solution for the breakdown voltage. Under these conditions, the ionization integral takes the form

$$\int_{-W}^{+W} 1.85 \times 10^{-35} \left[\frac{qG}{2\epsilon_s} (W^2 - x^2) \right]^7 dx = 1 \qquad (3.29)$$

From this equation, the depletion layer width at breakdown on each side of the junction can be derived:

$$W = 9.1 \times 10^5 \, G^{-7/15} \qquad (3.30)$$

The breakdown voltage of the linearly graded junction can be obtained by combining Eqs. (3.28) and (3.30):

$$V_{\text{LPP}} = 9.2 \times 10^9 \, G^{-2/5} \qquad (3.31)$$

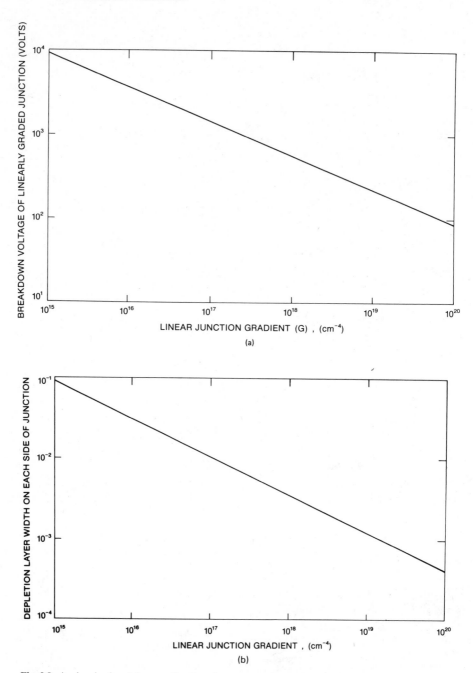

Fig. 3.8. Avalanche breakdown voltage and corresponding depletion layer width for a linearly graded junction.

The critical electric field for breakdown in these junctions can be obtained by combining Eqs. (3.26) and (3.30):

$$\mathscr{E}_c = 2.0 \times 10^4 \, G^{1/15} \tag{3.32}$$

The variation of these parameters with the grade constant is shown in Fig. 3.8. In general, the linearly graded junction offers higher breakdown voltage than the abrupt junction for the same range of doping levels because it can support voltage on both sides of the junction.

3.5. DIFFUSED JUNCTION DIODE

In practical devices, it is not feasible to fabricate either the abrupt junction or the linearly graded junction. By using alloying techniques to dope the semiconductor surface, it is possible to create a very sharp transition from the highly doped alloyed region into a homogeneously doped substrate. Alloying processes were used during early periods of device fabrication. This process is difficult to control and can lead to high stress in the semiconductor because of the presence of metal precipitates. (Note that device metallization relies on microalloying to avoid stress build up.) It is impractical to use alloying for the fabrication of modern power devices that require high-resolution patterns to achieve efficient power ratings. Modern device fabrication relies on either using diffusion with vapor transport to introduce dopants or by using ion implantation to introduce the dopant at the surface followed by high-temperature heat treatment to remove the lattice damage and redistribute the dopant. When the dopant is introduced into the semiconductor under conditions that create a constant surface doping concentration, the doping profile has the complementary error function form

$$N(x) = N_0 \, \text{erfc}\left(\frac{x}{d}\right) - N_B \tag{3.33}$$

where N_0 is the dopant surface concentration and N_B is the background doping level. Often the junctions are fabricated by ion implantation to introduce a precise number of dopants into the semiconductor followed by a thermal cycle to diffuse them to the appropriate depth. This process can be performed with great accuracy to fabricate devices with small features. Under these conditions, the doping profile takes a Gaussian shape:

$$N(x) = N_0 \, e^{-(x^2/d^2)} - N_B \tag{3.34}$$

In both these cases, the doping concentration rises rapidly on one side of the junction. Near the junction, where the concentration of the diffused dopant impurity approaches the background doping level, however, the net impurity density varies gradually as in the case of the linearly graded junction. Deeper into the semiconductor, compensation by the diffused impurity becomes negligible and the dopant density is homogeneous as in the case of the abrupt junction. Consequently, the diffused junction can be treated as being intermediate to the abrupt and linearly graded cases. It is not possible to obtain a closed-form analytical solution for the more complex profiles given in Eqs. (3.33) and (3.34) for the complementary error function and Gaussian cases.

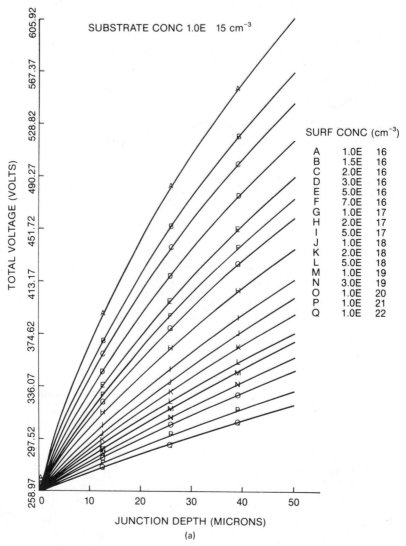

Fig. 3.9. Avalanche breakdown voltage and the depletion layer widths on the two sides of a diffused junction with a background doping level (lower side) of $1 \times 10^{15}/cm^3$. The upper side is heavily doped. (Figure is continued on next two pages.)

By using numerical techniques to solve the ionization integral, it is possible to calculate the breakdown voltage and the depletion layer widths on both sides of diffused junctions. A detailed compendium of breakdown voltage solutions has been published in Ref. 5. A typical example for the case of a background doping of $1 \times 10^{15}/cm^3$ is reproduced in Fig. 3.9. In general, the breakdown voltage of most practical junctions used to fabricate power devices is relatively insensitive to the surface concentration, and the depletion layer width on the

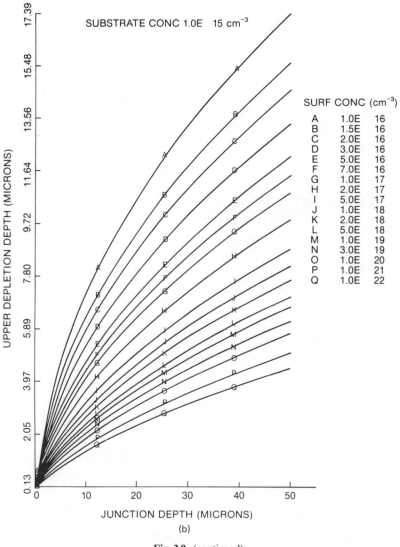

Fig. 3.9. (continued)

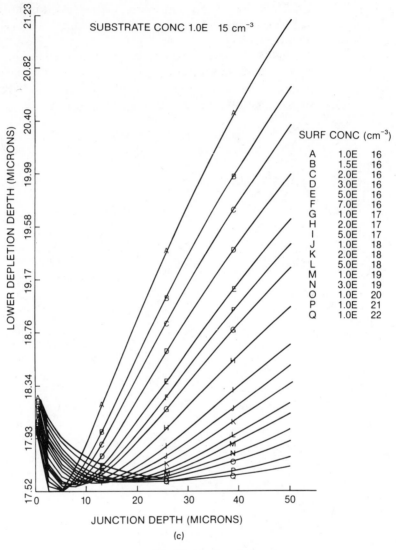

SUBSTRATE CONC 1.0E 15 cm^{-3}

SURF CONC (cm^{-3})

A	1.0E	16
B	1.5E	16
C	2.0E	16
D	3.0E	16
E	5.0E	16
F	7.0E	16
G	1.0E	17
H	2.0E	17
I	5.0E	17
J	1.0E	18
K	2.0E	18
L	5.0E	18
M	1.0E	19
N	3.0E	19
O	1.0E	20
P	1.0E	21
Q	1.0E	22

LOWER DEPLETION DEPTH (MICRONS)

JUNCTION DEPTH (MICRONS)

(c)

Fig. 3.9. (continued)

lightly doped side is much greater than on the diffused side. The impact of
changes in surface concentration and junction depth on the breakdown voltage
and depletion width can be better understood by using Fig. 3.10. In this figure,
two types of junction have been selected and their breakdown voltages and
depletion widths on both sides of the junction plotted as a function of the background doping concentration. The first type of diffused junction selected for
these plots has a high surface concentration of $10^{20}/cm^3$ as limited by dopant

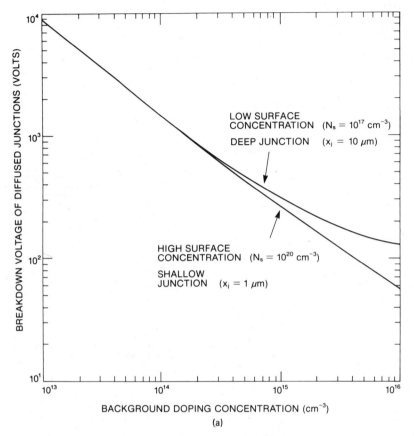

BACKGROUND DOPING CONCENTRATION (cm^{-3})

(a)

Fig. 3.10. Avalanche breakdown voltage and depletion layer widths on opposite sides of a diffused junction for background doping levels ranging from 10^{13} to 10^{16}/cm^3. (Continued on next page.)

solid solubility in silicon and a shallow depth of 1 μm. It is representative of an abrupt junction fabricated by use of diffusion techniques. The second type of diffused junction selected for the plots has a relatively low surface concentration of 10^{17}/cm^3 and a relatively large depth of 10 μm. It is representative of a linearly graded junction fabricated by using diffusion techniques. The spread between the curves in Fig. 3.10 provides an estimate of the variation in breakdown voltage that can be achieved by altering the diffusion profile.

3.6. JUNCTION TERMINATION

In previous sections, it has been assumed that the junctions are semi-infinite and no edge effects were considered. For practical devices, it becomes necessary to consider edge effects to obtain a realistic design. In fact, the edge termination

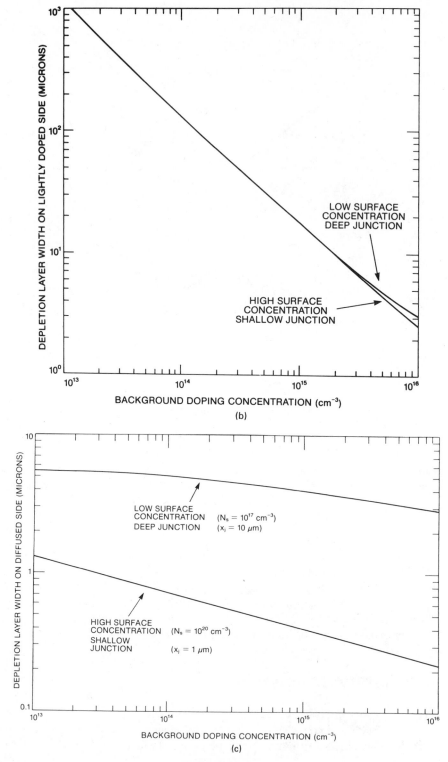

Fig. 3.10. (continued)

limits the breakdown voltage of practical devices to below the "ideal" limits set by the semi-infinite junction analysis. If the junction is poorly terminated, its breakdown voltage can be as low as 10–20% of the ideal case. This severe degradation in breakdown voltage can seriously compromise the device design and lead to reduced current ratings as well.

The study of device terminations is an intensively researched topic that has resulted in the development of a large variety of innovative designs. Early work on large-area power rectifiers and thyristors was based on shaping the edges of the devices by using lapping, sawing, and sandblasting techniques. These methods can be classified as beveled-edge terminations. A variant to this approach utilizes the removal of silicon using chemical etches. These terminations are known as *etched contours*.

With the advent of planar diffusion technology, it became practical to fabricate a large number of devices on a single wafer by selective diffusion of impurities through a silicon dioxide masking layer. This process creates cylindrical junctions at the mask edges and spherical junctions at the sharp corners of the diffusion window. The impact of junction curvature had to be taken into account during high-voltage device design. When it was discovered that this curvature results in a severe reduction in the breakdown voltage, methods to reduce the depletion layer curvature were investigated. Among these, the use of floating field rings and equipotential field plates has been found to be very useful. More recently, the widespread use of ion implantation for power device fabrication has allowed the tailoring of the charge at the edges of devices for improving the electric field distribution. This technique has been called *junction termination extension*.

In this section, each of the commonly used device termination techniques is discussed. Both the theoretical background for each approach and its practical aspects are considered. These techniques are then compared from the point of view of application to various types of power devices.

3.6.1. Planar Diffused Terminations

In the case of very high current rectifiers and thyristors where single devices are made from an entire wafer, it is common practice to use beveling techniques to terminate the edges. This approach to device termination is discussed later. For lower current devices more typical of the cases dealt with in this book, a large number of devices are fabricated in each wafer. In these cases, it is impractical to use the beveling approach. The most widely used edge terminations for these devices is based on the use of planar diffusion technology.

Planar diffusion is based on the ability to selectively introduce dopants into the semiconductor surface by using either an oxide layer as a mask against dopant diffusion or by using a photoresist layer to block the dopant during ion implantation. When the junction is fabricated by diffusing the dopant in a local region, it can be treated as a parallel plane inside the diffusion window.

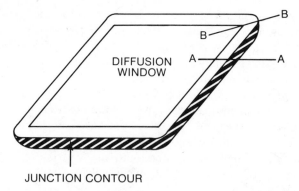

JUNCTION CONTOUR

Fig. 3.11. Planar junction formed by diffusion through a rectangular diffusion window.

However, the dopant also diffuses laterally at the edges of the diffusion window as illustrated in Fig. 3.11. The lateral diffusion of dopants at the edges of diffusion windows has been analyzed and shown to extend to about 85% of the vertical depth x_j. For purposes of breakdown analysis, an approximation can be made that the lateral diffusion is equal to the junction depth. From this assumption, it can be concluded that the junction takes a cylindrical contour at the edges of the diffusion window and a spherical contour at the corners of the diffusion window. The depletion layer of the junction follows these contours. The electric field distribution at the edges differs from the parallel lines representative of the parallel-plane portion. Since charge balance between the two sides of the junction must be established, the junction curvature leads to electric field crowding as illustrated in Fig. 3.12. The higher electric field at the junction edges leads to larger impact ionization at the edges. Consequently, the break-

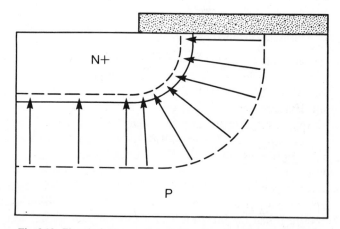

Fig. 3.12. Electric field crowding at the edge of a cylindrical junction.

down of planar diffused junctions can be expected to occur at the edges rather than in the parallel-plane portion.

3.6.1.1. Cylindrical Junction.

Consider the edge of the planar diffused junction shown in Fig. 3.11. A cross section of this edge at $A–A$ is illustrated in Fig. 3.13. Here it is assumed that the lateral diffusion is equal to the vertical junction depth. The resulting cylindrical junction has a radius r_j. The depletion layer extends from this junction on the lightly doped side to a radius r_d. For the case of a highly doped diffused region, the depletion layer width on the diffused side can be neglected. Inside the depletion layer, Poisson's equation can be set up in cylindrical coordinates:

$$\frac{1}{r}\frac{d}{dr}\left(r\frac{dV}{dr}\right) = -\frac{1}{r}\frac{d}{dr}(r\mathscr{E}) = \frac{qN_A}{\epsilon_s} \qquad (3.35)$$

where the potential distribution $V(r)$ and the electric field distribution $\mathscr{E}(r)$ are defined along a radius vector r extending into the depletion layer. The lightly doped side of the junction has homogeneous doping concentration N_A. Integration of this equation once with the boundary condition that the electric field must drop to zero at the edge of the depletion layer (i.e., at r_d) gives the electric field distribution:

$$\mathscr{E}(r) = \frac{qN_A}{2\epsilon_s}\left(\frac{r_d^2 - r^2}{r}\right) \qquad (3.36)$$

As in the case of the parallel-plane junction, the maximum electric field occurs at the metallurgical junction:

$$\mathscr{E}_m(r_j) = \frac{qN_A}{2\epsilon_s}\left(\frac{r_d^2 - r_j^2}{r_j}\right) \qquad (3.37)$$

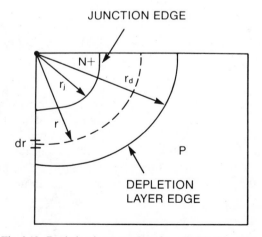

Fig. 3.13. Depletion layer at edge of a cylindrical junction.

This maximum electric field for the cylindrical junction can be substantially larger than the electric field in the parallel-plane portion. This can be illustrated easily for the case where the junction radius of curvature r_j is much smaller than the depletion layer width r_d. The maximum electric field is given approximately by

$$\mathscr{E}_{m,CYL}(r_j) \simeq \frac{qN_A}{2\epsilon_s} \frac{r_d^2}{r_j} \tag{3.38}$$

where the subscript CYL refers to the cylindrical region of the junction. The corresponding maximum electric field in the parallel-plane portion can be derived from Eq. (3.13):

$$\mathscr{E}_{m,PP}(r_j) = \frac{qN_A}{\epsilon_s} r_d \tag{3.39}$$

Taking a ratio of these expressions, we obtain

$$\frac{\mathscr{E}_{m,CYL}}{\mathscr{E}_{m,PP}} \simeq \frac{r_d}{2r_j} \tag{3.40}$$

This equation, valid for shallow junctions with small radii of curvature, indicates that the peak electric field at the cylindrical junction edge is substantially greater than in the parallel-plane portion. It also demonstrates that the effect of the cylindrical edge termination on the electric field becomes worse with increasing reverse bias applied to the junction. Avalanche breakdown is then confined to the edge because of the very strong dependence of the impact ionization coefficient on the electric field.

The potential distribution at the cylindrical junction termination can be obtained by integrating Eq. (3.36):

$$V(r) = \frac{qN_A}{2\epsilon_s} \left[\left(\frac{r_j^2 - r^2}{2} \right) + r_d^2 \ln\left(\frac{r}{r_j} \right) \right] \tag{3.41}$$

Using the boundary condition that the potential at $r = r_d$ must equal the applied reverse bias voltage, the depletion layer width can be derived from this equation.

To analyze the breakdown voltage of the cylindrical junction, it is necessary to solve the ionization integral by use of the electric field distribution given by Eq. (3.36). A closed-form analytical solution cannot be derived by using Eq. (3.36). However, a closed-form analytical solution can be obtained by utilizing an approximation to the electric field distribution that is based on the assumption that the impact ionization occurs primarily at the high electric field region close to the metallurgical junction [6]. This assumption allows the approximation

$$\mathscr{E}(r) = \frac{K}{r} \tag{3.42}$$

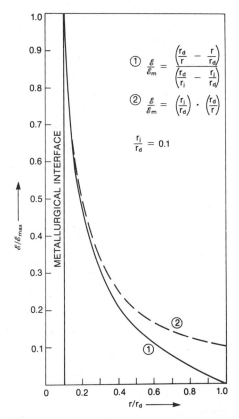

Fig. 3.14. Comparison of hyperbolic electric field distribution to the actual electric field distribution of a cylindrical junction.

This hyperbolic variation in the electric field distribution is compared with the actual field distribution in Fig. 3.14. The important point to note is the similarity in the field distribution near the metallurgical junction for the two cases. However, the hyperbolic field distribution implies a depletion layer that extends to infinity. Thus the ionization integral must also be evaluated from r_j to infinity when Eq. (3.42) is used. Solving the ionization integral, using the ionization coefficient given by Eq. (3.3), an expression for the critical electric field at breakdown for the cylindrical junction can be derived:

$$\mathscr{E}_{c,CYL} = \left(\frac{3.25 \times 10^{35}}{r_j} \right)^{1/7} \tag{3.43}$$

It is useful to normalize this parameter to the critical electric field at breakdown for the parallel-plane case. Equations (3.20) and (3.43) yield

$$\frac{\mathscr{E}_{c,CYL}}{\mathscr{E}_{c,PP}} = \left(\frac{3W_{c,PP}}{4r_j} \right)^{1/7} \tag{3.44}$$

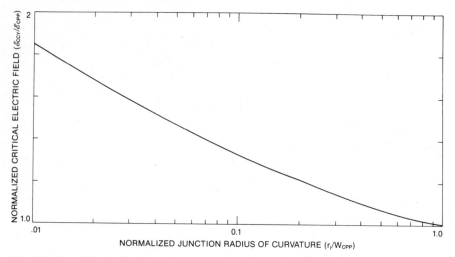

Fig. 3.15. Normalized critical electric field for a cylindrical junction as a function of the normalized radius of curvature.

This relationship is plotted in Fig. 3.15. It provides a general expression for the critical electric field of cylindrical junctions irrespective of the specific background doping level.

Using the critical electric field at breakdown for the cylindrical junction, an expression for the breakdown voltage can be derived. It is useful to normalize the breakdown voltage of the cylindrical junction to the breakdown voltage of the parallel-plane case to obtain a general expression that is independent of the specific background doping level:

$$\frac{BV_{CYL}}{BV_{PP}} = \left\{ \frac{1}{2} \left[\left(\frac{r_j}{W_c} \right)^2 + 2 \left(\frac{r_j}{W_c} \right)^{6/7} \right] \ln \left[1 + 2 \left(\frac{W_c}{r_j} \right)^{8/7} \right] - \left(\frac{r_j}{W_c} \right)^{6/7} \right\} \quad (3.45)$$

This expression is plotted in Fig. 3.16. It has been shown [6] that the breakdown voltages obtained by using this equation are in good agreement with those calculated by numerical techniques [7] for a wide range of radii of curvature and background doping level, as long as the radius of curvature is smaller than the depletion width at breakdown for the parallel-plane case and the background doping level is below $1 \times 10^{16}/\text{cm}^3$. This covers most of the practical junctions used for power device fabrication.

Numerical solutions for the avalanche breakdown of cylindrical junctions have been obtained, including the effect of the diffusion profile [8]. It has been shown that normalization of the critical electric field and breakdown voltage to the corresponding values for a semi-infinite diffused junction also results in condensation of all the data into a single curve. This curve is shown by the dashed lines in Fig. 3.16. To obtain this curve, the radius of curvature was normalized to the depletion layer width at breakdown on the lightly doped side

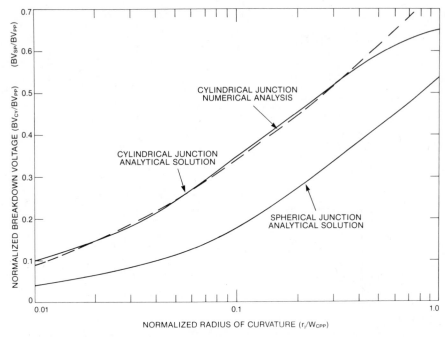

Fig. 3.16. Normalized breakdown voltage of cylindrical and spherical junctions as a function of the normalized radius of curvature.

of the junction for the parallel-plane case. An empirical expression that fits the curve is [8]

$$\frac{BV_{CYL}}{BV_{PP}} = \left[0.871 + 0.125 \ln\left(\frac{r_j}{W_{c,PP}}\right) \right]^2 \qquad (3.46)$$

From Fig. 3.16, it can be seen that Eq. (3.45), which was obtained from the closed-form analytical solution, provides nearly identical results.

From the preceding analysis and Fig. 3.16, it can be seen that the avalanche breakdown voltage of planar diffused junctions is always smaller than for the semi-infinite (parallel-plane) case. As the radius of curvature (i.e., the diffusion depth of the junction) increases, the breakdown voltage increases and begins to approach the parallel-plane case. The reason for this can be pictorially illustrated by comparing the electric field distribution of a shallow junction with that of a deep junction for the same background doping level. These cases will both have the same depletion widths on the lightly doped side for a given applied reverse-bias voltage as shown in Fig. 3.17. However, the electric field lines show much greater crowding, indicating a higher local electric field for the shallow junction because the positive charge in the N^+ diffused region, on which the field lines terminate, is located at a point close to the surface instead of being

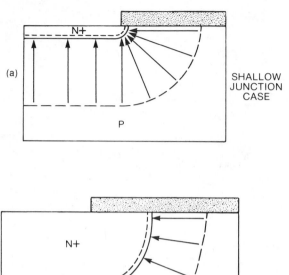

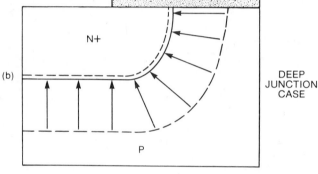

Fig. 3.17. Electric field crowding at the edges of (*a*) shallow and (*b*) deep junction.

spread out along a wide junction contour as in the case of the deep junction. The enhanced crowding of the electric field lines for the shallow junction is responsible for its lower breakdown voltage compared to the deeper junction. From Fig. 3.17 it can also be deduced that the crowding of the electric field lines is not determined by the absolute junction depth, but by its size relative to the depletion layer width. This is why normalization of the junction depth to the depletion layer width is effective in producing the elegant general expressions for breakdown of cylindrical junctions provided in Eqs. (3.45) and (3.46). It can be further concluded that shallower junctions can be used for the fabrication of low-voltage devices whereas high-voltage devices will require the extra processing steps that are usually necessary to obtain deep junctions.

For typical high-voltage devices, the depletion layer widths range from 20 to 50 μm, whereas practical planar process technology limits the junction depths to less than 10 μm. The ratio (r_j/W) for practical devices is typically less than 0.5 and may range down to less than 0.1. It is apparent from Fig. 3.16 that a cylindrical junction termination would yield breakdown voltages of less than half of the parallel-plane case under these circumstances. Methods to overcome this problem by using field rings and field plates are described later in the chapter.

3.6.1.2. Spherical Junction.   A planar junction formed by diffusion through a rectangular window is shown in Fig. 3.11. A cylindrical junction forms at the edges illustrated by the cross section taken at *A–A*. At the corners of the diffusion window the junction takes the form of one-quarter of a spheroid with a radius of curvature equal to the junction depth, if the lateral diffusion is assumed to be equal to the diffusion depth. The crowding of the electric field lines at the corners can be intuitively expected to be worse than at the edges because the field lines approach a point from three dimensions instead of a line from two dimensions. The breakdown voltage of this region of the planar junction can be analyzed by solution of Poisson's equation in spherical coordinates:

$$\frac{1}{r^2}\frac{d}{dr}\left(r^2\frac{dV}{dr}\right) = -\frac{1}{r}\frac{d}{dr}(r^2\mathscr{E}) = \frac{qN_A}{\epsilon_s} \tag{3.47}$$

Integrating this equation and using the boundary condition that the electric field is zero at the edge of the depletion layer (r_d), the electric field distribution can be derived:

$$\mathscr{E}(r) = \frac{qN_A}{3\epsilon_s}\left(\frac{r_d^3 - r^3}{r^2}\right) \tag{3.48}$$

The maximum electric field for the spherical junction occurs at the metallurgical interface and is given by

$$\mathscr{E}_m(r_j) = \frac{qN_A}{3\epsilon_s}\left(\frac{r_d^3 - r_j^3}{r_j^2}\right) \tag{3.49}$$

As in the case of the cylindrical junction, this maximum electric field is substantially greater than in the parallel-plane portion. If a junction with small radius of curvature compared with the depletion width is considered, the electric field can be approximated by

$$\mathscr{E}_{m,SP}(r_j) \simeq \frac{qN_A}{3\epsilon_s}\frac{r_d^3}{r_j^2} \tag{3.50}$$

Normalization of this expression to the maximum electric field in the parallel-plane portion of the junction gives

$$\frac{\mathscr{E}_{m,SP}}{\mathscr{E}_{m,PP}} = \frac{1}{3}\left(\frac{r_d}{r_j}\right)^2 \tag{3.51}$$

The maximum electric field at the spherical portion of the junction is thus substantially greater than in the parallel-plane portion. This difference in the electric field increases as the reverse bias increases because the depletion width increases. Avalanche breakdown at the corners of the planar diffusion will supersede breakdown at the parallel-plane portion.

A comparison with the cylindrical portion of the junction is also useful. Use of Eqs. (3.38) and (3.50) gives

$$\frac{\mathscr{E}_{m,SP}}{\mathscr{E}_{m,CYL}} = \frac{2}{3}\left(\frac{r_d}{r_j}\right) \tag{3.52}$$

Thus the maximum electric field in the spherical portion of the planar diffused junction is larger than that at the cylindrical portion. Consequently, the planar junction will break down first at the corners where the spherical junctions are formed.

The breakdown voltage of a planar diffused junction formed using a rectangular diffusion window can be obtained by integrating Eq. (3.48):

$$V(r) = \frac{qN_A}{3\epsilon_s}\left[\left(\frac{r_j^2 - r^2}{2}\right) + r_d^3\left(\frac{1}{r_j} - \frac{1}{r}\right)\right] \tag{3.53}$$

The depletion layer width can be obtained from this equation by using the boundary condition that the voltage at $r = r_d$ must be equal to the applied potential across the junction. To obtain the avalanche breakdown voltage, it is necessary to solve the ionization integral by use of the electric field distribution given by Eq. (3.48). A closed-form analytical solution has been derived [6] by

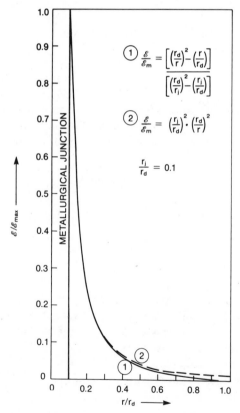

Fig. 3.18. Comparison of the electric field distribution used to obtain analytical solution (dashed lines) for the breakdown of spherical junctions with the actual (solid line) electric field distribution.

using the approximation

$$\mathcal{E}(r) = \frac{K}{r^2} \tag{3.54}$$

This is based on the fact that the impact ionization is confined close to the metallurgical junction. The electric field distribution described by Eq. (3.54) is compared with the actual distribution in Fig. 3.18. As in the case of the cylindrical junction, the similar field distribution near the metallurgical junction, where the impact ionization is predominant, allows the use of this approximation. Performing the ionization integral from r_j to infinity by use of Eq. (3.54) gives the critical electric field for breakdown of spherical junctions:

$$\mathcal{E}_{c,SP} = \left(\frac{7.0 \times 10^{35}}{r_j} \right)^{1/7} \tag{3.55}$$

Normalization of this field to the critical electric field at breakdown for the parallel-plane case gives

$$\frac{\mathcal{E}_{c,SP}}{\mathcal{E}_{c,PP}} = \left(\frac{13}{8} \frac{W_c}{r_j} \right)^{1/7} \tag{3.56}$$

This relationship is plotted in Fig. 3.19. It provides a general expression for the critical electric field at breakdown for the spherical junction irrespective of

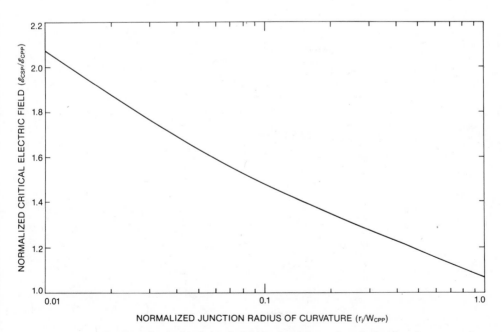

Fig. 3.19. Normalized critical electric field for breakdown of a spherical junction.

the background doping level. The breakdown voltage of spherical junctions can be derived from the critical electric field at breakdown. Normalization of this voltage to the parallel-plane case gives a general expression that is independent of the background doping level:

$$\frac{BV_{SP}}{BV_{PP}} = \left(\frac{r_j}{W_c}\right)^2 + 2.14\left(\frac{r_j}{W_c}\right)^{6/7} - \left[\left(\frac{r_j}{W_c}\right)^3 + 3\left(\frac{r_j}{W_c}\right)^{13/7}\right]^{2/3} \qquad (3.57)$$

This expression is plotted in Fig. 3.16 for comparison with the cylindrical junction. In general, the breakdown voltage of the spherical junction is found to be about a factor of 2 times lower than that of the cylindrical junction. The creation of spherical junctions at device terminations must, therefore, be avoided. Since the spherical junction is caused by the presence of sharp corners, as at the edges of the rectangular diffusion window in Fig. 3.11, such junctions can be avoided by rounding the devices edges. However, this results in loss in the active junction area by

$$\text{Area loss} = (4 - \pi)R^2 \qquad (3.58)$$

where R is the radius of curvature of the corner. To obtain the full benefits of rounding the corners during mask design, it is essential to make the radius of curvature R several times larger than the depletion width at breakdown.

3.6.2. Floating Field Rings

It is apparent from the previous section that the fabrication of devices using planar technology could lead to a serious degradation in the breakdown voltage as a result of high electric fields at the edges. An elegant approach to reducing the electric field at the edges is by using floating field rings [9]. The term floating field rings was coined because they consist of diffused regions that are isolated

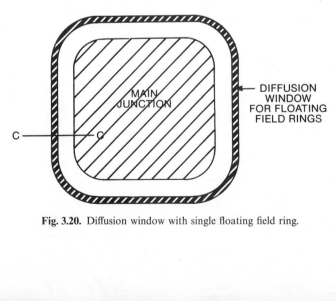

Fig. 3.20. Diffusion window with single floating field ring.

from the main junction but located close to it. These regions can assume a potential intermediate to that of either side of the *P–N* junction. Their potential is established by the depletion layer extending from the main junction. These floating field rings are almost invariably fabricated simultaneously with the main junction because this can be achieved in the same processing step during device fabrication by creating an extra diffusion window in the mask that surrounds the main junction as illustrated in Fig. 3.20. It is important to retain a uniform spacing between the main junction and the floating field ring.

When the floating field ring is fabricated simultaneously with the main junction, their diffusion depths will be equal. A cross section of this structure taken along line *C–C* is illustrated in Fig. 3.21*a*. A cross section of a junction without the floating field ring is shown in Fig. 3.21*b* for comparison. It can be seen that the electric field crowding responsible for the low breakdown voltage of cylindrical junctions is reduced by the presence of the floating field ring. During reverse-bias operation of the main junction at low voltages, its depletion layer is small and does not extend to the floating field ring. The floating field ring then retains the potential of the lightly doped *P* region. As the reverse bias on the main junction increases, its depletion layer widens until it punches through

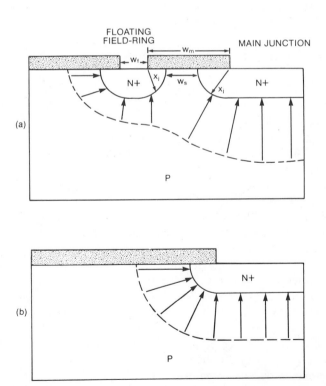

Fig. 3.21. Comparison of the electric field crowding for a planar junction (*a*) with and (*b*) without a floating field ring.

to the floating field ring. The voltage at which this occurs is related to the minimum spacing W_s between the floating field ring and the main junction. If the lateral diffusion of the junction is assumed to be equal to the vertical depth, the mask dimension W_m used for the location of the floating field ring relative to the main junction can be calculated from the punch-through voltage V_{PT} by using the relationship

$$W_m = 2x_j + W_s = 2x_j + \sqrt{\frac{2\epsilon_s V_{PT}}{qN_A}} \qquad (3.59)$$

Once the punch-through occurs, the potential on the floating field ring tracks the main junction potential. As an extreme case, if a floating field ring is considered that is infinitesimally small in width [27], it will not have any effect on the voltage distribution in the depletion layer. Its potential can then be related to the main junction potential by using Eq. (3.14):

$$V_{ffr} = \frac{qN_A}{\epsilon_s} \left(W_d W_s - \frac{W_s^2}{2} \right) \qquad (3.60)$$

where W_d is the depletion width of the main junction, which is related to the main junction voltage V_a by Eq. (3.15), and the subscript ffr denotes floating field ring. Combination of Eqs. (3.15) and (3.60) yields

$$V_{ffr} = \sqrt{\frac{2qN_A}{\epsilon_s} W_s^2 V_a} + \frac{qN_A}{2\epsilon_s} W_s^2 \qquad (3.61)$$

The floating field ring potential varies as the square root of the applied potential on the main junction for this case. In the case of actual floating field rings that distort the electric field distribution, it has been found that the floating field ring potential varies as the 0.65th power of the reverse bias on the main junction.

Although the floating field ring can be intuitively expected to increase the breakdown voltage by reducing the electric field crowding, its exact location relative to the main junction is crucial to its effectiveness in increasing the breakdown voltage. If the floating field ring is placed too far away from the main junction, it will have little effect on the depletion layer curvature at the main junction. Breakdown will then occur at the main junction without a substantial increase from the cylindrical case. If the floating field ring is placed too close to the main junction, its potential will be nearly equal to that of the main junction. Breakdown will then occur at the field ring at nearly the same voltage as without the field ring. However, optimal placement of the floating field ring can result in nearly a twofold increase in the breakdown voltage [10].

The breakdown voltage of planar junctions with optimally placed floating field rings has been analyzed numerically. An example of the effect of the floating field ring location on the breakdown voltage is shown in Fig. 3.22. Here the breakdown voltage has been normalized to the parallel-plane case and the field ring spacing has been normalized to the depletion layer width on the lightly doped side of the junction. It can be seen that the optimal floating field ring

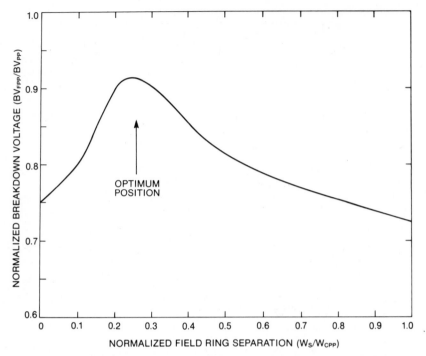

Fig. 3.22. Effect of floating field ring position relative to the main junction on its breakdown voltage.

location for this case is about 0.25 times the depletion layer width at breakdown on the lightly doped side for the parallel-plane case. An important point to note in Fig. 3.22 is that the breakdown voltage predicted by the numerical analysis shows a sharply pointed distribution around the optimal spacing. This implies that, to obtain the full benefits of a single floating field ring, it is essential to precisely control the mask dimension W_m and the junction depth of the diffusion. In general, this can be easily achieved during device design and fabrication.

The optimal spacing indicated in Fig. 3.22 is based on the assumption that there is no charge at the surface of the semiconductor over the lightly doped region. The presence of surface charge has a strong influence on the depletion layer spreading at the surface because this charge complements the charge due to the ionized acceptors inside the depletion layer. Examples of the depletion layer shape for the case of positive, zero, and negative surface charge are provided in Fig. 3.23 for a planar junction. Positive surface charge causes the depletion layer at the surface of the lightly doped side to extend further, whereas negative charge will tend to retard the depletion layer. The opposite effect will apply to a junction with a lightly doped N-type region. For the cylindrical junction illustrated in Fig. 3.23, the presence of positive surface charge will cause a decrease in the crowding of the electric field lines and result in raising the

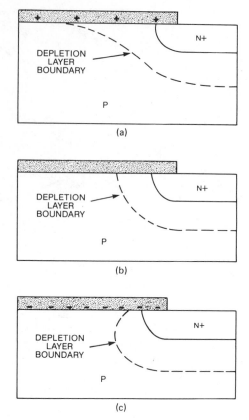

Fig. 3.23. Influence of surface charge on depletion layer spreading at the edge of a planar junction: (*a*) positive charge; (*b*) zero charge; (*c*) negative charge.

breakdown voltage, whereas negative surface charge will have the opposite effect. Thus positive charge is beneficial for the simple planar junction with a lightly doped *P* region. In the case of junctions with field rings, however, surface charge of either polarity can lower the breakdown voltage because it affects the punch-through voltage to the floating field ring. This alters the optimal spacing and results in lowering the breakdown voltage because of the sharply peaked breakdown voltage distribution around the optimum spacing. If the surface charge is precisely known, it is possible to analyze for the optimal floating field ring location, including the effect of this charge, and then design the mask spacing W_m. In practice, the charge at the surface of thermally oxidized silicon is positive and in the range of 10^{10} to $10^{12}/\text{cm}^2$. Even with a well-controlled, clean fabrication sequence for a complex device such as a power MOSFET, the surface charge will vary from wafer to wafer and even across a wafer by $\pm 1 \times 10^{11}/\text{cm}^2$. This represents an inherent practical limitation to designing the breakdown of devices by use of single floating field rings.

Under the assumption that the floating field ring is optimally located with respect to the main junction, the breakdown voltage has been found to be enhanced by nearly a factor of 2 over the simple cylindrical planar diffused junction. Through use of the same normalization scheme adopted in the earlier sections for cylindrical and spherical junctions, it has been found that the breakdown voltage of all junctions with floating field rings can be represented by a single line as shown in Fig. 3.24. The breakdown voltage of cylindrical junctions without the floating field ring is included in this figure for comparison. The impact of the floating field ring is the greatest for high voltage devices fabricated by use of shallow junctions. For these junctions with small $r_j/W_{c,PP}$, where the breakdown voltage is less than 30% of the parallel-plane case, the addition of

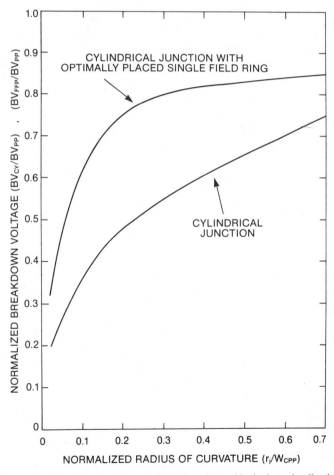

Fig. 3.24. Normalized breakdown voltage of planar junctions with single optimally placed floating field ring.

the floating field ring can raise the breakdown voltage by a factor of nearly 2. For deep junctions, where the curvature effects are small to begin with, the addition of the floating field ring will raise the breakdown voltage by a much smaller amount. Since the floating field ring can take up valuable space on the periphery of the device and reduce the utilization of chip area, its use is more commonplace in small devices operating at low voltages (< 1500 V).

The optimal spacing of floating field rings was discussed in the preceding section. Another design parameter that is important is the width W_f of the window through which the floating field region is diffused into the semiconductor. If the width of the window W_f is very small, the floating field ring can become ineffective in reducing the depletion layer curvature even when it is at the optimal location. This is illustrated in Fig. 3.25a. To be fully effective in raising the breakdown voltage, it is necessary to make the width of the floating field ring comparable to the depletion layer width. Making the floating field ring width much larger than the depletion layer width, as shown in Fig. 3.25b, is not advisable because it does not improve the breakdown voltage but results in wasting space at the edge of chip.

The floating field ring concept, with its extensions using multiple rings or field plates, represents the most widespread device termination technique for the small field controlled devices discussed in this book. Its design is crucial

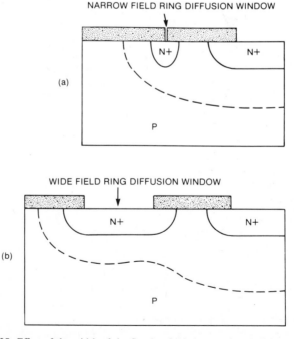

Fig. 3.25. Effect of the width of the floating field ring on electric field crowding.

to achieving the high-performance ratings that have been reported for power MOSFETs.

3.6.3. Multiple Floating Field Rings

Since a single field ring reduces depletion layer curvature and the electric field crowding, it can be expected that several floating field rings working in conjunction with each other may raise the breakdown voltage even closer to the parallel-plane case. As in the case of the single floating field ring, multiple floating field rings are generally fabricated with the main junction by designing the mask with multiple windows surrounding the main junction.

Two design philosophies exist for designing the multiple floating field ring termination. In one case the spacing between individual floating field rings is varied together with its width. The floating field ring spacing and its width should both decrease with increasing distance from the main junction. This provides a gradual extension of the depletion layer away from the main junction as illustrated in Fig. 3.26. The field rings further away from the junction can be made narrower because the depletion layer depth below them becomes progressively smaller, thus saving space at the device periphery. However, this approach is based on the assumption that the surface space charge is precisely known, and the field ring spacings are designed by including the effect of this space charge. If the surface space charge in the actual device is more positive than that assumed during device design, the inner field rings can become ineffective, transferring all the voltage to the outer field rings and resulting in premature breakdown at the outer field ring. In the ideal case, these field rings will all share the applied voltage equally, producing avalanche breakdown at the outer edges of all the floating field rings simultaneously.

In the second design approach, all the floating field rings are made narrow and are equally spaced. Because of the smaller width of these floating field rings and their closer spacing, more of them can be accommodated within a given

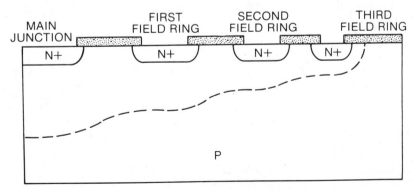

Fig. 3.26. Multiple field ring termination with gradually decreasing field ring width and spacing.

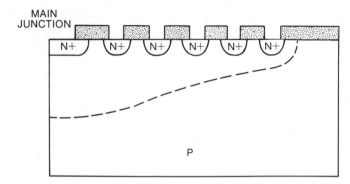

Fig. 3.27. Multiple field ring termination with equal width and spacing.

edge area as shown in Fig. 3.27. This design produces a much finer gradation in the depletion layer at the edge of the device. The design of this termination is also easier because it is essentially based on use of the minimum design rules when laying out the windows and spacings that control the field ring spacing and width. This termination is also sensitive to the surface charge, but the presence of a larger number of rings reduces the impact of surface charge variations when compared to the earlier multiple field ring design.

Through application of these multiple floating field ring designs, the breakdown voltage can be raised arbitrarily close to the parallel-plane case. The only limitation to applying this approach lies in the space taken up at the edge of the chip. In practical devices, it is usual to use three floating field rings. Designs with up to 10 floating field rings have been utilized where it is important to achieve close to the parallel-plane junction breakdown.

3.6.4. Beveled-Edge Terminations

With the application of solid-state thyristors to power control, there has been a constant need to develop higher-voltage devices. It was apparent during the early stages of thyristor development that it was necessary to design edge terminations that promote bulk breakdown. Since these high-voltage, large-area devices were manufactured by using gallium and aluminum diffusions to obtain highly graded deep junctions, planar termination techniques were not applicable. It was instead necessary to bring the junction to the edges of the chip and provide adequate surface passivation to avoid premature surface breakdown.

The simplest approach was to make the device edge perpendicular to the wafer surface. This can be done by either sawing or scribing and breaking the wafer. This approach creates considerable surface damage at the edges of the chip which is difficult to passivate. Furthermore, it was discovered that cutting the edge of the wafer at an angle improved the breakdown voltage.

Consider the case of a junction whose area at the edge decreases when proceeding from the heavily doped side to the lightly doped side. This edge contour

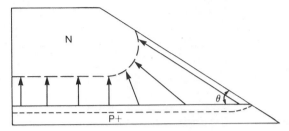

Fig. 3.28. Positive-bevel-edge contour.

is known as a *positively beveled junction* and is illustrated in Fig. 3.28. The depletion layer shape and corresponding electric field distribution are also indicated in this figure. To maintain the charge balance on the opposite sides of the junction, the depletion layer on the lightly doped side of the junction is forced to expand near the surface. This expansion of the depletion layer causes a reduction in the electric field crowding. Since the depletion layer width along the surface is much larger than in the bulk, it can be concluded that the electric field along the surface will be much smaller than that in the bulk. This is an ideal design for the termination of junctions because it ensures bulk breakdown prior to surface breakdown if the surface electric field is sufficiently low. It should be noted that even if the surface electric field is lower than in the bulk, surface breakdown may precede bulk breakdown because the ionization coefficients at the surface are generally larger than in the bulk for the same electric field strength because of the presence of defects at the surface.

If the edge of the junction is cut in the opposite direction so that the area of the junction increases when proceeding from the highly doped side toward the lightly doped side, the contour is known as a *negative-beveled junction*. The depletion layer shape for this case is illustrated in Fig. 3.29. The establishment of charge balance on the opposite sides of the junction causes the depletion layer at the surface of the lightly doped side to decrease while the depletion layer on the heavily doped side expands. If the diffused side of the junction is heavily doped, the depletion layer shrinkage on the lightly doped side will have

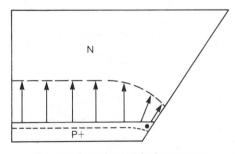

Fig. 3.29. Negative-bevel-edge contour.

the dominant influence. Since the junction potential is being supported across a narrower depletion layer at the surface, the electric fields at the surface can be expected to be higher than those in the bulk. Surface breakdown will precede bulk breakdown in the negatively beveled junction, degrading its breakdown characteristics. Negatively beveled contours are undesirable and should be avoided during device processing.

An exception to this reasoning occurs if the junction is highly graded and a very shallow negative-bevel angle is created. Now, the surface electric field can be reduced as illustrated in Fig. 3.30. Here, the gradual change in doping level on the diffused (p^+) side of the junction and the large amount of material removed from the diffused side force the depletion layer to expand considerably along the surface on the diffused side of the junction. Note that the depletion layer on the lightly doped side can become pinned to the metallurgical junction edge at the surface under these circumstances. The expansion of the depletion layer on the diffused side of the junction lowers the surface electric field. Negative-bevel-edge contours are used primarily in devices containing two back-to-back junctions, such as power thyristors. In these cases, a positive bevel is used for the reverse blocking junction and the negative bevel for the forward blocking junction.

Although beveled-junction terminations are widely used for the fabrication of high-voltage, large-area rectifiers and thyristors, they are infrequently used for small-area devices such as those discussed in this book. The treatment of the edges of individual devices by the lapping and polishing technology used to prepare beveled-edge terminations is economical only when the pellet size is large. As the current ratings of the devices increase, it is conceivable that this device termination approach may become important. With this in mind, the following sections treat the beveled junctions in further detail.

3.6.4.1. Positively Beveled Junction. The positively beveled junction is one in which the junction area decreases when proceeding from the more highly doped side to the lightly doped side. Two-dimensional computer analysis of the posi-

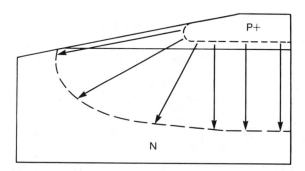

Fig. 3.30. Shallow negatively beveled junction with highly graded diffusion profile.

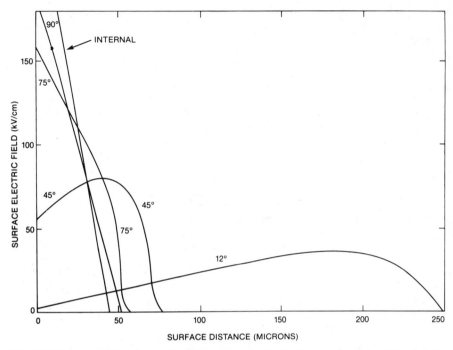

Fig. 3.31. Electric field distribution along the surface of a positively beveled junction. (After Ref. 11, reprinted with permission from the IEEE © 1964 IEEE.)

tively beveled junction [11] has confirmed the surface field reduction described with the aid of Fig. 3.28. As an example, plots of the electric field distribution along the surface of a typical junction for various bevel angles are shown in Fig. 3.31. From this analysis it has been found that the maximum electric field along the surface is always lower than the maximum electric field in the bulk for all positive-bevel angles. The point at which the maximum electric field occurs moves away from the metallurgical junction into the lightly doped side as the positive-bevel angle is reduced. In fact, it can be seen from Fig. 3.31 that when the positive-bevel angle becomes small, as in the case of the 12° positive-bevel angle, the depletion layer on the diffused side of the junctions becomes pinned at the metallurgical junction. In this case the electric field approaches zero where the metallurgical junction meets the surface.

The maximum electric field at the surface of the positively beveled junction decreases with decreasing bevel angle θ. As an example, a plot of the calculated maximum electric field as a function of bevel angle is shown for a specific case in Fig. 3.32. The variation in the electric field from the surface toward the bulk has also been examined. It has been found that the electric field decreases monotonically when going from the bulk toward the surface. Thus the positively beveled junction exhibits the ideal characteristics for achieving bulk breakdown.

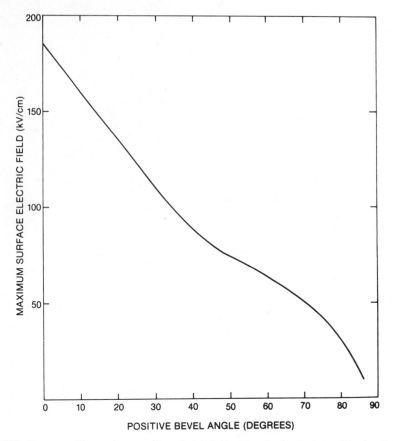

Fig. 3.32. Decrease of the maximum surface electric field with reduction in the positive-bevel angle.

It is worth noting that this is the only device termination technique that has been developed thus far that can allow achieving bulk breakdown at the avalanche breakdown voltage of the parallel-plane junction.

Although the preceding analysis indicates that bulk breakdown will occur for a positively beveled junction if the impact ionization coefficients at the surface are assumed to be equal to those in the bulk, in practice it is important to reduce the maximum surface electric field to at least 50% lower than the maximum electric field in the bulk because of the presence of surface defects that enhance impact ionization. The maximum surface electric field of a variety of positively beveled junctions with different background doping levels and diffusion profiles has been analyzed, and it has been found that all these cases can be represented by a single curve by using a normalization scheme similar to that adopted for the planar junctions [12], that is, by normalizing to the electric field in the parallel-plane portion of the junction. This curve is provided in Fig. 3.33. From this curve, it can be concluded that positive-bevel angles ranging from 30 to 60° are adequate for ensuring bulk breakdown. Shallower positive-

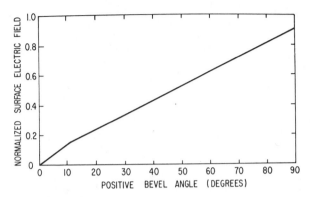

Fig. 3.33. Normalized surface electric field as a function of the positive-bevel angle.

bevel angles are not recommended because of the wastage of space at the edge of the chip.

The reduced surface electric field of positively beveled junctions makes them relatively stable and easy to passivate. The fabrication of the positive-bevel angle in high-voltage, large-area devices is usually performed by grit blasting by use of an abrasive powder emanating from a nozzle placed at the appropriate angle to the wafer edge while the wafer is being rotated. The crystal damage caused by the grit blasting is subsequently removed by using a chemical etch just prior to coating the wafer edge with the passivant. In the case of a large number of small-area devices fabricated on a single wafer, a positive bevel can be achieved by using a V-shaped diamond saw blade to cut through the wafer from the lightly doped side. Again, a chemical etch is essential after the saw cut to remove the residual damage. Such positively beveled pellets are difficult to handle during packaging because of their sharp edges. Furthermore, the passivation of these pellets is difficult since it must be done after the die mount down process, which can contaminate the beveled surface.

3.6.4.2. Negatively Beveled Junction.   In a negatively beveled junction, the area of the junction decreases when proceeding from the lightly doped side toward the diffused (highly doped) side. It was pointed out earlier with the aid of Fig. 3.30 that the surface electric field will be reduced below the parallel-plane case for a negatively beveled junction if the angle is small and the diffusion is highly graded. The electric field distribution along the surface of a negatively beveled junction is shown in Fig. 3.34 for several bevel angles. It can be seen that as the bevel angle decreases, the depletion layer becomes pinned on the lightly doped side at the $P-N$ junction metallurgical interface while the depletion layer on the diffused side continues to grow larger, thus lowering the maximum surface electric field. The maximum in the surface electric field occurs on the diffused side of the junction for these small bevel angles. The variation of maximum electric field as a function of the negative-bevel angle over the full range

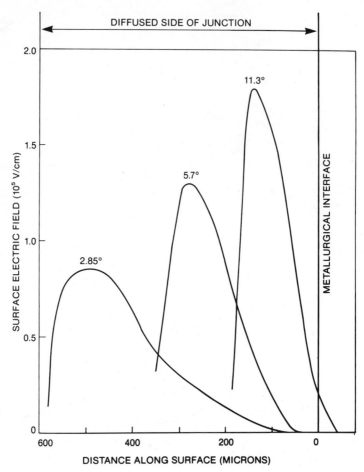

Fig. 3.34. Examples of surface electric field distribution for several negatively beveled junctions. (After Ref. 13, reprinted with permission from the IEEE © 1973 IEEE.)

from 0 to 90° is illustrated in Fig. 3.35 for a specific case (the same as in Fig. 3.31). Note that the maximum surface electric field is higher than that in the bulk at large negative-bevel angles. This occurs due to the reduction in depletion layer width on the lightly doped side as illustrated in Fig. 3.29 that cannot be made up on the diffused side. It is only when the negative-bevel angle is made very small that the maximum surface electric field becomes less than that in the bulk.

Despite the reduction of the maximum surface electric field to less than the bulk electric field, the avalanche breakdown voltages of negatively beveled junctions do not equal the breakdown voltage of the parallel-plane junction. The reason for this is the occurrence of a peak in the electric field beneath the surface [13]. The electric field distribution proceeding into the bulk from the surface

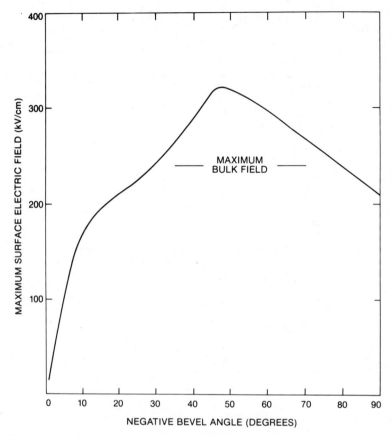

Fig. 3.35. Variation of the maximum surface electric field with negative-bevel angle. (After Ref. 11, reprinted with permission from the IEEE © 1964 IEEE.)

of the negatively beveled junction is shown for a specific case in Fig. 3.36. In all cases, the electric field is found to exhibit a value higher than that in the bulk. This implies that the breakdown voltage of negatively beveled junctions will always be lower than that for the parallel-plane junction.

Numerical calculations of the breakdown voltage of a variety of diffused junctions with a broad range of surface concentrations and junction depths have been performed as a function of the background doping and the negative-bevel angle [14]. It has been found that the breakdown voltage of all these cases can be represented by a single curve by use of a normalization scheme similar to that used for planar diffused junctions. In this case, a bevel parameter ϕ called the *effective-bevel angle* must be defined as

$$\phi = (0.04)\theta\left(\frac{W_{\mathrm{L}}}{W_{\mathrm{H}}}\right)^2 \qquad (3.62)$$

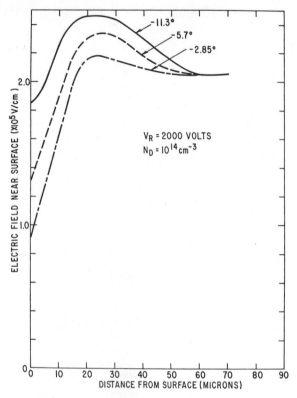

Fig. 3.36. Electric field distribution near the surface of a negatively beveled junction. (After Ref. 13, reprinted with permission from the IEEE © 1973 IEEE.)

where θ is the actual bevel angle in degrees and W_L and W_H are the depletion layer widths on the lightly and highly doped sides of the junction, respectively. A curve representing the breakdown voltage of positively beveled junctions normalized to the parallel-plane case is provided in Fig. 3.37a. To achieve a breakdown voltage approaching the parallel-plane case, it is necessary to use very shallow negative-bevel angles in the range of 2 to 6° to reduce the surface electric field as shown in Fig. 3.37b. It is also necessary to achieve a highly graded junction that has a small ratio (W_L/W_H). For a high-voltage (3000-V) thyristor, a negative-bevel angle of 2 to 4° is typically used with a diffusion gradient that achieves a ratio (W_L/W_H) of 4 to 5. This results in a breakdown voltage of close to 90% of the parallel-plane case.

Negative-bevel angles are formed by using a lapping process followed by chemical etching to remove the surface damage. This technique requires consumption of a large amount of space on the device edges because of the shallow

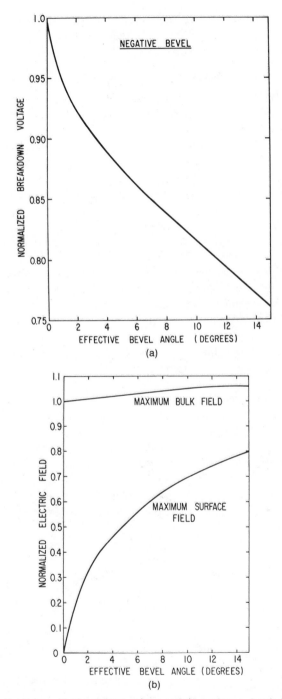

Fig. 3.37. Normalized plots of (*a*) breakdown voltage and (*b*) maximum electric field for negatively beveled junctions. (After Ref. 14, reprinted with permission from the IEEE © 1976 IEEE.)

bevel angles required to reduce the surface electric field. The application of negative bevels is generally confined to large-area, high-current devices.

3.6.5. Etch Contour Terminations

The concept of using chemical etches to remove surface damage and improve the breakdown voltage or reduce surface leakage current has been known since the early days of high-voltage device development. A moat etch such as that illustrated in Fig. 3.38 was typically used to terminate devices. This moat was used to separate adjacent devices as well. The etching was done at wafer level by using masking material such as photoresist or even the aluminum metallization itself. The scientific application of etching to reduce surface electric field by the precise removal of silicon from selective areas of the high-voltage $P-N$ junction near the surface was developed only recently [15, 16].

Chemical etches used for forming these etch contours are mixtures of nitric, hydrofluoric, and acetic acid. The composition and temperature of the acid must be carefully controlled for regulation of the etch rate. In some etch contours, the depth must be controlled to within 0.1 μm. Such precision is often difficult to achieve with good uniformity across an entire wafer by using chemical etching. With the advent of dry etch technologies, such as plasma and reactive ion etching, much greater precision of etch depth and uniformity is achievable. Furthermore, these dry etching processes are less sensitive to doping concentration in the silicon than chemical etches.

3.6.5.1. Positive-Bevel-Etch Contours.   As in the case of the positive-bevel angle described in Section 3.6.4 by use of lapping techniques, an effective positive-bevel angle can be achieved by using a chemical etch to remove more material from the lightly doped side of the junction. In fact, by using etching techniques it is possible to selectively remove material from only the lightly doped side of the junction without removing material from the diffused or highly doped side. Examples of a positive-bevel etch contour for a parallel-plane junction and a planar diffused junction are given in Fig. 3.39. In both cases, the selective removal of

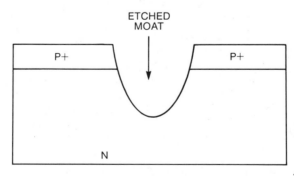

Fig. 3.38. Moat etch contour.

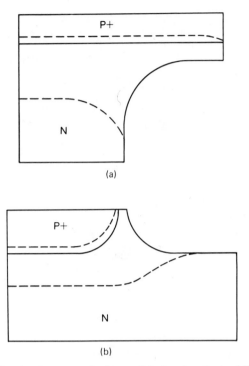

Fig. 3.39. Positive-bevel-etch contours for (*a*) parallel-plane junction and (*b*) planar junction.

material on the lightly doped side of the junction at the surface forces the depletion layer on the lightly doped side to expand. This can result in a significant reduction in the maximum surface electric field.

Numerical analysis of the electric field distribution at the surface and in the bulk for the planar junction with the etch contour illustrated in Fig. 3.39*b* have been performed to optimize the location of the etched region [15]. A typical set of examples of the effect of changing the etch contour location and depth with respect to the planar junction is provided in Fig. 3.40. It should be noted that the maximum surface electric field will occur on the etch contour sidewall if the etch contour intersects the diffused region of the junction. If the etch contour is confined to the lightly doped side, the maximum surface electric field will occur on the top unetched surface. This transition of the maximum electric field location is accompanied by a minimum in the peak surface electric field. In general, this minimum occurs when the etch contour just intersects the planar junction at the surface. This change in electric field distribution can be intuitively understood by first considering the case of an etch contour that intersects the junction in its parallel-plane portion. This contour is similar to a positive-bevel contour where the peak electric field occurs on the beveled surface. Now consider the case where the etch contour is located remote from the planar junction. In

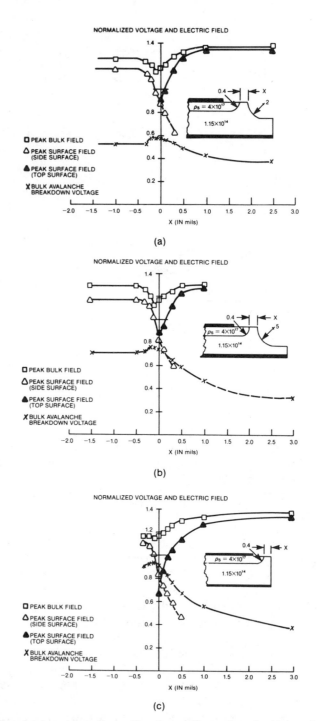

Fig. 3.40. Effect of etch contour radius and location relative to planar junction on the electric field and breakdown voltage. (After Ref. 15, reprinted with permission from the IEEE © 1977 IEEE.)

this case, the peak surface electric field will occur at the top surface because the etch contour does not perturb the junction. When the etch contour is placed near the planar junction edge, therefore, the location of the peak surface electric field can be shifted as indicated by the modeling results shown in Fig. 3.40.

It is worth pointing out that the highest breakdown voltage as determined by the bulk electric field does not occur when the peak surface electric field is at its minimum. The highest bulk breakdown voltage occurs when the peak bulk electric field is at its lowest value. This is found to take place when the etch intersects the diffused region. Unfortunately, this peak bulk electric field is always greater than in the case of the parallel-plane junction.

The breakdown voltages of planar junctions with positive-bevel-etch contours approach those of the parallel-plane case within 90%. Their most significant advantage is the markedly lower surface electric field, which greatly facilitates surface passivation and breakdown stability. Furthermore, this termination is less sensitive to surface charge when compared to the field ring termination. To achieve these results, it is important to control the etch contour location in within a few micrometers. It can also be concluded from Fig. 3.40 that the breakdown voltage will increase as the radius of curvature of the etch contour grows. Thus a precisely located vertical cut near the planar junction will provide the best results.

3.6.5.2. Negative-Bevel-Etch Contours.

As in the case of the negatively beveled junction, more material is removed from the highly doped side of a negative-bevel-etch contour than from the lightly doped side. Examples of this type of device termination are provided in Fig. 3.41. In both cases, the removal of material from the diffused side of the junction forces the depletion layer to expand, causing a reduction in the maximum surface electric field. Since the depletion layer extends to much smaller distances on the diffused side of the junction, the control of the etch depth for the negative-bevel-etch contour must be much more precisely accomplished when compared to the positive-bevel-etch contour.

Numerical analysis of the negative-bevel-etch contour for the planar junction as well as experimental measurements have been carried out [16]. The very precise control of the etch depth demanded by the negative-bevel-etch contour is illustrated in Fig. 3.42. It can be seen that the breakdown voltage exhibits a pronounced peak and that the etch depth must be controlled to within ± 1 μm at a depth of 35 μms; that is, this process would require control and uniformity in etching to within $\pm 3\%$. If this is achieved, the breakdown voltage can approach within 90% of the ideal case.

3.6.6. Junction Termination Extension

In the previous sections on bevel and etch contours, the breakdown voltage was enhanced by removal of material (and hence charge) from either the heavily doped or lightly doped side of the junction. A similar result can be achieved by

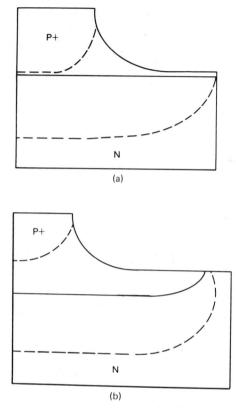

Fig. 3.41. Negative-bevel-etch contours for (*a*) parallel-plane junction and (*b*) planar junction.

the complementary process of adding precisely controlled charge to the junction at the surface. This approach has been called *junction termination extension* [17]. This termination is contingent on the ability to introduce charge at the surface near the junction with an accuracy of better than 1% by using ion implantation. The control over the charge introduced by ion implantation is significantly superior to that achievable by the etching techniques and can be performed with much better uniformity.

Examples of the application of the junction termination extension technique to the parallel-plane and planar junctions are illustrated in Fig. 3.43. The charge required to obtain high breakdown voltages must be accurately controlled. If the dose of the implanted charge is too low, it will have little influence on the electric field distribution and the maximum in the electric field will occur at point *A* as in the case of the termination without the junction termination extension. If the dose of the implanted charge is too high, the junction will have been simply extended to point *B*. Since the depth of the ion-implanted region is small, the extension region has a very small radius of curvature. The termination will then break down as a result of the high electric field at point *B* as deter-

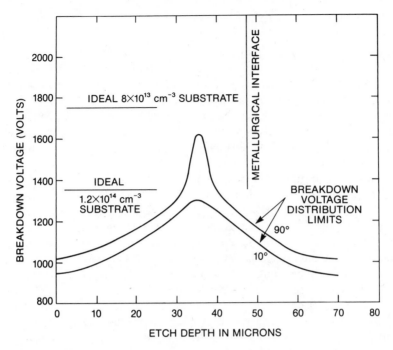

Fig. 3.42. Impact of etch depth on breakdown voltage of a negative-bevel-etch contour. (After Ref. 16, reprinted with permission from the IEEE © 1976 IEEE.)

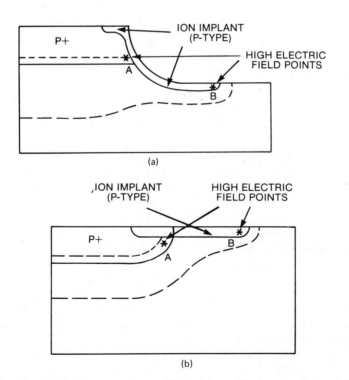

Fig. 3.43. Junction termination extension applied to (*a*) parallel-plane junction and (*b*) planar junction.

mined by planar junction breakdown (see Section 3.6.1). To obtain a reduction in the electric field and an increase in the breakdown voltage, it is essential to control the implanted charge so that the implanted region becomes completely depleted under reverse bias. This occurs when the dose of the implanted charge ranges from 60 to 80% of the charge obtained by taking the product of the dielectric constant for silicon and the maximum electric field at breakdown for the parallel-plane case. With this charge, breakdown voltages to within 95% of the parallel-plane case can be achieved.

Although this termination technique is very promising for achieving close to the ideal parallel-plane junction breakdown voltage, two problems have been encountered. First, the presence of surface charge over the implanted region can strongly alter the electric field distribution. The optimum ion-implant doses are in the range of 10^{11} to 10^{12} charges/cm^2, which are comparable to surface charges arising from typical passivation processes. This can cause high electric fields to arise at either points A or B and wide variations in breakdown voltage from device to device across a wafer. Second, the high electric fields near the surface at points A and B can cause excessive leakage current flow. When applied with adequate control over the implant dose, however, the junction termination extension approach offers the possibility for achieving nearly ideal parallel-plane junction breakdown voltage by using a process that is compatible with modern device processing [28].

3.6.7. Field Plates

In previous sections it was shown that the electric field at the surface of a planar diffused junction is higher than in the parallel-plane junction because of depletion layer curvature effects. The depletion layer curvature can be controlled by altering the surface potential. The simplest method for achieving this is by placing a metal field plate at the edge of the planar junction as illustrated in Fig. 3.44. By altering the potential on the field plate, the depletion layer shape can be adjusted. When a positive bias is applied to the metal field plate with respect to the N-type substrate, it will attract electrons to the surface and cause the depletion layer to shrink as illustrated by case A. If a negative bias is applied to the field plate, it will drive away electrons from the surface, causing the depletion layer to expand as illustrated in case C. The latter phenomenon can be expected to increase the breakdown voltage. This has been observed experimentally [18]. It has been found that the breakdown voltage of the diode with field plate is related to the field plate potential V_{FP} by

$$V_D = mV_{FP} + \text{constant} \tag{3.63}$$

where $m \simeq 1$. The value for m is closer to unity for small oxide thicknesses. With sufficient bias on the field plate, the breakdown voltage of the planar diode can be made to approach the parallel-plane value.

In devices it is impractical to provide a separate bias to control the potential on the field plate. Instead, the field plate is created by merely extending the

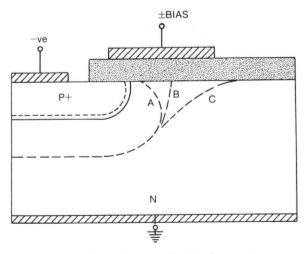

Fig. 3.44. Planar junction with field plate at edge.

junction metallization over the oxide as shown in Fig. 3.45. The presence of the field plate at the diffusion region potential forces the depletion layer to extend at the surface beyond the edge of the field plate. This reduces the depletion layer curvature and reduces the electric field at point A. However, a high electric field can occur at the edge of the field plate at point B.

Numerical analysis of the potential distribution for a planar junction with a field plate tied to its highly doped side has been performed for a variety of substrate doping concentrations and oxide thicknesses [19]. The analysis shows a high electric field arising at the edge of the field plate if the oxide thickness is small. The calculated value of the maximum ionization integral near the edge of the field plate is shown in Fig. 3.46. In this figure the ionization integral has been normalized to the value in the parallel-plane portion of the junction. The

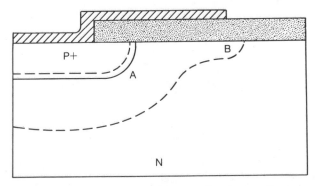

Fig. 3.45. Planar junction with field plate formed by extending the metallization over an oxide at the junction edge.

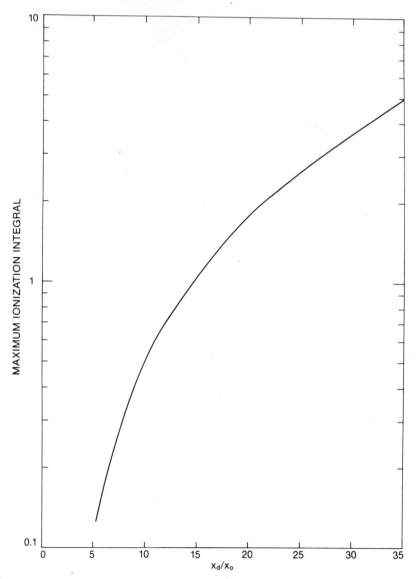

Fig. 3.46. Maximum ionization integral at the edge of a planar junction with field plate. (After Ref. 19, reprinted with permission from *Solid State Electronics*.)

term X_d is the depletion layer width, and X_o is the oxide thickness. To avoid breakdown at the edge of the field plate, it is necessary to use an oxide thickness sufficiently large to obtain X_d/X_o less than 12. However, the results in Fig. 3.46 were obtained without including the effect of curvature at the metallurgical junction. As the oxide thickness increases, the influence of the field plate on the junction curvature becomes smaller, and breakdown can now occur at the

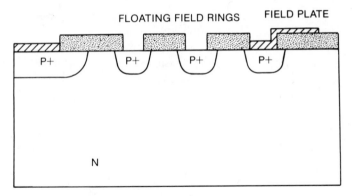

Fig. 3.47. Device termination combining floating field rings with a field plate.

metallurgical junction as a result of a high electric field at point A as in the case of a planar junction without the field plate. An optimum oxide thickness is, therefore, dictated to balance the peak electric fields that occur at points A and B.

An alternative to this compromise is to tailor the oxide thickness from small values near the junction to thicker values near the edge of the field plate. This requires a complex device fabrication process that is often unwarranted by the gain in breakdown voltage. A more common practice is to use the field plate in conjunction with floating field rings to achieve high breakdown voltage as illustrated in Fig. 3.47. The field plate is tied to the last floating field plate and extends beyond its edge. Field plates must not be placed between the main junction and the floating field rings or between adjacent floating field rings because they will transfer the potential between these regions and destroy the ability of the floating field rings to enhance the breakdown voltage.

3.7. OPEN-BASE TRANSISTOR BREAKDOWN

The breakdown voltage of $P-N$ junctions with various junction terminations was considered in the previous sections. In these cases, the maximum achievable breakdown voltage is that of a parallel-plane $P-N$ junction. However, many power devices contain back-to-back $P-N$ junctions with floating middle regions. The maximum voltage blocking capability of these devices is limited by the open-base transistor breakdown of the parasitic transistor formed in these structures [20]. The influence on the breakdown voltage of the resistivity and the spacing between the $P-N$ junctions in these devices is discussed in the paragraphs that follow in this section.

The open-base transistor structure is shown in the inset of Fig. 3.48 with one of the $P-N$ junctions reverse-biased. This region of the device then acts as the collector and the other junction, as the emitter. When the width of the

N-base layer is larger than the depletion width W of the reverse-biased collector junction, the breakdown voltage of the collector junction is limited by avalanche breakdown. Avalanche breakdown occurs when the multiplication factor of the junction becomes infinitely large. The variation of the avalanche breakdown voltage with the resistivity of the N base is shown in Fig. 3.48 as the avalanche limit. The avalanche breakdown voltage decreases with increasing N-base doping level.

When the emitter junction is brought close to the collector junction, open-base transistor breakdown occurs as a result of the injection of electrons from the emitter as the collector depletion layer approaches the emitter junction. If the minority carrier diffusion length in the N base is extremely small, the collector breakdown occurs when its depletion layer punches through. This breakdown voltage for the case of abrupt junctions is given by the expression

$$BV_{\text{PT}} = \frac{qN_{\text{D}}}{2\epsilon_{\text{s}}} W_n^2 \tag{3.64}$$

where W_n and N_{D} are the width and doping level of the N base, respectively. For a given N-base width, the punch-through breakdown voltage decreases with increasing N-base resistivity. Examples of the variation in the punch-

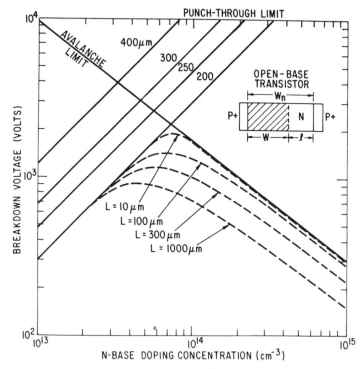

Fig. 3.48. Breakdown voltage of an open-base transistor structure.

through breakdown voltage with N-base doping are shown in Fig. 3.48 for N-base widths ranging from 200 to 400 μm.

Depending on the N-base width and doping, the device breakdown voltage may be limited by either avalanche or punch-through breakdown. It is important to design the device parameters to obtain the peak breakdown voltage for each base width and resistivity. In practice, the peak breakdown voltage, indicated by the point of intersection of the avalanche and punch-through limits, is not achievable because of the finite diffusion length for minority carriers in the N base. With a finite diffusion length, open-base transistor breakdown occurs when the product of the multiplication factor and the base transport factor becomes equal to unity. The base transport factor is given by

$$\alpha_T = \frac{1}{\cosh(l/L_p)} \tag{3.65}$$

where l is the width of the undepleted N base and L_p is the diffusion length for holes. Combining this with the increase in the multiplication factor with reverse junction voltage, the variation of the breakdown voltage with N-base resistivity can be calculated. Some examples are shown in Fig. 3.48 by the dashed lines. It should be noted that larger hole diffusion lengths tend to lower the punch-through-limited breakdown voltage and, consequently, decrease the maximum achievable open-base transistor breakdown. Since the breakdown voltages are measured at low current densities, a small low-level minority carrier lifetime is desirable for achieving larger breakdown voltages. However, a low minority carrier lifetime will degrade the forward conduction characteristics of bipolar devices.

The presence of back-to-back $P-N$ junctions imposes new design criteria for the termination of high-voltage devices. The terminations discussed in earlier sections are oriented to lowering the maximum surface field for a single junction. To achieve a similar surface field reduction at both junctions, the terminations must be combined. These types of termination are discussed next.

3.7.1. Negative–Positive-Bevel Combination

The termination of back-to-back $p-n$ junctions using a combination of negative- and positive-bevel contours is illustrated in Fig. 3.49. The depletion layer positions on both sides of junction J1 under reverse blocking conditions for the upper junction are shown by the dashed lines. Note the extension of the depletion layer on the diffused (P^+) side of the junction near the surface as a result of the negative bevel and its compression on the lightly doped (N) side. The design of the breakdown voltage at the termination in this case can be carried out as described in Section 3.6.4.2 on negative-bevel contours.

When junction J2 is reverse-biased, the depletion layer acquires the shape indicated by the dotted lines in Fig. 3.49. In this case, the depletion layer expands on the lightly doped (N) side of the junction near the surface as a result of the positive bevel. The design of this termination is not as straightforward as

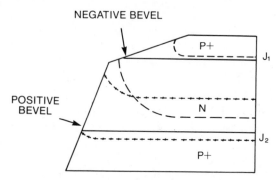

Fig. 3.49. Combination of the positive- and negative-bevel contours for termination of a *P–N–P* open-base transistor structure.

in the case of the simple positive-bevel angle described in Section 3.6.4.1 because of the possibility of premature breakdown at the surface if the depletion layer were to punch through to the forward-biased junction J1. The tendency for punch-through breakdown at the surface becomes greater as the positive-bevel angle θ is reduced because the depletion layer extends further on the lightly doped sided as the positive-bevel angle decreases. Very small positive-bevel angles should not be used for this reason despite the advantage of reduced surface fields. In practice, the positive-bevel angles used for the case of back-to-back *P–N* junctions range from 45 to 60°.

Because of the large area consumed at the edge by the shallow negative bevel (typically 3°) required to achieve high breakdown voltages, the negative–positive-bevel combination is used only for large-area devices. This termination also relies on highly graded diffusions that are applicable to extremely high voltage devices with breakdown voltages of over 2000 V.

3.7.2. Double Positive-Bevel Contours

Since the positive-bevel contour produces the most effective surface field reduction among all device termination techniques and offers the opportunity to achieve the breakdown voltage of the parallel-plane junction, the combination of positive bevels at both the back-to-back junctions is certainly attractive. Two techniques for achieving this type of edge termination are illustrated in Fig. 3.50. In one case (Fig. 3.50*a*), a circular groove is formed at the edge of the wafer usually by means of lapping methods, whereas in the other case (Fig. 3.50*b*) a groove formed by grit blasting by using a nozzle is combined with the conventional positive-bevel contour. In both cases, a positive-bevel angle θ is formed locally at each of the junctions. When either junction is reverse-biased, the depletion layer expands on the lightly doped side of the junction as illustrated by the dashed lines for junction J1 and the dotted lines for junction J2.

The design of this termination is, however, complicated by the interaction between the upper and lower terminations. In the case shown in Fig. 3.50*b*, the

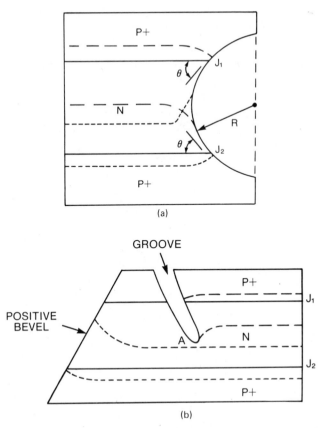

Fig. 3.50. Double positive-bevel contours: (a) circular edge double positive bevel; (b) groove edge double positive bevel.

groove must be sufficiently deep so that the depletion layer on the lightly doped side does not extend beyond its tip if a truly positive bevel is to be realized at junction J1. In addition, the lightly doped N region must be thick enough that the depletion layer of junction J2 does not intersect the tip of the groove when junction J2 is reverse-biased. Thus, unlike the negative–positive-bevel combination in which the N region can be thin and the depletion layers can overlap as shown in Fig. 3.49, the double positive bevel of the termination shown in Fig. 3.50b requires a thicker N region. This can degrade other device performance parameters such as forward voltage drop and switching speed.

In the case of the double positive-bevel termination shown in Fig. 3.50a, the depletion layers can be designed to overlap as illustrated by the dashed and dotted lines. However, this has the consequence of reducing the effective positive-bevel angle for both junctions. A comparison of the surface field for the double positive bevel and the single positive bevel has been obtained by numerical analysis [21]. The electric field distribution along the surface for the two cases is provided in Fig. 3.51. The actual surface contour modeled is also provided since it was not the idealized circular shape shown in Fig. 3.50a. Important points to note are the significantly higher surface electric fields observed for the double positive-bevel case and the presence of field maxima located at regions of discontinuity in the contour geometry. It can be seen that such discontinuities are highly undesirable and should be prevented during the contour formation or subsequently removed by use of isotropic etching techniques.

Investigation of a variety of surface contour geometries by numerical analysis indicates that it is very difficult to reduce the maximum surface electric field to below 60% of the bulk for the double positive bevel. For example, the

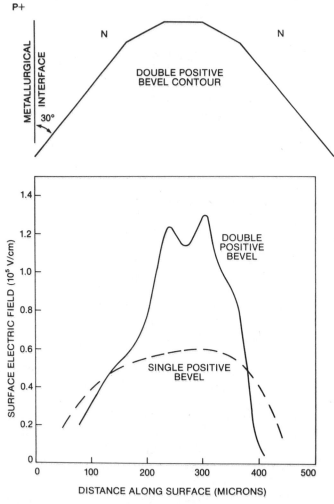

Fig. 3.51. Surface electric field distribution for a double positive-bevel contour. (After Ref. 21, reprinted with permission from the IEEE © 1974 IEEE.)

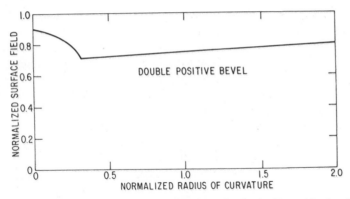

Fig. 3.52. Normalized maximum surface electric field for a circular double-positive bevel as a function of its radius of curvature. (After Ref. 12, reprinted with permission from the IEEE © 1978 IEEE.)

impact of changing the radius of curvature R of the circular double positive bevel, shown in Fig. 3.50a, is provided in Fig. 3.52 [12]. In this figure, the surface electric field has been normalized to the peak electric field at the parallel-plane portion and the radius of curvature normalized to the depletion layer depth on the lightly doped side of the parallel-plane portion. The maximum surface electric field achievable by use of the double positive-bevel contour is significantly higher than the typical maximum surface field (40% of the bulk) achievable using negative-bevel contours. The higher surface electric fields obtained with the double positive-bevel contour gives rise to greater difficulty in device passivation. However, the space consumed at the edge of the wafer is much smaller than that for the negative-bevel angle. Furthermore, it is worth pointing out that the double positive bevel still offers the possibility of achieving bulk breakdown at the parallel-plane portion of the junction as in the case of the simple positive bevel because no field maximum occurs underneath the surface. This is not achievable with the use of the negative-bevel contour, despite the lower surface electric field, because of the occurrence of an electric field maximum in the bulk near the surface whose value exceeds the maximum electric field in the bulk at the parallel-plane portion of the junction as discussed in Section 3.6.4.2.

3.8. HIGH-VOLTAGE SURFACE PASSIVATION

The passivation of the surface of discrete power devices plays an important role in the fabrication of these devices because it determines the ability of the device to withstand high surface electric fields. In addition, the surface charge induced by the passivant can strongly influence the electric fields in the bulk. Methods of passivation extend from the application of rubberlike coatings on the beveled edges of the device to the chemical vapor deposition of inorganic films.

The most commonly used passivation process for high-voltage, large-area devices with bevel contoured surfaces is by means of rubberized coatings or organic polymers [22]. These techniques consist of the initial removal of surface damage resulting from the mechanical surface beveling techniques by use of chemical etching immediately followed by the application of an organic material such as polyimide in a solution of dimethylacetamide. The passivant is then cured by heating in a nitrogen ambient to temperatures near 250°C. This technique of device passivation has been applied for the fabrication of devices with breakdown voltages exceeding 5000 V to obtain stable device characteristics. To ensure long-term stability, it is still necessary to enclose these devices in a hermetically sealed package. This problem can be overcome by using coatings such as silicon nitride that are impervious to the migration of ions and moisture.

Silicon dioxide and silicon nitride films applied to surfaces of devices have been observed to provide stable operation of high-voltage devices with both positively beveled [22] and negatively beveled [23] surfaces. It has been found that the cleaning procedure for the surface of the device prior to the growth of the silicon dioxide layer by thermal oxidation is a critical factor in obtaining sharp breakdown characteristics. The best results are obtained by using the following steps: (1) 30-sec dip in 40% hydrofluoric acid, (2) water rinse, (3) perchlorate etch for 10 sec to remove surface oxides, (4) water rinse and dry by spinning, and (5) immediate loading in a furnace for oxidation. The oxide growth temperature has been found to strongly influence the device leakage current, with the best results obtained by growth at 800°C followed by slow cooling of the furnace (2°C/min) from 800 to 720°C. In addition, the use of oxidation tubes and boats made of pure silicon to hold the wafers is recommended for obtaining low leakage currents. Furthermore, the use of wet oxidation is acceptable only if preceded by a dry oxidation cycle. Since these silicon dioxide films are not impervious to moisture or ion migration, they are generally covered by a film of silicon nitride obtained by using chemical vapor deposition techniques. The successful application of this process has been reported for devices with breakdown voltages of up to 3,000 V [22, 23].

The silicon dioxide–silicon nitride "sandwich" has also been used for the passivation of planar junctions. However, superior results can be obtained in these junctions by using semi-insulating polycrystalline silicon (SIPOS) films. Polycrystalline silicon films with high resistivities can be obtained by the addition of oxygen or nitrogen during the deposition [24]. The resulting films have a resistivity of about 10^{10} $\Omega\cdot$cm compared with 10^5 $\Omega\cdot$cm for undoped polysilicon and 10^{14} $\Omega\cdot$cm for silicon dioxide. As discussed earlier, the breakdown voltage of planar junctions is lower than that of parallel-plane junctions as a result of higher electric fields in the curved portion of the junction. These fields can be decreased by application of the SIPOS layer on the junction surface to result in higher breakdown voltages. In this method of passivation, the negative potential of the P-diffused region of the junction is transmitted to the surface region of the N-type substrate by current flow in the SIPOS film. This results in an extension of the surface depletion layer that lowers the surface electric field. The improvement in the breakdown voltage of the junction depends on the conductivity of the SIPOS film, with larger voltages being achieved at lower oxygen doping levels.

These films, however, have a lower resistivity than does silicon dioxide, which results in an additional junction leakage current. An optimum oxygen content of 15–35% is found to produce the best device characteristics. The oxygen-doped SIPOS films have low charge densities and can be applied for the passivation of both P- and N-type substrates but are not impervious to moisture or ionic contamination. Consequently, it is necessary to coat them with a film of SIPOS containing nitrogen. A layer of silicon dioxide is also added on top of the nitrogen-doped SIPOS layer to increase the breakdown strength of the composite dielectric film. Power bipolar transistors with collector breakdown voltages of up to 10,000 V have been fabricated with the SIPOS passivation technique.

Another technique that has been used to passivate high-voltage junctions in power devices is by the electrophoretic deposition of zinc borosilicate glasses containing additives such as Ta_2O_5 or Sb_2O_3 [25]. This technique is capable of providing thick glass coatings in selective areas with expansion coefficients matched to that of silicon. The process consists of the preparation of a suspension of glass particles in a medium such as acetone or isopropyl alcohol. The glass particles are then deposited on the silicon substrates by passage of current through the suspension with the silicon as the cathode. After the deposition of the glass, it is first densified to remove the solvent and then fired at temperatures between 800 and 900°C in an oxygen-containing ambient. The addition of Ta_2O_5 to the glass has been found to improve junction stability, and the addition of Sb_2O_3 to the glass has been found to improve the dielectric breakdown strength. Successful application of this technique has been demonstrated with junction breakdown voltages of over 1000 V.

3.9. COMPARISON OF TERMINATIONS

A large variety of high-voltage device termination techniques have been described above. These techniques are compared in Table 3.1 on the basis of the typical breakdown voltage achievable with the use of each technique and the surface electric field reduction [26]. It can be seen that the positive-bevel contour offers the best results. However, the selection of a specific technique for any particular application is dependent on several other important factors.

First, the approach depends on the size of the device. In the case of small devices, beveling techniques are not practical. The most commonly used field termination method for small dies is the use of planar junctions with field rings. This allows the achievement of up to 80% of the ideal breakdown voltage of the parallel-plane junction. This is adequate for most bipolar devices. In the case of power MOSFET and JFET devices, where their on-resistance is strongly influenced by the breakdown voltage, the junction termination extension offers

TABLE 3.1. High-Voltage Device Termination Techniques

Technique	Typical Breakdown Voltage (%)[a]	Peak Surface Electric Field (%)[b]	Typical Device Size	Device Types	Remarks
Planar junction	50	80	Small (<100 mils)	BJT, MOSFET	Seldom used for high-voltage devices
Planar junction with field ring	80	80	Medium (≤1 in.)	BJT, MOSFET, SCR	Well suited for a large number of devices per wafer
Planar junction with field plate	60	80	Medium (≤1 in.)	BJT, MOSFET	Usually used in conjunction with field ring
Positive bevel	100	50	Large (>1 in.)	Rectifier, SCR	Well suited for single device per wafer
Negative bevel	90	60	Large (>1 in.)	SCR	Well suited for single device per wafer
Double positive bevel	100	80	Large (>1 in.)	SCR	Well suited for single device per wafer only
Positive etch contour	90	60	All	BJT, MOSFET, SCR	Well suited for a large number of devices per wafer
Negative etch contour	80	60	All	BJT, MOSFET, SCR	Well suited for a large number of devices per wafer
Junction termination extension	95	80	All	BJT, MOSFET, SCR	Well suited for both single devices and a large number of devices per wafer; high leakage current; passivation sensitive

[a] As percentage of parallel-plane case.
[b] As percentage of bulk.

the best promise, albeit at higher leakage currents. Alternately, the positive etch contour or passivation with SIPOS can be used to achieve close to 90% of the ideal breakdown voltage.

Second, the planar junction with field ring and the planar junction with junction termination extension can be used to simultaneously process a large number of small devices on each wafer. In contrast, the beveling methods require handling individual devices and are not cost-effective for small devices. For large-area devices, fabricated from individual silicon wafers, for instance, the beveling techniques are the most promising because of the nearly ideal breakdown voltages achievable with this technique. The beveling technique ensures a large surface electric field reduction that makes surface passivation less critical. Today most high-current rectifiers and thyristors are fabricated by using combinations of positive- and negative-bevel contours. Devices with breakdown voltages of up to 6500 V and current handling capability of thousands of amperes are commercially available.

In summary, the choice of the edge termination technique should be made based on the die size, the sensitivity of the other electrical characteristics of the device on the breakdown voltage, and the ease of surface passivation of the edges. Until recently, the termination of high-voltage devices was considered an art rather than a science. The development of two-dimensional numerical solution techniques now allows the design of edge terminations with a high degree of confidence.

REFERENCES

1. R. Van Overstraeten and H. DeMan, "Measurements of the ionization rates in diffused silicon $p-n$ junctions," *Solid State Electron.*, **13**, 583–608 (1970).

2. N. R. Howard, "Avalanche multiplication in silicon junctions," *J. Electron. Control*, **13**, 537–544 (1962).

3. W. Fulop, "Calculation of avalanche breakdown of silicon $p-n$ junctions," *Solid State Electron.*, **10**, 39–43 (1967).

4. R. A. Kokosa and R. L. Davies, "Avalanche breakdown of diffused silicon $p-n$ junctions," *IEEE Trans. Electron Devices*, **ED-13**, 874–881 (1966).

5. M. S. Adler and V. A. K. Temple, *Semiconductor Avalanche Breakdown Design Manual*, GE Technology Marketing Operation, Schenectady, NY, 1979.

6. B. J. Baliga and S. K. Ghandhi, "Analytical solutions for the breakdown voltage of abrupt cylindrical and spherical junctions," *Solid State Electron.*, **19**, 739–744 (1976).

7. S. M. Sze and G. Gibbons, "Effect of junction curvature on breakdown voltage in semiconductors," *Solid State Electron.*, **9**, 831–845 (1966).

8. V. A. K. Temple and M. S. Adler, "Calculation of the diffusion curvature related avalanche breakdown in high voltage planar $p-n$ junctions," *IEEE Trans. Electron Devices*, **ED-22**, 910–916 (1975).

9. Y. C. Kao and E. D. Wolley, "High voltage planar $p-n$ junctions," *Proc. IEEE*, **55**, 1409–1414 (1967).

10. M. S. Adler, V. A. K. Temple, A. P. Ferro, and R. C. Rustay, "Theory and breakdown voltage for planar devices with a single field limiting ring," *IEEE Trans. Electron Devices*, **ED-24**, 107–113 (1977).

11. R. L. Davies and F. E. Gentry, "Control of electric field at the surface of $p-n$ junctions," *IEEE Trans. Electron Devices*, **ED-11**, 313–323 (1964).

12. M. S. Adler and V. A. K. Temple, "Maximum surface and bulk electric fields at breakdown for planar and beveled devices," *IEEE Trans. Electron Devices*, **ED-25**, 1266–1270 (1978).

13. J. Cornu, "Field distribution near the surface of beveled $p-n$ junctions at high voltage devices," *IEEE Trans. Electron Devices*, **ED-20**, 347–352 (1973).

14. M. S. Adler and V. A. K. Temple, "A general method for predicting the avalanche breakdown voltage of negative beveled devices," *IEEE Trans. Electron Devices*, **ED-23**, 956–960 (1976).

15. V. A. K. Temple, B. J. Baliga, and M. S. Adler, "The planar junction etch for high voltage and low surface fields in planar devices," *IEEE Trans. Electron Devices*, **ED-24**, 1304–1310 (1977).

16. V. A. K. Temple and M. S. Adler, "The theory and application of a simple etch contour for near ideal breakdown voltage in plane and planar $p-n$ junctions," *IEEE Trans. Electron Devices*, **ED-23**, 950–955 (1976).

17. V. A. K. Temple, "Junction termination extension, a new technique for increasing avalanche breakdown voltage and controlling surface electric fields in $p-n$ junctions," *IEEE International Electron Devices Meeting Digest*, Abstract 20.4, pp. 423–426 (1977).

18. A. S. Grove, O. Leistiko, and W. W. Hooper, "Effect of surface fields on the breakdown voltage of planar silicon $p-n$ junctions," *IEEE Trans. Electron Devices*, **ED-14**, 157–162 (1967).

19. F. Conti and M. Conti, "Surface breakdown in silicon planar diodes equipped with field plate," *Solid State Electron.*, **15**, 93–105 (1972).

20. A. Herlet, "The maximum blocking capability of silicon thyristors," *Solid State Electron.*, **8**, 655–671 (1965).

21. J. Cornu, S. Schweitzer, and O. Kuhn, "Double positive bevel: A better edge contour for high voltage devices," *IEEE Trans. Electron Devices*, **ED-21**, 181–184 (1974).

22. R. R. Verderber, G. A. Gruber, J. W. Ostrowski, J. E. Johnson, K. S. Tarneja, D. M. Gillott, and B. J. Coverton, "SiO_2/Si_3N_4 passivation of high power rectifiers," *IEEE Trans. Electron Devices*, **ED-17**, 797–799 (1970).

23. R. E. Blaha and W. R. Fahrner, "Passivation of high breakdown voltage $p-n-p$ structures by thermal oxidation," *J. Electrochemi. Soc.*, **123**, 515–518 (1976).

24. T. Matsushita, T. Aoki, T. Ohtsu, H. Yamoto, H. Hayashi, M. Okayama, and Y. Kawana, "Highly reliable high voltage transistors by use of the SIPOS process," *IEEE Trans. Electron Devices*, **ED-23**, 826–830 (1976).

25. K. Miwa, M. Kanno, S. Kawashima, S. Kawamura, and T. Shibuya, "Glass passivation of silicon devices by electrophoresis," *Denki Kagaku*, **40**, 478–484 (1972).

26. B. J. Baliga, "High voltage device termination techniques—a comparative review," *IEE Proc.*, **129**, 173–179 (1982).

27. S. K. Ghandhi, *Semiconductor Power Devices*, Wiley, New York, 1977.

28. R. Stengl and U. Gosele, "Variation of lateral doping—a new concept to avoid high voltage breakdown of planar junctions", *IEEE International Electron Devices Meeting Digest*, Abstract 6.4, pp. 154–157 (1985).

PROBLEMS

3.1. What is the background doping level required to achieve a breakdown voltage of 60, 200, 600, 1200 and 2000 V for an abrupt parallel-plane junction? What are the corresponding depletion layer thicknesses at breakdown?

3.2. Consider a punch-through diode with an N-base doping level of $5 \times 10^{13}/cm^3$ and a thickness of 20 μm. What is its breakdown voltage? What is the background doping level and depletion layer thickness for the corresponding non-punch-through case?

3.3. A P^+-N junction with a cylindrical junction termination is produced by planar diffusion into an N-type region with a doping concentration of $1 \times 10^{14}/cm^3$. Determine the junction depth required to obtain a breakdown voltage of 800 V.

3.4. What is the depth of the junction for the parameters given in Problem 3.3 when a single optimally spaced floating field ring is used?

3.5. What is the background doping level required to achieve a breakdown voltage of 800 V when junction termination extension is used?

3.6. Determine the positive-bevel angle required to reduce the surface electric field to less than 50% of the bulk value. Calculate the percentage area consumed at the edge of a device 5 cm in diameter for a wafer thickness of 40 mils (0.1 cm).

3.7. Calculate the breakdown voltage of an open-base P^+-N-P^+ transistor with N-base doping and width identical to that used in Problem 3.2. Compare this case to the punch-through P^+-N-N^+ structure.

3.8. What is the doping concentration and N-base thickness required for the open-base P^+NP^+ transistor to obtain the same breakdown voltage as the punch-through P^+NN^+ structure given in Problem 3.2? Assume that the minority carrier diffusion length is very small.

3.9. What are the depletion layer widths on the two sides of a $P-N$ junction formed by diffusion of a P-type region with a surface concentration of $10^{17}/cm^3$ to a depth of 10 μm into an N-type region of doping $1 \times 10^{14}/cm^3$?

3.10. What is the negative-bevel angle required to obtain a breakdown voltage of 1200 V?

4 POWER JUNCTION FIELD-EFFECT TRANSISTORS

The basic concept of the junction field-effect transistor (JFET) was described by Shockley in 1952 [1]. These devices were intended for low-voltage signal processing applications. Their primary feature in this regard was a high input impedance and faster inherent switching speed. With the advent of MOS technology, the use of JFETs in integrated circuits has been relegated to only a few special applications such as in current limiters and operational amplifiers. Its kin, the metal–semiconductor field-effect transistor (MESFET), has been used for more applications such as microwave amplification and very-high-speed signal processing using gallium arsenide as the base material. In these cases, the lack of an MOS technology for gallium arsenide has resulted in the use of MESFETs.

The extension of silicon JFET technology to power control at high operating voltages occurred in the 1970s. Prior to this, the development of high-voltage JFETs with substantial current handling capability was considered as part of the "blue skies" department [2]. The reasons for the delay in the evolution of JFET technology can be traced to processing difficulties rather than fundamental theoretical limitations. Improvements in process technology, coupled with the conception of several novel device structures, have now allowed the fabrication of devices with blocking voltages exceeding 100 V.

When compared with high-voltage, power bipolar junction transistors, JFETs have several advantages as power switching devices. They have a higher input impedance than do bipolar transistors and a negative temperature coefficient for the drain current that prevents thermal runaway, thus allowing the coupling of many devices in parallel to increase the current handling capability. Furthermore, the JFET is a majority carrier device with a higher inherent switching speed because of the absence of minority carrier recombination, which limits the speed of bipolar transistors. This also eliminates the stringent lifetime control requirements that are essential during the fabrication of high-speed bipolar transistors. Another advantage of these devices is the absence of second breakdown. This results in a much larger safe operating area in comparison to bipolar transistors.

4.1. BASIC STRUCTURES AND OPERATION

Discrete, high-voltage power devices with large current carrying capability are invariably fabricated with a vertical current conduction path. The vertical current flow in discrete power devices is favored with respect to lateral current flow for several reasons. First, the internal structure of power devices with vertical current flow paths can be designed so that electric field crowding can be minimized. The prevention of field crowding within the internal structure is extremely important for achieving nearly ideal breakdown characteristics within the limits set by the device termination. The design of device terminations to achieve within 90% of the parallel plane junction breakdown was discussed extensively in Chapter 3.

Second, in a vertical device design, one of the high current carrying terminals is placed at the top of the wafer and the other, at the bottom. The control electrode of these three terminal active power switching devices is also placed on the top surface for easy access during packaging. When compared with a lateral device structure, the placement of one of the large, high current carrying terminals at the bottom of the chip results in significant reduction in chip area.

Many structures for high-voltage JFETs have been developed based on various processing innovations. These structures can be classified into two basic types: (1) the surface gate device structure and (2) the buried gate device structure. Typical examples of these JFET structures are illustrated in Fig. 4.1. Fundamentally, the JFET consists of a bar of semiconductor material whose resistance can be controlled by the application of a reverse-bias voltage to a gate region. For the cases illustrated in Fig. 4.1, a $P–N$ junction gate region has been shown. A metal–semiconductor (Schottky) junction can also be utilized to achieve the same function.

In the absence of a gate bias, that is, with the gate short-circuited to the source, the current flow between drain and source is limited by the resistance of the lightly doped N-type region between these current carrying terminals. The N-type region consists of two portions. The region between the junction gates is called the *channel* and the region below the junction gates is called the *drift region*. The resistances of both these regions add together to determine the total resistance to current flow between the drain and source terminals.

With the application of a reverse gate bias with respect to the source (negative voltage to the gate terminal for the n-channel devices shown in Fig. 4.1), a depletion layer forms around the gate junctions and extends out into the channel. Since the depletion region is devoid of free carriers, the resistance of the channel region increases with the application of higher reverse gate bias voltages. By use of the gate junction depletion layer, the resistance of the JFET can be altered by changing the reverse gate bias voltage. Thus the JFET is a voltage-controlled device, in contrast to a bipolar transistor, which is a current-controlled device. In the JFET, the gate bias supply needs to provide only a small displacement current to modulate the depletion region width. The resulting high input impedance is an important feature of these devices.

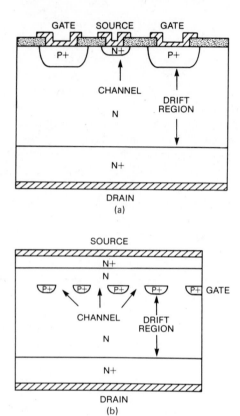

Fig. 4.1. Structures of high-voltage JFETs with gates located (*a*) at the surface and (*b*) buried below the source region.

The resistance of the JFET is controlled by changing the channel conductivity. For low-voltage signal processing applications, JFETs are designed with minimum drift region width to minimize their parasitic resistance. In a JFET designed for operation at high drain voltages it is essential to include a wide drift region. During the application of large drain voltages (positive for the *n*-channel devices shown in Fig. 4.1) in the forward blocking mode, the drain voltage is supported across the gate depletion layer. This depletion layer extends down from the gate junction toward the drain. The drift region doping concentration and thickness must be designed to support the drain–gate voltage, that is, the sum of the absolute values of the drain and gate voltages, because they combine to determine the total reverse-bias voltage across the gate *P–N* junction. It should also be noted that the gate–source structure must be capable of supporting the highest reverse gate–source voltage. This imposes stringent requirements on the processing of surface gate devices.

In the forward blocking mode, the depletion layers from the gate junctions extend through the entire channel. The applied reverse gate bias sets up a potential barrier in the channel. For current to flow between drain and source, electrons must surmount this potential barrier. As the drain voltage increases, the potential barrier is lowered and electron injection becomes easier. The maximum blocking voltage capability of high-voltage JFETs is determined by the current flow across the channel potential barrier.

4.2. BASIC DEVICE CHARACTERISTICS

Low-voltage, lateral JFETs exhibit pentodelike output current–voltage characteristics; that is, the drain current saturates at high drain voltages. This characteristic is typical for these long-channel devices in which the length of the gate along the current flow path is much larger than the channel width. An example of this type of device structure is shown in Fig. 4.2. At small drain voltages, the depletion layer width is determined by the gate voltage. It extends with uniform width across the gate junction as illustrated by the dashed lines in Fig. 4.2a. When the drain voltage becomes comparable to the gate voltage

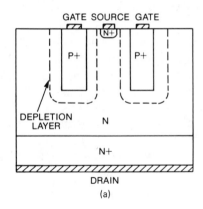

(a)

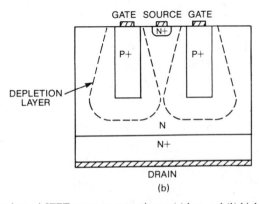

(b)

Fig. 4.2. Long-channel JFET structure operating at (a) low and (b) high drain voltage.

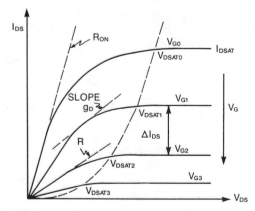

Fig. 4.3. Pentodelike characteristics of a long-channel JFET.

in the presence of substantial vertical drain–source current flow, a large drop in voltage occurs along the channel. The depletion layer now takes the form shown in Fig. 4.2b. The depletion layer progressively widens from the source toward the drain. The channel width is then smaller on the drain side. This increases the resistance to current flow. On the basis of these considerations, it can be concluded that the resistance of a JFET will increase with increase in gate and drain voltages. The resulting pentodelike characteristics are shown in Fig. 4.3. Note that at higher drain voltages the characteristics show current saturation. The saturated drain current is a strong function of the gate bias voltage. Current saturation occurs when the adjacent gate depletion layers intersect in the channel.

In addition to the lowest resistance to current flow indicated by the dotted lines marked R_{on} for the case of zero gate bias (V_{GO}), an important figure of

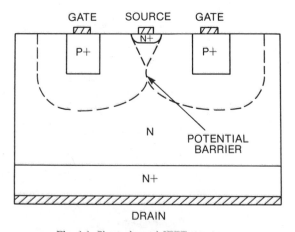

Fig. 4.4. Short-channel JFET structure.

merit for the JFET is the rate of change in drain current with gate bias voltage. This parameter is known as *transconductance*. It is defined as

$$g_{\mathrm{m}} = \frac{\Delta I_{\mathrm{DS}}}{\Delta V_{\mathrm{GS}}}\bigg|_{V_{\mathrm{G}}} \tag{4.1}$$

where ΔI_{DS} is the change in drain current for a change $\Delta V_{\mathrm{GS}} = (V_{\mathrm{G1}} - V_{\mathrm{G2}})$ in gate bias voltage. The saturated drain current ($I_{\mathrm{D,sat}}$) is another parameter that must be considered during device design.

The pentodelike characteristics for JFETs are typically observed in long-channel devices. In these devices, the drain potential cannot penetrate into the channel. If the gate structure is designed with a small channel length compared with the channel width, the JFET characteristics are significantly altered. In the short-channel structure, the potential barrier established in the channel by the reverse gate bias extends over only a small vertical distance. The depletion layer shape for such a device is shown in Fig. 4.4. As the drain voltage is increased, the drain potential penetrates into the channel and lowers the potential barrier. To illustrate this, the two-dimensional potential distribution in the channel is shown in Fig. 4.5. At low drain voltages, the potential barrier established by the gate voltage is pronounced and extends throughout the channel

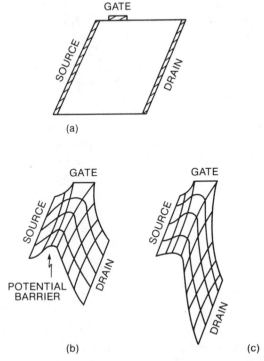

Fig. 4.5. Two-dimensional view of the potential distribution between drain and source of the JFET. Note the reduction in the channel potential barrier height when the drain voltage increases from *b* to *c*.

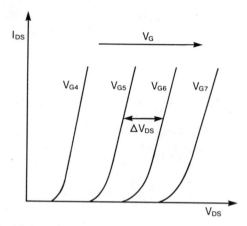

Fig. 4.6. Triodelike characteristics of a short-channel JFET.

as shown in Fig. 4.5*b*. When the drain voltage is increased as shown in Fig. 4.5*c*, the potential barrier is pulled down by the drain potential in regions of the channel away from the gate region. Electron injection can now occur in these regions. Since the injection of carriers across the potential barrier varies exponentially with the barrier height, the drain current exhibits a rapid increase once the potential barrier is reduced. The resulting triodelike characteristics, that is, drain current continuously increasing with increasing drain voltage, of JFETs with short channel length are illustrated in Fig. 4.6.

An important parameter for devices with triodelike characteristics is the voltage blocking gain. The AC blocking gain can be defined as

$$g_B = \frac{\Delta V_{DS}}{\Delta V_{GS}}\bigg|_{I_{DS}} \tag{4.2}$$

where ΔV_{DS} is the incremental increase in the blocking voltage for an increase $\Delta V_{GS} = (V_{G6} - V_{G5})$ in gate voltage. It is crucial to achieve a large blocking gain in high-voltage JFETs for two reasons: (1) a higher blocking gain reduces the gate drive voltage that must be supplied by the gating circuit to achieve any desired forward blocking capability, and (2) a higher blocking gain reduces the total voltage that must be supported by the gate junction. The total reverse bias across the gate junction is given by

$$V_{RJ} = (V_{DS} + V_{GS}) = V_{DS}\left(1 + \frac{1}{G_B}\right) \tag{4.3}$$

where $G_B = V_{DS}/V_{GS}$ is the DC blocking gain. A large blocking gain reduces the breakdown voltage that must be designed for the gate junctions.

The first high-voltage JFET devices had blocking gains of less than 5, thus limiting their blocking capability to below 100 V. Improvements in gate structure have allowed increasing the blocking gain to over 20. This has allowed the fabrication of devices capable of blocking up to 500 V.

The preceding discussion described long-channel devices with pentodelike characteristics and short-channel devices with triodelike characteristics. Neither of these cases provides optimum power JFET characteristics. The pentodelike characteristics are conducive to achieving high current handling capability, whereas the triodelike characteristics provide high forward blocking capability. In practical power JFET designs, an intermediate channel length must be utilized, resulting in mixed pentode–triode characteristics. These devices exhibit pentodelike characteristics at high drain currents and low gate bias voltages and triodelike characteristics at low drain currents and large gate bias voltages.

4.3. DEVICE ANALYSIS

The design of high-voltage JFETs with large current handling capability requires an understanding of current transport in the triode and pentode regimes of operation. The analysis must include the changes in resistance of the channel, including current saturation at high drain currents. The analysis of the forward conduction characteristics in this pentode regime must account for the resistance of the drift region that is in series with the channel resistance. For analysis of the triode regime that determines the foward blocking capability, it is necessary to understand the parameters controlling the channel potential barrier and how they influence the blocking gain.

4.3.1. Forward Conduction

The fundamental structure of the junction gate FET consists of a thin layer of semiconductor sandwiched between gate regions of opposite conductivity type as shown in Fig. 4.7 for an *n*-channel device with P^+ gate regions. In this figure

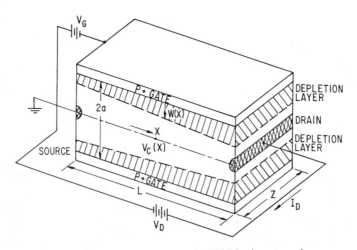

Fig. 4.7. Fundamental structure of a JFET in the gate region.

only the channel region of the JFET is considered. The drift region is considered later in the analysis. The current flow between the drain and source contacts is carried by majority carriers in the channel region. When a reverse bias is applied to the gate regions with respect to the source, the gate junction depletion layer extends into the channel region, reducing its cross-sectional area. This allows the modulation of the drain current by means of the applied gate voltage.

To prevent forward biasing of the gate junction, the n-channel JFET shown in Fig. 4.7 must be operated with positive drain voltages. In the presence of drain current flow, a resistive voltage drop occurs along the channel. Since the gate and drain voltages are additive in reverse biasing the junction, the resistive voltage drop causes the junction reverse bias to vary along the channel. The gate junction depletion layer will be narrowest at the end of the gate near the source and widest at the end of the gate near the drain. These depletion layer widths can be determined by using expressions for reverse-bias junction voltages of V_G and $(V_G + V_D)$. Because of the vertical channel geometry used for high-voltage JFETs, the channel regions are invariably homogeneously doped and it is not necessary to consider more complex doping profiles. In the case of a uniformly doped channel with doping level N_D per cubic centimeter, the depletion layer widths at the two ends of the channel are given by

$$W_S = \sqrt{\frac{2\epsilon_s}{qN_D}(V_G + V_{bi})} \qquad (4.4)$$

and

$$W_D = \sqrt{\frac{2\epsilon_s}{qN_D}(V_D + V_G + V_{bi})} \qquad (4.5)$$

where the absolute values of the gate (V_G), drain (V_D), and built-in (V_{bi}) potentials must be used in the equations. The junction built-in potential is given by

$$V_{bi} = \frac{kT}{q}\ln\left(\frac{N_D N_A}{n_i^2}\right) \qquad (4.6)$$

where N_A and N_D are the doping concentration on opposite sides of the abrupt metallurgical junction. The preceding expressions are based on an abrupt approximation for the depletion layer edge.

4.3.1.1. Field-Independent Mobility Analysis.

The current flow in the channel of the JFET can be derived analytically by making two additional assumptions: (1) the mobility is constant (the decrease in mobility with increasing electric field described in Chapter 2 is considered later), and (2) the magnitude of the electric field along the channel (i.e., the longitudinal field) is much smaller than the electric field normal (i.e., the transverse field) to the channel—this is termed the *gradual channel approximation*. This allows a one-dimensional calculation of the depletion width of the gate junction after taking into account the local potential in the channel. This approximation is a good one for devices with large length to width (L/a) ratios before channel pinch-off occurs.

Under the gradual channel approximation, Poisson's equation can be used to compute the gate depletion layer width using the local channel potential. Assuming an abrupt gate junction, the depletion width at location x along the channel is given by

$$W(x) = \sqrt{\frac{2\epsilon_s}{qN_D} (V_C(x) + V_G + V_{bi})} \tag{4.7}$$

where V_C is the local potential in the channel at location x. The channel cross section at x is $2[a - W(x)]$. From this cross section and the conductivity of $(q\mu N_D)$, an expression for the drain current is obtained:

$$I_D = 2[a - W(x)](q\mu N_D)Z\left(\frac{dV}{dx}\right) \tag{4.8}$$

where a is half the width of the undepleted channel. Thus

$$I_D\, dx = 2[a - W(x)](q\mu N_D)Z\, dV \tag{4.9}$$

Differentiation of Eq. (4.7) yields

$$dV = \frac{qN_D}{\epsilon_s} W\, dW \tag{4.10}$$

Substitution for dV in Eq. (4.9) yields

$$I_D\, dx = 2\frac{(q^2\mu N_D^2)}{\epsilon_s} Z(a - W)W\, dW \tag{4.11}$$

Integration from $x = 0$ to $x = L$ leads to

$$I_D L = \frac{q^2\mu N_D^2 Z}{3\epsilon_s} a^3 \left[\frac{3}{a^2}(W_D^2 - W_S^2) - \frac{2}{a^3}(W_D^3 - W_S^3)\right] \tag{4.12}$$

where W_D and W_S are the depletion layer widths at the drain $(x = L)$ and source $(x = 0)$ ends of the channel, respectively. Using Eqs. (4.4) and (4.5) for these boundary values, it can be shown that

$$I_D = 2a(q\mu N_D)\frac{Z}{L}\left\{V_D - \frac{2}{3a}\left(\frac{2\epsilon_s}{qN_D}\right)^{1/2}[(V_D + V_G + V_{bi})^{3/2} - (V_G + V_{bi})^{3/2}]\right\} \tag{4.13}$$

This expression provides an explicit relationship between the drain current and the gate–drain voltages. The output current–voltage characteristics of the JFET in the pentode regime can be obtained from this equation.

The basic I–V characteristics described by Eq. (4.13) were illustrated in Fig. 4.3. These characteristics can be divided into three segments.

Segment 1. The first segment occurs at very low drain voltages. In this segment, the drain current increases linearly with increasing drain voltage. In the linear region of the output characteristics, the JFET exhibits an resistive characteristic, with the resistance a function of the gate bias voltage. This resistance can be obtained by differentiating Eq. (4.13) with respect to the drain voltage:

$$\frac{1}{R} = \frac{dI_\mathrm{D}}{dV_\mathrm{D}}$$

$$= 2(q\mu N_\mathrm{D})\frac{Z}{L}\left\{a - \sqrt{\frac{2\epsilon_\mathrm{s}}{qN_\mathrm{D}}(V_\mathrm{D} + V_\mathrm{G} + V_\mathrm{bi})}\right\} \tag{4.14}$$

Since the drain voltage V_D is much smaller than the gate voltage V_G in the linear region, it follows that

$$\frac{1}{R} = 2q\mu N_\mathrm{D}\frac{Z}{L}(a - W_\mathrm{S}) \tag{4.15}$$

The resistance in the linear region given by Eq. (4.15) is simply equal to that of semiconductor region of width $2(a - W_\mathrm{S})$ where W_S is related to the gate bias voltage by Eq. (4.4). The lowest value of the resistance is an important parameter because it determines the power dissipation in the JFET during current conduction in the on state. The lowest resistance occurs when the gate bias is reduced to zero. This value is called the on-resistance. It is given by the expression

$$R_\mathrm{on} = \frac{L}{2q\mu N_\mathrm{D}Z}\left[a - \sqrt{\frac{2\epsilon_\mathrm{s}}{qN_\mathrm{D}}V_\mathrm{bi}}\right]^{-1/2} \tag{4.16}$$

Note that this expression applies to the channel region only and excludes the additional resistance contributed by the drift region.

Segment 2. As the drain voltage increases, the depletion layer width W_D on the drain end of the channel increases in size. Ultimately, the depletion layer width W_D becomes equal to half the channel width a. At this point the depletion layers of the adjacent gate junctions join together and pinch off the channel. The condition for pinch-off can be represented by

$$V_\mathrm{p} = (V_\mathrm{D} + V_\mathrm{G} + V_\mathrm{bi}) = \frac{qN_\mathrm{D}a^2}{2\epsilon_\mathrm{s}} \tag{4.17}$$

where V_p is called the pinch-off voltage. The drain voltage at which pinch-off occurs decreases linearly with increasing gate bias.

Segment 3. As illustrated in Fig. 4.3, the drain current increases with increasing drain current until the pinch-off condition is reached. At this point the drain current becomes constant. The saturated drain current can be obtained from

Eq. (4.13) by substituting for the drain voltage at pinch-off using Eq. (4.17):

$$I_{D,sat} = 2a(q\mu N_D)\frac{Z}{L}\left\{\frac{qN_Da^2}{6\epsilon_s} - (V_G + V_{bi}) + \frac{2}{3a}\left(\frac{2\epsilon_s}{qN_D}\right)^{1/2}(V_G + V_{bi})^{3/2}\right\}$$

(4.18)

The saturated drain current can be calculated from this equation by using the known device parameters. Alternately, it can be expressed in terms of the pinch-off voltage:

$$I_{D,sat} = \frac{a^3q^2\mu N_D^2}{3\epsilon_s}\frac{Z}{L}\left\{1 - 3\left(\frac{V_G + V_{bi}}{V_p}\right) + 2\left(\frac{V_G + V_{bi}}{V_p}\right)^{3/2}\right\}$$

(4.19)

The maximum value of the saturated drain current occurs when the gate voltage is zero. If the built-in junction potential is neglected, the following condition will result:

$$I_{D,sat,max} = \frac{a^3q^2\mu N_D^2}{3\epsilon_s}\frac{Z}{L}$$

(4.20)

Transconductance. Another important JFET characteristic is its transconductance, defined by Eq. (4.1). The transconductance provides a measure of the control of the drain current by the gate voltage. A large transconductance is desireable to minimize the gate drive and provide high power gain. A large transconductance also results in higher frequency response for these devices as discussed later in this chapter. The transconductance can be derived from the drain current Eq. (4.13) by differentiating with respect to the gate bias voltage:

$$g_m = \frac{dI_D}{dV_G}\bigg|_{V_D}$$

$$= 2a(q\mu N_D)\frac{Z}{L}\left(\frac{2\epsilon_s}{qN_D}\right)^{1/2}[(V_D + V_G + V_{bi})^{1/2} - (V_G + V_{bi})^{1/2}]$$

(4.21)

The transconductance can be expressed in terms of the depletion layer widths

$$g_m = 2a(q\mu N_D)\frac{Z}{L}(W_D - W_S)$$

(4.22)

The maximum transconductance occurs in the saturated current region of the device characteristics:

$$g_{m,max} = 2a(q\mu N_D)\frac{Z}{L}(a - W_S)$$

(4.23)

From the expressions for the drain current and transconductance derived previously, it can be concluded that the current handling capability can be maximized by using a large channel width a and short channel length L. It is also necessary to maximize the channel width Z by using a highly interdigitated device design. Since the drain current and transconductance are linearly dependent on the free carrier mobility, n-channel devices are favored over p-channel devices because, for silicon, the electron mobility is three times larger than that for holes.

4.3.1.2. Field-Dependent Mobility Analysis.   The assumption of a constant mobility for the free carriers in the channel, irrespective of the drain voltage applied to the JFET, is valid as long as the electric field strength along the channel remains below 1×10^3 V/cm. The drain current saturation then occurs as a result of channel pinch-off. This holds true for devices with channels that have a small width and large length. However, in devices designed for high current handling capability it becomes necessary to use short and wide channels to reduce the channel resistance. In these devices, the electric field along the channel can exceed 1×10^3 V/cm prior to pinch-off, and the decrease in the mobility with increasing electric field must be accounted for in computing their output characteristics.

A comparison of the output characteristics of a typical high voltage JFET calculated with and without taking into account the variation of the mobility with electric field is provided in Fig. 4.8. Note that the field dependence of the mobility has its strongest influence at higher drain currents. At high gate voltages where only low drain currents are observed, the longitudinal electric field along the channel remains small up to pinch-off. When the gate bias voltage is reduced, the channel width increases allowing large drain currents to flow. The larger drain current flow is accompanied by a corresponding increase in the longitudinal electric field in the channel. Now the field dependence of the mobility strongly impacts the output characteristics of the device.

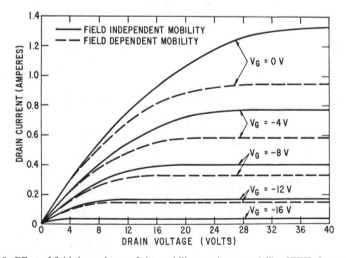

Fig. 4.8. Effect of field dependence of the mobility on the pentodelike JFET characteristics.

At low gate bias voltages, the field dependence of the mobility has several important effects on the characteristics. First, it results in a reduction in the slope of the output characteristics near the origin. Although the field dependence of the mobility does not alter the on-resistance because this resistance is defined for low drain voltages, it results in an increase in the power dissipation because of the larger forward drop across the JFET at higher drain currents. Second, the field dependent mobility results in drain current saturation occurring at lower drain voltages. Furthermore, it produces a reduction in the magnitude of the saturated drain current.

In the case of the device characteristics illustrated in Fig. 4.8, channel pinch-off takes place well before velocity saturation occurs. In Chapter 2, it was pointed out that when the electric field exceeds 3×10^4 V/cm, the carrier velocity for both electrons and holes becomes constant. For devices designed with very wide channels, the pinch-off effect can become weak. In these devices, the output characteristics will essentially follow the velocity–field characteristics for the free carriers. The drain current will then be given by

$$I_D = 2(a - W_S)qN_D Z\mu(\mathscr{E}_L)\mathscr{E}_L \tag{4.24}$$

where $\mathscr{E}_L$ is the longitudinal electric field given by

$$\mathscr{E}_L = \frac{V_D}{L} \tag{4.25}$$

For the case of n-channel devices, the electric field dependence of the mobility at room temperature is given by Eq. (2.5). Using this expression in Eq. (4.24):

$$I_D = 2(a - W_S)qN_D Z\frac{9.85 \times 10^6 \mathscr{E}_L}{[1.04 \times 10^5 + \mathscr{E}_L^{1.3}]^{0.77}} \tag{4.26}$$

If channel pinch-off is negligible, the drain current will saturate when the drift velocity for the free carriers saturates. The saturated drain current is then given by

$$I_{D,sat} = 2(a - W_S)qN_D Z v_{sat} \tag{4.27}$$

where v_{sat} is the saturated drift velocity. The saturated drift velocity for electrons and holes is discussed in Chapter 2.

When the drain output characteristics are controlled by the velocity–field characteristics, the transconductance differs from that determined by channel pinch-off. When the velocity–field characteristics determine the output characteristics, the transconductance can be derived from Eq. (4.24):

$$g_m = \frac{2\epsilon_s Z}{W_S}\mu(\mathscr{E}_L)\mathscr{E}_L \tag{4.28}$$

When the drain current saturates due to saturation of the drift velocity, the transconductance becomes

$$g_m = \frac{2\epsilon_s Z}{W_S}v_{sat} \tag{4.29}$$

The transconductance determined by the velocity–field characteristics is the highest achievable value because it omits the effect of channel pinch-off. In practice, channel pinch-off results in a significantly lower transconductance for power JFET devices.

4.3.1.3. Drain Output Conductance.   The drain output conductance is defined as the rate of change of the drain current with drain voltage.

$$g_D = \frac{dI_D}{dV_D}\bigg|_{V_G} \tag{4.30}$$

The drain output conductance is the dynamic slope of the output characteristics shown in Fig. 4.3. An expression for this device parameter can be derived from Eq. (4.13) by differentiating with respect to the drain voltage:

$$g_D = 2(q\mu N_D)\frac{Z}{L}\left\{a - \sqrt{\frac{2\epsilon_s}{qN_D}(V_D + V_G + V_{bi})}\right\} \tag{4.31}$$

Alternately, by using the expression for the depletion layer width at the drain end of the channel:

$$g_D = 2(q\mu N_D)\frac{Z}{L}(a - W_D) \tag{4.32}$$

The maximum value for the drain output conductance occurs near the origin and is simply the reciprocal of the device resistance R. As the drain voltage increases, the channel pinch-off condition is approached with the depletion width W_D at the drain end tending to equal half the channel width a. The drain output conductance then becomes zero; that is, the slope of the output characteristics in the pentode regime becomes zero.

In practical devices, a finite drain output conductance is always observed. It can be postulated that the observed increase in drain current with increasing drain voltage beyond the pinch-off (or current saturation) point can arise from either impact ionization or due to the thermal generation of carriers. However, the electric field strength is typically well below 10^5 V/cm in the channel at pinch-off, resulting in negligible impact ionization, and the observed increase in drain current is significantly larger than that accountable by space-charge carrier generation.

The Shockley gradual channel theory presented earlier predicts that the drain current saturates when the pinch-off point is reached. This pinch-off point is assumed to occur when the gate depletion layers punch through in the channel near the drain. Since no mobile charge is assumed to exist in the depletion layer and the electric field is well below that required to create the impact ionization current, the flow of drain current is inconsistent with depletion layer punch-through in the channel. To circumvent this difficulty, the saturated current regime has been modeled by making the assumption that the channel is not completely depleted under these conditions and that a finite neutral region exists in the channel to support the drain current flow [3].

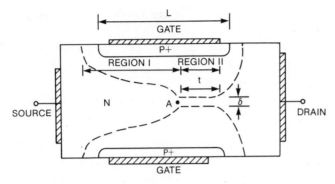

Fig. 4.9. Junction Field-effect transistor structure under pinch-off with a finite residual channel at the drain end.

Consider the channel region of a JFET under pinch-off conditions. If a finite residual channel is assumed to exist after pinch-off, the depletion layer profile will have the form shown in Fig. 4.9. The width of the residual channel is δ and its length is t. The channel can now be regarded as consisting of two regions: region I, in which the depletion layer width of the adjacent gate junctions is less than half the channel width a, and region II, in which the depletion layer widths are essentially equal to half the channel width a since the residual channel δ is much smaller than a. In region I, the gradual channel approximation holds and the voltage distribution can be analyzed by using the one-dimensional Poisson equation. In region II, a more complex two-dimensional Poisson's equation must be analyzed. The boundary between these regions at point A must have a potential equal to the pinch-off voltage.

Within region II, it is assumed that the width of the undepleted channel δ adjusts itself with increasing drain voltage to maintain current continuity. This requires that δ decreases with increasing drift velocity if the carrier density is assumed to remain constant. When the drift velocity saturates, the width attains its smallest value given by

$$\delta = \frac{I_D}{2ZqN_D v_s} \tag{4.33}$$

The length of region II has been derived using the two-dimensional Poisson's equation for the channel [3]:

$$t = \frac{2a}{\pi} \sinh^{-1}\left[\frac{\pi(V_D + V_G - V_P)}{2a\mathscr{E}_0}\right] \tag{4.34}$$

In deriving this equation it has been assumed that the velocity for free carriers increases linearly with increasing electric field until an electric field of $\mathscr{E}_0$ is reached. The free-carrier velocity is then assumed to remain constant at electric field strengths above $\mathscr{E}_0$. For silicon, $\mathscr{E}_0$ is 2×10^4 V/cm for electrons and 3×10^4 V/cm for holes. It should be noted that the length of region II increases with increasing drain voltage; that is, the pinch-off point A moves closer to the source with increasing drain voltage. This reduces the series resistance and produces an increase in drain current with increasing drain voltage in the pinch-off region of operation. The drain output conductance can then be derived from Eq. (4.34):

$$g_D = \frac{L\mathscr{E}_0}{I_D} \left\{ 1 + \left[\frac{\pi(V_D + V_G - V_P)}{2a\mathscr{E}_0} \right]^2 \right\}^{1/2} \tag{4.35}$$

This expression is valid for long channel devices. Numerical solutions of current transport in long-channel devices have confirmed the movement of the pinch-off point toward the source with increasing drain voltage [4].

4.3.1.4. Space-Charge-Limited Current Flow.

In the previous section, the JFET channel was divided into two regions. In region I, the current flow was described by using the gradual channel approximation. In region II, the drain current is controlled by space-charge-limited current flow. Space-charge limited current flow can be described with the aid of Fig. 4.10, where a bar of lightly doped N-type semiconductor is illustrated without gate electrodes. The bar can

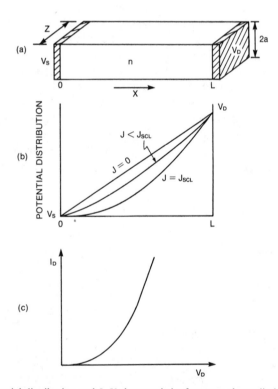

Fig. 4.10. Potential distribution and $I–V$ characteristics for space-charge-limited current flow.

be treated as an insulator under the condition that the mobile electron (free-carrier) concentration is much greater than the background charge arising from the presence of ionized donors. Poisson's equation for this case can be written as

$$\frac{d^2V}{dx^2} = -\frac{d\mathscr{E}}{dx} = \frac{qn}{\epsilon_s} \tag{4.36}$$

The current flow between the contacts can then be obtained from

$$J = \sigma\mathscr{E} = -q\mu n\mathscr{E} \tag{4.37}$$

where J is the current density and σ is the conductivity. Substitution for n in Eq. (4.36):

$$\mathscr{E} \, d\mathscr{E} = \frac{J}{\epsilon_s\mu} \, dx \tag{4.38}$$

Integration of this equation gives

$$\mathscr{E}^2(x) = \frac{2J}{\epsilon_s\mu} x + \mathscr{E}^2(0) \tag{4.39}$$

where $\mathscr{E}(0)$ is the electric field at the source electrode. The solution to this equation when the current density tends to zero is simply a linear variation in the potential from source to drain. When the current density increases, the presence of the mobile charge depresses the potential near the source as illustrated in Fig. 4.10. The current density at which the electric field $\mathscr{E}(0)$ at the source becomes zero is called the *space-charge-limited current density*. This current density is given by

$$J_{scl} = \frac{9aZ\epsilon_s\mu}{4L^3} V_D^2 = \frac{I_D}{2aZ} \tag{4.40}$$

where a is half the width of the semiconductor bar and Z is its depth. The quadratic increase in current with increasing drain voltage is shown at the bottom of Fig. 4.10. Note the contrast between the continuously increasing current in the space-charge limited condition and the saturating current described earlier in the pentode regime of JFET operation. The continuous quadratic increase in drain current with increasing drain voltage, known as *Child's law*, is reminiscent of vacuum-tube triodes. The operation of JFETs in this mode is described by triodelike characteristics.

4.3.1.5. Triode–Pentode Transition.

The pentodelike saturating drain current characteristics of JFETs occur when the channel length L is large and the channel width $2a$ is small. In these devices, the pinched-off region, region II, is confined to a portion close to the drain. The pinch-off point A in Fig. 4.9 does not approach the source and the drain potential does not penetrate the channel significantly. As the channel length L is reduced and the channel width $2a$ is increased, the penetration of the drain potential becomes prominent. In these devices, the channel consists predominantly of region II and the devices exhibit triodelike characteristics described by space-charge-limited current flow resulting from the action of the drain field on the source.

Analysis of the generalized characteristics of JFETs has been performed for the case when the mobility is assumed to be constant independent of the electric

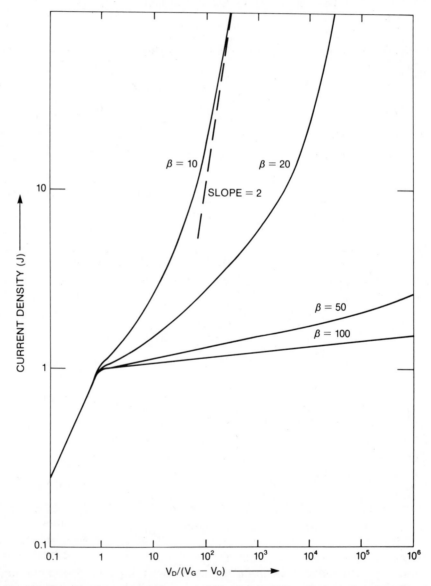

Fig. 4.11. Transition from pentodelike to triodelike characteristics with decreasing channel aspect ratio. (After Ref. 6, reprinted with permission from *Solid State Electronics*.)

field [5, 6]. The solution for the current–voltage characteristics is given by the transcendental equation [6]

$$\frac{\beta}{J} = \alpha - \ln(\alpha) - 1 \tag{4.41}$$

where

$$\beta = \frac{\pi L}{a} \tag{4.42}$$

$$J = -\frac{I_D}{(\mu \epsilon_s Z / aL)(V_G - V_0)^2} \tag{4.43}$$

$$V_0 = -\frac{qaN}{2\epsilon_s L} \tag{4.44}$$

and

$$\alpha = \left[\frac{e^{-\beta}}{1 - e^{-\beta}} \right] \left[\beta + \frac{(\eta^2 - 1)}{(j/\beta)} \right] \tag{4.45}$$

with

$$\eta = \frac{V_D}{(V_G - V_0)} - 1 \tag{4.46}$$

For large values of β, that is, long-channel devices, pentodelike characteristics are observed. As β decreases, the characteristics become increasingly triodelike. To observe this transition from pentodelike to triodelike characteristics, examine the plots of the variation of the current density J with the normalized drain voltage shown in Fig. 4.11. It can be seen that pentodelike characteristics are preserved when β exceeds 50. As β decreases, the characteristics become more triodelike until, for the case of $\beta = 10$, the characteristics display a slope of 2 at high current densities corresponding to space-charge limited current flow as described by Child's law.

4.3.1.6. Effect of the Drift Region.

The preceding analyses of current transport during forward conduction focused on the channel region of the JFET. High-voltage JFETs contain a drift region in addition to the channel region. The drift region is designed to support the high applied voltage between drain and source during forward blocking. To provide a high blocking voltage capability, the drift region must be lightly doped. Its resistance has a strong influence on the on-resistance of high-voltage JFETs because the drain current must flow through the drift region.

To analyze the influence of the drift region on drain current flow, consider the cross section of an element of a vertical channel, high-voltage JFET illustrated in Fig. 4.12. In this figure, the depletion region is shown by the dashed lines and it is assumed that the current transport occurs exclusively in the neutral undepleted channel region and then spreads through the drift region as

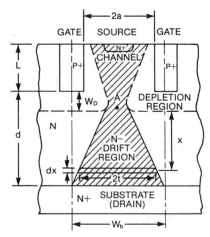

Fig. 4.12. High-voltage, vertical-channel JFET structure illustrating current conduction region (shaded).

illustrated. In this elemental JFET cross section, W_b is the repeat distance of each conducting channel including the gate regions. The analysis is performed for the case of a rectangular gate geometry. Although this may seem idealized, this device structure has actually been physically realized by using preferential etching techniques. If the drain voltage is negligibly small compared with the gate bias (i.e., at very low drain currents), the channel cross section becomes equal in area along its entire length L with an undepleted channel width of $2(a - W_D)$, where the gate junction depletion width W_D is given by

$$W_D = \sqrt{\frac{2\epsilon_s}{qN_D}(V_G + V_{bi})} \tag{4.47}$$

The resistance of the channel region can then be obtained from

$$R_{ch} = \frac{(L + W_D)}{q\mu_n N_D Z(2a - 2W_D)} \tag{4.48}$$

Note that the depletion layer spreading toward the drain adds to the effective channel length.

The resistance contributed by the drift region must include the effect of current spreading as illustrated in Fig. 4.12. The resistance of a segment dx located at a distance x from the channel is given by

$$R(x) = \frac{dx}{q\mu_n N_D Z(2t)} \tag{4.49}$$

where $2t(x)$ is the width of the current flow path at point x in the drift region. Assuming that the current spreading occurs by the linear expansion of the width $2t(x)$ from the channel to the repeat spacing, the following expression for $t(x)$

is obtained:

$$t(x) = (a - W_D) + \left[\frac{W_b}{2(d - W_D)} - \frac{(a - W_D)}{(d - W_D)}\right]x \qquad (4.50)$$

Substitution for $t(x)$ in Eq. (4.49) and integration from the channel edge A to the drain end of the drift region gives the drift region resistance:

$$R_D = \frac{(d - W_D)}{q\mu_n N_D Z[W_b - (2a - 2W_D)]} \ln\left(\frac{W_b}{2a - 2W_D}\right) \qquad (4.51)$$

The total on-resistance of the high-voltage JFET is the sum of the channel resistance given by Eq. (4.48) and the drift region resistance given by Eq. (4.51):

$$R_{on} = \frac{(L + W_D)}{q\mu_n N_D Z(2a - 2W_D)} + \frac{(d - W_D)}{q\mu_n N_D Z[W_b - (2a - 2W_D)]} \ln\left(\frac{W_b}{2a - 2W_D}\right) \qquad (4.52)$$

The lowest on-resistance occurs when the depletion width W_D is the smallest. This is achieved by operating at zero gate bias where the depletion width is created by only the built-in potential across the gate junction. From a design point of view, the on-resistance can be minimized by reducing the channel length L and increasing the channel width $2a$. However, this conflicts with achieving high blocking voltage, as is discussed later in the chapter.

If an ideal device is considered, in which an infinite number of infinitesimally small channels of zero length are used, the on-resistance becomes equal to that of a bar of semiconductor defined by the drift region with a length d equal to the depletion layer width at breakdown and a doping concentration equal to the background doping at the desired breakdown voltage. These parameters were related to the breakdown voltage in Chapter 3. Using those relationships, it can be shown that:

$$N = 2 \times 10^{18}(BV_{PP})^{-4/3} \qquad (4.53)$$

and

$$d = 2.58 \times 10^{-6}(BV_{PP})^{7/6} \qquad (4.54)$$

The on-resistance per unit area for this ideal case is known as the specific on-resistance. By use of the preceding equations for the doping level and width of the drift region, an expression for specific on-resistance as a function of the breakdown voltage can be derived:

$$R_{on,sp} = \frac{d}{q\mu N} \qquad (4.55)$$

$$= 8.06 \times 10^{-6}\frac{(BV_{PP})^{2.5}}{\mu} \qquad (4.56)$$

From this expression, it can be seen that the on-resistance will decrease inversely with increasing mobility. Consequently, the on-resistance of n-channel JFETs will be lower than that for p-channel JFETs. By use of the mobility values for electrons and holes at room temperature for low doping levels, the specific on-resistance for the two cases can be obtained:

$$R_{\substack{\text{on,sp} \\ n\text{-channel}}} = 5.93 \times 10^{-9}(BV_{PP})^{2.5} \tag{4.57}$$

and

$$R_{\substack{\text{on,sp} \\ p\text{-channel}}} = 1.63 \times 10^{-8}(BV_{PP})^{2.5} \tag{4.58}$$

These expressions indicate a very sharp increase in the specific on-resistance with increasing breakdown voltage, that is, increasing forward blocking capability for the JFETs. The sharp increase in on-resistance arises from a combination of the lower drift region doping and the increase in its thickness required to support higher breakdown voltages. The variation of the specific on-resistance for n- and p-channel JFETs is plotted as a function of the breakdown voltage

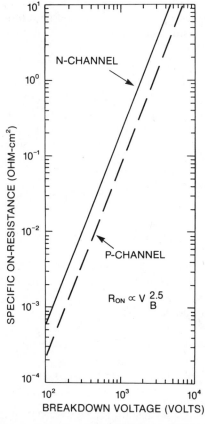

Fig. 4.13. Specific on-resistance of n- and p-channel JFETs as a function of the breakdown voltage.

in Fig. 4.13. These lines represent the lowest possible on-resistances that can be achieved with silicon power JFETs.

In practical power JFETs, the ideal specific on-resistance cannot be achieved for two reasons. The first reason is that the contribution to the on-resistance from the channel cannot be made negligible. To obtain a reasonable blocking gain, the channel region must be made long and narrow. Typically, the channel contribution results in an increase in on-resistance by 25–50% over the ideal case. The second and more important reason is that the gate junction must support not only the drain voltage but the gate bias voltage as well. Even if the devices were to be operated right up to the breakdown voltage, therefore, the maximum drain voltage and hence the forward blocking capability are less than the breakdown voltage. The reduction in the drain blocking voltage V_{DB} compared to the gate junction breakdown voltage BV_{PP} is dependent on the blocking gain G_B defined by Eq. (4.3):

$$V_{DB} + V_G = V_{DB}\left(1 + \frac{1}{G_B}\right) = BV_{PP} \tag{4.59}$$

For the case of n-channel devices, the specific on-resistance can then be related to the drain blocking voltage and the blocking gain:

$$R_{on,sp} \atop {}_{n\text{-channel}} = 5.93 \times 10^{-9}(V_{DB})^{2.5}\left(1 + \frac{1}{G_B}\right)^{2.5} \tag{4.60}$$

Depending on the technology used to fabricate the devices, DC blocking gains ranging from 2 to 100 have been achieved. The increase in specific on-resistance due to a finite DC blocking gain is illustrated in Fig. 4.14. It can be seen that a severe increase in the on-resistance occurs when the DC blocking gain drops below 5. Early device technology produced gate structures limited to DC blocking gains of 2–5. This restricted the development of high-voltage power JFETs. With the development of gate structures capable of achieving DC blocking gains of over 10, the fabrication of devices with blocking voltage capability in excess of 100 V has now become feasible without suffering from extremely high on-resistances.

The presence of the drift region not only adds to the on-resistance of the JFET, but also alters the output drain characteristics at higher drain currents. The current-voltage characteristics of the device, including the effect of the drift region, can be derived by coupling the channel and drift regions by use of the voltage at point A. In the presence of large drain current flow, the voltage drop along the channel becomes appreciable and the gate depletion layer becomes wider near the end of channel closest to the drain. The depletion layer width at the end of the channel (point A in Fig. 4.12) is given by

$$W_A = \sqrt{\frac{2\epsilon_s}{qN_D}(V_A + V_G + V_{bi})} \tag{4.61}$$

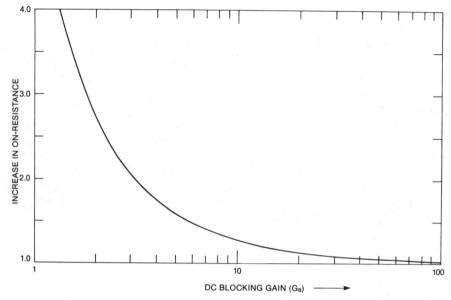

Fig. 4.14. Impact of the DC blocking gain on the specific on-resistance of JFETs.

Assuming a linear change in the cross section of the channel from the source end to point A, the resistance of the channel can be derived in terms of the depletion layer widths at the two ends:

$$R_{ch} = \frac{1}{2q\mu N_D Z}\left(\frac{L + W_A}{W_A - W_S}\right)\ln\frac{a - W_S}{a - W_A} \qquad (4.62)$$

In this expression, the depletion width at the source end (W_S) is independent on the gate bias alone:

$$W_S = \sqrt{\frac{2\epsilon_s}{qN_D}(V_G + V_{bi})} \qquad (4.63)$$

The resistance of the drift region is also dependent on the potential at point A because this determines the width of the current flow path into the drift region. By modification of the previously derived equation [Eq. (4.51)] to include the potential at point A, the drift region resistance can be obtained:

$$R_D = \frac{1}{q\mu N_D Z}\left[\frac{(d - W_A)}{W_b - (2a - 2W_A)}\right]\ln\frac{W_b}{2a - 2W_A} \qquad (4.64)$$

To analyze the output characteristics, the drain voltage must be related to the drain current:

$$V_D = V_A + I_D R_D \qquad (4.65)$$

The potential at point V_A is related to the drain current by the current flow expressions derived for the channel using the gradual channel approximation:

$$I_D = 2a(q\mu N_D)\frac{Z}{(L + W_A)}$$

$$\left\{V_A - \frac{2}{3a}\left(\frac{2\epsilon_s}{qN_D}\right)^{1/2}[(V_A + V_G + V_{bi})^{3/2} - (V_G + V_{bi})^{3/2}]\right\} \quad (4.66)$$

Since the depletion width at point A is dependent on the drain current, a closed-form analytical solution for the output characteristics cannot be derived. It is necessary to use an iterative process to solve these equations and derive the output characteristics. In performing this analysis, the change in the mobility with electric field should be taken into account.

4.3.1.7. Punch-through Structure. In the analysis of the specific on-resistance for the ideal case, it was assumed that the drift region had a doping concentration and thickness corresponding to the parallel-plane breakdown case. The drift region should not be thicker than the ideal case because it simply adds an extra resistance to the device. However, an alternate device structure can be postulated based on achievement of the same breakdown (and blocking voltage) capability by use of a punch-through structure. A comparison of the electric field distribution in the drift region for the two cases is provided in Fig. 4.15. The punch-through structure has a lower doping level N_{DP} and a smaller thickness W_p to support the same voltage as the ideal parallel-plane case. The width of the punch-through device can be related to the doping for the punch-through

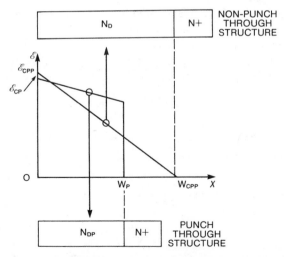

Fig. 4.15. Comparison of electric field distribution for the punch-through JFET structure with the non-punch-through structure.

case and the ideal non-punch-through device parameters:

$$W_p = W_{Cp}\left\{1 - \sqrt{1 - \left(\frac{N_p}{N_I}\right)^{3/4}}\right\}$$ (4.67)

The specific on-resistance for the punch-through device is given by

$$R_{on,sp,pt} = \frac{W_p}{q\mu_n N_{DP}}$$ (4.68)

The ratio of the specific on-resistance for the punch-through case to that of the ideal non-punch-through case can then be obtained by using Eqs. (4.67) and (4.68):

$$\frac{R_{on,sp,pt}}{R_{on,sp,ideal}} = \left(\frac{N_I}{N_{DP}}\right)\left\{1 - \sqrt{1 - \left(\frac{N_{DP}}{N_I}\right)^{3/4}}\right\}$$ (4.69)

A plot of this ratio is shown in Fig. 4.16 as a function of the doping level for the punch-through device normalized to the ideal doping level. Note that the specific on-resistance goes through a minimum when the doping level of the punch-through structure is reduced. Since a lower doping level also favors improved blocking characteristics, the use of a punch-through structure for high-voltage JFETs is favored, with the optimum doping level of 70% of that for the ideal case. The corresponding normalized drift region width can be obtained from Fig. 4.16. For the optimum doping, a drift region width of about 50% of the ideal case can be used for the punch-through device. The thinner drift region is another advantage of using a punch-through structure because it is easier to grow by using epitaxial technology.

4.3.1.8. Effect of Semiconductor Material.

In the preceding discussion it was assumed that the devices were fabricated by using silicon as the base material. The performance of these devices is limited by the properties of silicon. The impact of changes in these properties on the on-resistance is considered here [7].

In the general case, the avalanche breakdown voltage of any semiconductor is related to its doping concentration and its energy gap:

$$V_B = 60\left(\frac{E_g}{1.1}\right)^{3/2}\left(\frac{N_B}{10^{16}}\right)^{-3/4}$$ (4.70)

where E_g is the energy gap. For a fixed breakdown voltage, the doping concentration is dependent only on the energy gap:

$$N_B \propto E_g^2$$ (4.71)

Since the depletion layer width at breakdown is given by

$$W_D = \sqrt{\frac{2\epsilon_s V_B}{q N_B}}$$ (4.72)

for a fixed breakdown voltage, the depletion layer width also changes with energy gap:

$$W_D \propto E_g^{-1}$$ (4.73)

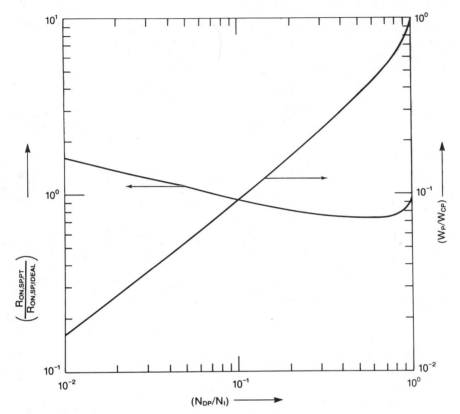

Fig. 4.16. Normalized plot of the on-resistance of the punch-through JFET structure as a function of the normalized doping level.

From these expressions, the specific on-resistance is given by

$$R_{\mathrm{on}} \propto W \mu^{-1} N_{\mathrm{B}}^{-1} \tag{4.74}$$

or

$$R_{\mathrm{on}} \propto \mu^{-1} E_{\mathrm{g}}^{-3} \tag{4.75}$$

Thus the specific on-resistance is a strong function of the energy gap. The variation of the on-resistance, normalized to that for silicon, with changes in the energy gap and mobility is shown in Fig. 4.17. It is evident that a high mobility and a large energy gap will result in a low on-resistance.

A review of the properties of elemental and compound semiconductors [7] indicates that the on-resistance for devices fabricated from gallium arsenide, aluminum arsenide, gallium phosphide, and indium phosphide will be lower than that for silicon devices. Of these semiconductors, the on-resistance of

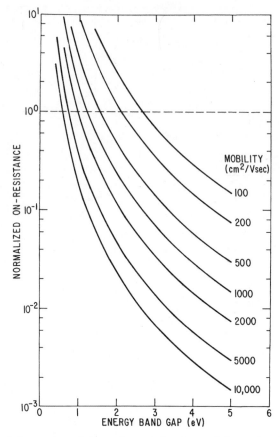

Fig. 4.17. Effect of changes in mobility and energy band gap of the semiconductor on the on-resistance of high-voltage JFETs.

gallium arsenide devices is the lowest with an improvement of a factor of 12.7 over silicon. An even greater reduction in the on-resistance is indicated for some ternary compound semiconductors such as gallium arsenide phosphide, gallium aluminum arsenide, and gallium indium phosphide. Although the fabrication of high-voltage devices from compound semiconductors has been attempted as discussed later in the chapter, most of the effort on power devices has been devoted to silicon because of its advanced technology.

4.3.2. Forward Blocking

Together with the current rating, the forward blocking capability defines the power controlling ability of a power device. The current handling capability is determined by the forward conduction characteristics (primarily the on-resistance) as discussed in the previous section, where both pentodelike and triodelike operations were considered. It was shown that at the high current levels typical of the forward conduction mode, the devices exhibit pentodelike characteristics.

In the forward blocking mode, the device must be operated at low current levels. Otherwise, excessive power dissipation would occur at high blocking voltages. When operated at high voltages and low currents, power JFETs exhibit triodelike characteristics. These characteristics are controlled by the presence of a potential barrier in the channel.

4.3.2.1. Channel Potential Barrier. When the gate voltage exceeds the pinch-off voltage V_p, the depletion layers of adjacent gate junctions intersect in the channel and a potential barrier is formed. The transport of electrons from source to drain is controlled by the injection of carriers over the potential barrier. This produces the triodelike characteristics.

The effect of changes in the gate and drain voltages on the potential barrier height are illustrated schematically in Fig. 4.18. In the case shown in Fig. 4.18*a*,

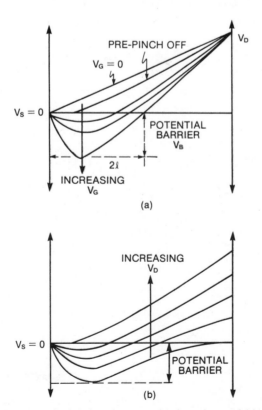

Fig. 4.18. Effect of gate and drain voltages on the channel potential barrier height.

the drain voltage is assumed to be fixed. As the gate voltage increases a potential barrier forms after the pinch-off condition is reached. Further increase in gate voltage results in a proportionate increase in the barrier height. The drain voltage has an opposite influence as shown by the case illustrated in Fig. 4.18b. Here the gate voltage is held constant. As the drain voltage increases, the barrier height is reduced by the penetration of the drain potential into the channel.

As may be expected, the geometric shape of the channel and its doping level influence the modulation of the potential barrier by the gate and drain voltages. As an example, the effect of changing the gate length is shown in Fig. 4.19. The potential distribution between drain and source shown in this figure was obtained by two-dimensional numerical analysis [8]. Note that for the case of a small gate length (along the source–drain direction), the drain potential has a very strong influence on the potential barrier. A rapid reduction in barrier height occurs with increasing drain voltage. In contrast, when the gate length is large, the barrier height remains nearly unaffected by the drain potential. These plots demonstrate that a longer gate length is required to obtain higher blocking voltage capability.

The numerical analysis also indicates that a smaller channel width will result in an increase in the potential barrier height. This can be intuitively expected because a narrower channel produces punch-through at lower gate voltages

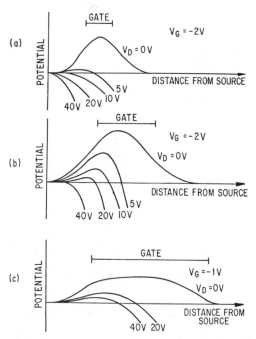

Fig. 4.19. Impact of increasing gate length on the channel potential barrier reduction by the drain voltage. (After Ref. 8, reprinted with permission from the IEEE © 1977 IEEE.)

and brings the gate electrodes closer to the center of the channel so that they can exert a stronger electrostatic field. In addition, a smaller channel width is found to reduce the influence of the drain potential. This results in higher blocking gains.

Although the channel length and width have the strongest influence on the potential barrier, other parameters that have a second-order effect are the channel doping concentration and the separation between the drain and gate regions, that is, the drift region width. When the channel doping level is reduced, the gate depletion layers extend more rapidly with increasing gate voltage. A higher potential barrier that extends over a longer distance along the channel is established in this case. Decreasing the doping concentration in the channel then produces a higher blocking gain. A higher blocking gain is also observed when the drain–source separation is increased. As may be intuitively expected, moving the drain electrode further from the channel lessens its ability to lower the barrier height.

4.3.2.2. Current Transport.

The current flow in a JFET operating in the triode regime can be limited by the injection of carriers across the channel potential barrier. This injection phenomenon is similar to that at the emitter–base junction of a bipolar transistor. It can be described by the equation

$$I_D = I_0 \, e^{-(qV_B/kT)} \tag{4.76}$$

where V_B is the height of the channel potential barrier.

To relate the drain current to the drain and gate voltages, it is necessary to perform a numerical analysis including the two-dimensional potential distribution. An approximate analytical solution has been derived [9] by considering only the central point in the channel and solving a one-dimensional current continuity equation. This is justified by the fact that the potential barrier is lowest at the central point in the channel and most of the source–drain current will flow through this point as a result of the exponential dependence of the current on the barrier height. With this assumption:

$$J_D = qnv + qD \, \frac{dn}{dx} \tag{4.77}$$

If the mobility is assumed to be constant, it follows that

$$v = \mu \mathscr{E}(x) \tag{4.78}$$

Using Einstein's relationship

$$\mu = \frac{q}{kT} \frac{1}{D} \tag{4.79}$$

and multiplying both sides of Eq. (4.77) with the factor $\exp[-qV(x)/kT]$:

$$J_D \, e^{-[qV(x)/kT]} = qD \, \frac{d}{dx} \{n \, e^{-[qV(x)/kT]}\} \tag{4.80}$$

Integration of this equation yields

$$J_D = \frac{qDN}{\int e^{-[qV(x)/kT]} \, dx} \tag{4.81}$$

where N is the doping level. For the case of a parabolic potential distribution in the channel

$$V(x) = \frac{V_B}{l^2} (x^2 - 2xl) \tag{4.82}$$

where V_B is the height of the potential barrier and $2l$ is the width of the barrier as shown in Fig. 4.18. Use of this potential distribution in Eq. (4.81) and solution for the drain current density gives [9]

$$J_D = \frac{qDN}{l} \sqrt{\frac{qV_B}{\pi kT}} e^{-(qV_B/kT)} \tag{4.83}$$

This expression differs from Eq. (4.76) in that the barrier height also affects I_0.

If it is further assumed that the potential barrier height varies linearly with the gate and drain voltages

$$V_B = \alpha V_G - \beta V_D \tag{4.84}$$

where α and β are constants, then

$$J_D = \frac{qDN}{l} \sqrt{\frac{q}{\pi kT} (\alpha V_G - \beta V_D)} \exp \left\{ -\left[\frac{q}{kT} (\alpha V_G - \beta V_D) \right] \right\} \tag{4.85}$$

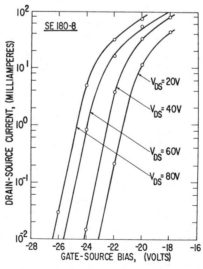

Fig. 4.20. Measured exponential increase in drain current with decreasing gate bias voltage in the triode regime. (After Ref. 10, reprinted with permission from the IEEE © 1980 IEEE.)

According to this expression, the drain current should increase nearly exponentially with decreasing gate voltage and increasing drain voltage. The deviation from a purely exponentially dependence arises from the term under the
square root in Eq. (4.85). The variation in drain current with gate voltage for
a typical device is shown in Fig. 4.20 for various fixed drain voltages [10]. The
curves exhibit a nearly exponential variation, especially at lower drain currents.
In contrast, the deviation from an exponential variation of the drain current is
stronger with respect to the drain voltage. This can be seen from Fig. 4.21, where
the drain current is plotted as a function of the drain voltage [10].

The difference between the operation of the device in the pentodelike and
triodelike regions can be clarified further by examining the mobile carrier distribution. The carrier distribution in the channel, for devices operating in both
the triodelike and pentodelike regions, has been obtained by numerical analysis
and is shown in Fig. 4.22 at two representative points in the $I-V$ characteristics
for each case [8]. In the triode region of operation, it can be seen that the mobile

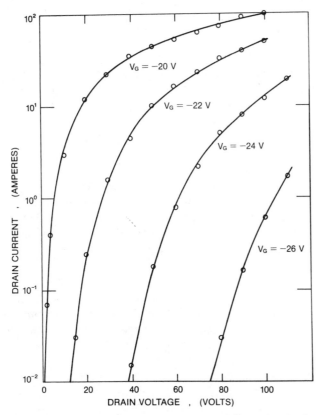

Fig. 4.21. Measured increase in drain current with increasing drain voltage in the triode regime.
(After Ref. 10, reprinted with permission from the IEEE © 1980 IEEE.)

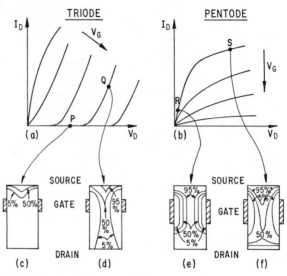

Fig. 4.22. Carrier distribution in the channel of JFETs operating in the triode and pentode regimes. (After Ref. 8, reprinted with permission from the IEEE © 1977 IEEE.)

carrier density vanishes in the channel at low drain voltages (Fig. 4.22c) and increases with increasing drain voltage (Fig. 4.22d). Thus the current transport in the triode region occurs by enhancement of the carrier density in the channel. In contrast, for the pentode region of operation, a neutral undepleted channel exists at low drain voltages as shown in Fig. 4.22e. As the drain voltage increases, this channel becomes depleted of carriers (Fig. 4.22f), leading to the observed saturation of the drain current.

Power JFETs exhibit triodelike characteristics when the gate length is short, allowing the drain potential to penetrate the gate region, whereas devices with long channels exhibit pentodelike characteristics. At intermediate gate lengths, the devices exhibit mixed characteristics consisting of the pentode shape at large drain currents and low gate voltages and triodelike characteristics at large gate voltages and low drain currents. Such characteristics are found to be the optimum for achieving a low drain on-resistance while retaining high blocking voltage capability. To exhibit such mixed characteristics, it is essential that the zero-bias gate depletion width be insufficient to pinch off the channel. If the zero-bias depletion width were sufficient to completely deplete the channel, the channel potential barrier would be established even at zero gate bias and exclusively triodelike characteristics would be observed.

For an abrupt gate junction with a doping concentration of N_A on the heavily doped side, the zero-bias depletion layer width can be obtained by using the following expression:

$$W_0 = \sqrt{\frac{2\epsilon_s kT}{q^2 N_D} \cdot \ln\left(\frac{N_A N_D}{n_i^2}\right)} \tag{4.86}$$

Assuming a doping level of $10^{20}/cm^3$ on the heavily doped side of the junction, the zero-bias depletion layer width at room temperature has been calculated and provided in Fig. 4.23 for reference. In this figure, the maximum depletion layer width as limited by avalanche breakdown is also shown. If the half-channel width a is smaller than the zero-bias depletion layer width, the channel will be pinched off without the application of a gate bias and a channel potential barrier established by the built-in potential of the gate junction. Consequently, purely triodelike characteristics will occur in this region. If the half-channel width a is greater than the maximum depletion layer width at avalanche breakdown, an undepleted channel region will exist for all gate bias voltages and no channel potential barrier will form. Consequently, these devices will exhibit purely pentodelike characteristics. To observe the mixed triode–pentode characteristics, it is necessary to design the devices with a half-channel width in the region between the zero bias and breakdown depletion layer widths.

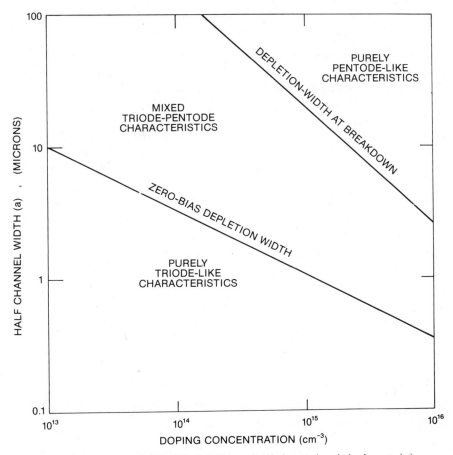

Fig. 4.23. Regimes of pentodelike, triodelike, and mixed pentode–triode characteristics.

4.3.2.3. Blocking Gain.   The blocking gain of a high-voltage JFET operated in the triode region is defined as the ratio of the drain voltage to the gate voltage at a specified leakage current. This parameter has also been called the *voltage amplification factor*. The blocking gain is a function of the length of the gate, the width of the channel, and the doping concentration in the channel. A longer gate inhibits the penetration of the drain field into the gate region and produces a larger blocking gain. The channel width and resistivity also influence the blocking gain because they determine the height of the channel potential barrier formed for a given gate bias. Since a narrower channel formed from higher-resistivity material favors a larger potential barrier height, a larger blocking gain is observed in these cases. In the case of high-voltage power devices, it becomes essential to obtain a large voltage amplification factor to retain a relatively low gate control voltage.

An expression for the blocking gain can be derived from the current transport equation for the triode regime. By differentiating Eq. (4.85) with respect to V_G and taking J_D as a constant, it can be shown that

$$G = \frac{dV_D}{dV_G} = \frac{\alpha}{\beta} \tag{4.87}$$

where α and β are parameters that determine the rate of variation of the channel potential barrier with gate and drain voltage, respectively, as described by Eq. (4.84). These parameters are a function of the device geometry and channel doping concentration.

Several analyses of the blocking gain have been performed by developing analytical solutions by use of the analogy between the JFET in the triode regime and a vacuum-tube triode [11, 12]. In addition, numerical analysis of the forward blocking characteristics has been undertaken for a variety of gate geometries [13].

By use of conformal mapping, the linear grid structure used in power JFETs can be transformed to cylindrical coordinates [12]. Poisson's equation for the device now takes the form

$$\frac{1}{r}\frac{d}{dr}[r\mathscr{E}(r)] = \frac{\rho_c(r)}{\epsilon_s} \tag{4.88}$$

where $\rho_c(r)$ is the charge density. Because of the conformal mapping, the charge density $\rho_c(r)$ is a function of r for the uniformly doped linear channel. The transformation is performed by using the function

$$x + jy = \frac{p}{2\pi}\ln(c) \tag{4.89}$$

where p is the pitch of a grid of radius r_g as illustrated in Fig. 4.24 and c is given by

$$c = re^{j\theta} \tag{4.90}$$

Using Eqs. (4.89) and (4.90), it can be shown that

$$r = e^{2\pi(x/p)} \tag{4.91}$$

$$\theta = \frac{2\pi y}{p} \tag{4.92}$$

An elemental volume in cylindrical coordinates is given by

$$dV_c = r \, d\theta \, dr \, dz \tag{4.93}$$

Using the preceding expressions, it can be shown that

$$dV_c = \left(\frac{2\pi}{p}\right)^2 r^2 \, (dx \, dy \, dz) \tag{4.94}$$

In this expression, $(dx \, dy \, dz)$ is the elemental volume in the original linear device structure shown in Fig. 4.24. If the linear device is uniformly doped, the charge distribution in the cylindrical coordinates becomes

$$\rho_c = \left(\frac{p}{2\pi}\right)^2 \frac{\rho_0}{r^2} = \left(\frac{p}{2\pi}\right)^2 \frac{qN_D}{r^2} \tag{4.95}$$

where N_D is the channel doping level. Using this charge distribution in Eq. (4.88) and performing an integration, an expression for the potential distribution can be derived.

$$V(r) = -\left(\frac{p}{2\pi}\right)^2 \frac{qN_D}{8\epsilon_s} [\ln(r)^2]^2 - A \ln(r^2) + C \tag{4.96}$$

where A and C are constants. These constants can be expressed in terms of the gate and drain voltages. From this solution, an expression for the blocking gain can be derived [12]:

$$G = \frac{G_0}{1 + [V_P/(V_G + V_D)]} \tag{4.97}$$

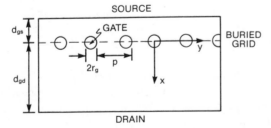

Fig. 4.24. Buried-grid structure used for blocking gain analysis.

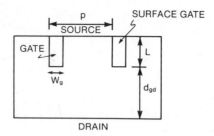

Fig. 4.25. Vertically walled grid structure used for blocking gain analysis.

where V_P is the punch-through voltage between gate and drain:

$$V_P = \frac{qN_D}{2\epsilon_s} d_{gd}^2 \qquad (4.98)$$

For the buried-grid structure shown in Fig. 4.24, it is found that

$$G_0 = \frac{(2\pi d_{gd}/p) - \ln[\cosh(2\pi r_g/p)]}{\ln[\coth(2\pi r_g/p)]} \qquad (4.99)$$

and for the case of a vertically walled device structure, as shown in Fig. 4.25:

$$G_0 = \frac{(2\pi d_{gd}/p) - \ln[2 + \frac{1}{2} e^{2\pi L/p}]}{\ln[1 + 4 e^{-(2\pi L/p)}]} \qquad (4.100)$$

The calculated blocking gains obtained by using these expressions are in good agreement with the measured data obtained for both buried-grid and vertically walled gate devices [12].

To achieve high blocking gains, it is necessary to have a channel length L comparable in size to the grid pitch p. For these cases, the expression in Eq. (4.100) can be simplified to

$$G_0 = \left[\frac{\pi}{2} \left(\frac{d_{gd}}{p} - \frac{L}{p} \right) + 0.17 \right] e^{2\pi L/p} \qquad (4.101)$$

A plot of the DC blocking gain for a number of vertically walled devices fabricated by using a recessed-gate process [14] is shown in Fig. 4.26 as a function of the ratio (L/p). This ratio is called the *channel aspect ratio* if the width of the grid (W_g) is negligible. Note that the blocking gain increases nearly exponentially with increasing aspect ratio; the slight deviation is due to the first term in Eq. (4.101).

The blocking gain increases with increasing drain voltage, as shown in Fig. 4.27 for vertically walled devices fabricated by using the recessed-gate process. Note that the DC blocking gain tends to reach a limiting value and remains nearly independent of the drain voltage at higher voltages. This behavior is consistent with the drain voltage dependence given by Eq. (4.97), because, when

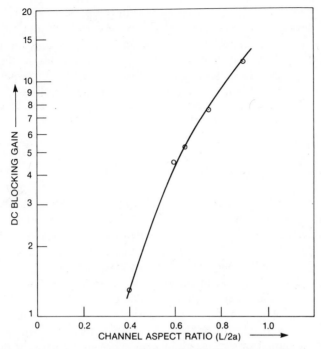

Fig. 4.26. Measured increase of the blocking gain of recessed gate JFETs with increasing channel aspect ratio. (After Ref. 14, reprinted with permission from the IEEE © 1982 IEEE.)

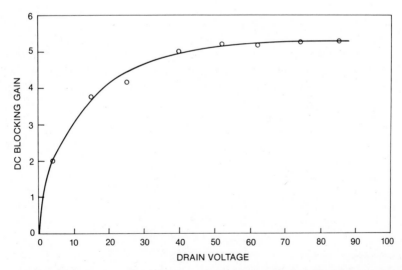

Fig. 4.27. Measured increase of the blocking gain of recessed-gate JFETs with increasing drain voltage. (After Ref. 14, reprinted with permission from the IEEE © 1982 IEEE.)

the drain voltage increases, the denominator in this expression tends to unity. In terms of device output characteristics, such as observed on a curve tracer, this behavior of the blocking gain is exhibited by the fact that the $I-V$ characteristics are bunched closer together at low gate bias voltages. As the gate bias increases, the characteristics become further spaced apart and eventually display a uniform spacing until the blocking voltage becomes limited by avalanche breakdown.

4.3.3. Frequency Response

The drain current of a power JFET is controlled by the gate voltage. In the idealized case with negligible leakage current, the gate current is zero in steady state. The power JFET then has an infinite input impedance in the steady-state conduction or blocking modes of operation. During the transition between these states, the depletion layer must be modulated by the applied gate bias. This requires the charging and discharging of an input gate capacitance. The input impedance of the JFET is determined by this capacitance.

A simple equivalent circuit for the JFET is shown in Fig. 4.28. In this equivalent circuit, g_m is the transconductance as described in Section 4.3.1, and the drain output conductance has been assumed to be zero. A maximum frequency of operation of the JFET can be defined as the frequency at which the input current becomes equal to the output current. The input current is determined by the input capacitance, and it increases as the frequency rises. In the simple equivalent circuit shown in Fig. 4.28, the output current remains constant. By equating the input and output currents, an expression for the maximum frequency of operation can be obtained:

$$f_m = \frac{g_m}{2\pi C_{in}} \qquad (4.102)$$

The transconductance g_m is dependent on the gate and drain voltage. For operation in the high current active region, pentodelike operation can be assumed, and from Section 4.3.1

$$g_m = 2a\mu \frac{Z}{L}(2\epsilon_s q N_D)^{1/2}[(V_D + V_G + V_{bi})^{1/2} - (V_G + V_{bi})^{1/2}] \qquad (4.103)$$

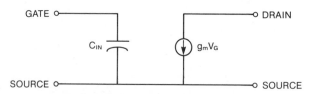

Fig. 4.28. Simple equivalent circuit for a JFET.

The input capacitance is also dependent on the gate and drain voltages since they determine the channel depletion width:

$$C_{\text{in}} = \frac{2\epsilon_s ZL}{W_D} \tag{4.104}$$

Substitution of these expressions in Eq. (4.102) yields

$$f_m = \frac{a\mu\epsilon_s}{\pi} \frac{Z}{L} (V_D + V_G + V_{bi})^{1/2} [(V_D + V_G + V_{bi})^{1/2} - (V_G + V_{bi})^{1/2}] \tag{4.105}$$

A typical set of frequency response curves is shown in Fig. 4.29 (solid lines) as a function of gate bias for various drain voltages. Note that the frequency response improves with increasing drain and gate voltages as the result of a reduction in the input capacitance.

In actual devices several parasitic elements must be considered during the analysis of the frequency response. The equivalent circuit with these elements

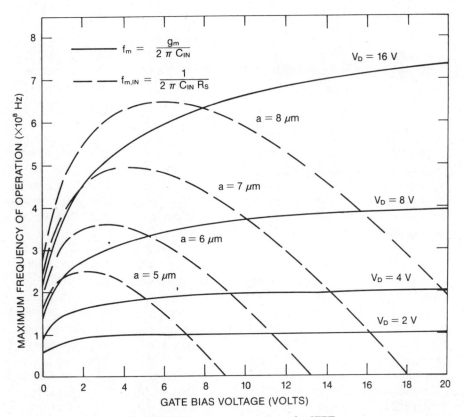

Fig. 4.29. Frequency-response curves for JFETs.

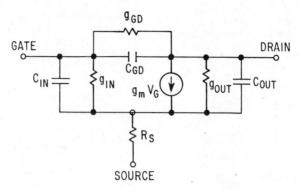

Fig. 4.30. Equivalent circuit of JFET including parasitic elements.

is shown in Fig. 4.30. The presence of a finite channel conductivity creates a source series resistance R_S, which results in a change in transconductance:

$$g'_m = \frac{g_m}{1 + g_m R_S} \qquad (4.106)$$

This change in the transconductance leads to a reduction in the frequency response as described in Eq. (4.102).

Because of the finite channel resistance, the frequency response can become limited by the charging rate of the input capacitance. This limit to the frequency response is given by

$$f_{m,in} = \frac{1}{2\pi C_{in} R_S} \qquad (4.107)$$

Both C_{in} and R_S are a function of the gate bias; C_{in} decreases with increasing gate voltage as the result of a widening of the gate depletion layer. In contrast, R_S increases with increasing gate voltage because of a reduction in the cross-sectional area of the channel. These parameters have an opposing influence on the frequency response. Some typical examples of the calculated variation in the frequency response as limited by the input impedance are shown in Fig. 4.29 by the dashed lines. It can be seen that the frequency response goes through a maximum as the gate voltage increases. This is due to the rapid decrease in the input capacitance at low gate voltages, which tends to increase the frequency response, followed at high gate voltages by the rapid increase in the source resistance, which tends to lower the frequency response [15]. The improvement in frequency response with increase in the channel width a is a direct consequence of the reduction in the source resistance due to the increased cross-sectional area of the channel.

In addition to these frequency response limitations, the presence of the long drift region in high-voltage JFETs results in a significant transit time for the carriers through it. Assuming that the field in the drift region is high, the transit time will be given by the source–drain spacing divided by the saturated drift velocity of the carriers. The source–drain spacing is given by the sum of the channel length and drift region length; the latter is a function of the breakdown voltage of the device. Using the expressions for the depletion layer width as a function of the breakdown voltage provided in Chapter 3 and the saturated drift velocity for electrons in silicon, an expression for the transit time limited frequency response, as a function of the breakdown voltage, can be derived:

$$f_T = \frac{6.11 \times 10^{11}}{(1 + L/d)(BV_{pp})^{7/6}} \tag{4.108}$$

where L is the channel length and d is the drift region thickness. This expression shows that the frequency response will decrease rapidly with increase in the device breakdown voltage. In addition, to maintain a high frequency response, it is important to minimize the length of the channel. This conflicts with achieving a high blocking gain.

In general, the frequency response of JFETs is very high in comparison to other power devices. The calculated values shown in Fig. 4.29 indicate that devices operating at hundreds of megahertz are feasible. In fact, devices have been developed with cut-off frequencies exceeding 1 GHz [16, 17]. These devices have blocking voltages of less than 100 V. For operation directly off the 120-V AC line, it is desirable to use devices with blocking voltages of over 400 V. Because of the degradation in frequency response with increasing operating voltage, these devices have a frequency response in the range of 0.5–0.8 GHz [14]. By reducing the parasitics with the use of higher-resolution process technology and using gallium arsenide in place of silicon, it should be possible to extend the frequency response of these higher voltage devices to above 1 GHz.

4.4. BIPOLAR OPERATION

When originally conceived, the JFET was intended to be operated with reverse gate bias. This mode of operation ensures unipolar current conduction, which results in a very high operating frequency. This is an attractive feature of low-voltage JFETs. As the operating voltage increases, the series resistance of the drift region rises and the on-resistance of JFETs becomes very large. This severely limits the current handling capability. An approach to lowering the on-resistance is to modulate the conductivity of the drift region by the introduction of free carriers. The injection of free carriers into the drift region can be accomplished by forward-biasing the gate junction during current conduction. The gate–source current flow produces a large excess electron and hole concentration in the channel region between the gate and the source. In addition, the free carriers diffuse into the drift region and significantly reduce its resistance as well. It has been found that the on-resistance of high-voltage JFETs can be reduced by over two orders of magnitude by forward-biasing the gate [18]. This is known as the *bipolar mode of operation of the power JFET*.

4.4.1. Simple On-Resistance Analysis

In a simple analysis of the bipolar operation of the JFET, it can be assumed that the gate current flow serves to supply the recombination in the N-base region and in the end regions of the P–i–N diode formed between gate and source. Using high-level forward-biased P–i–N diode theory [19], it can be shown that

$$J_G = \left[\frac{qD_{np}\bar{n}_p}{L_{np}\tanh(W_p/L_{np})} \right]\left(\frac{n^*}{n_i} \right)^2 + \left(\frac{2q\bar{n}a}{\tau_{HL}} \right) + \left[\frac{qD_{pn}\bar{p}_n}{L_{pn}\tanh(W_n/L_{pn})} \right]\left(\frac{n^{**}}{n_i} \right)^2$$

(4.109)

where J_G is the gate current density; n^* is the carrier concentration at the gate P^+–N junction; n^{**} is the carrier concentration at the N–N^+ junction; $\bar{n}$ is the average carrier concentration in the N-base region; D_{np} and L_{np} are the electron diffusion coefficient and diffusion length in the P^+-gate region, respectively; D_{pn} and L_{pn} are the hole diffusion coefficient and diffusion length in the N^+-source region, respectively; $\bar{n}_p$ and $\bar{p}_n$ are the equilibrium minority carrier concentrations in the P^+-gate and N^+-source regions, respectively; W_p and W_n are the thickness of the P^+-gate and N^+-source regions, respectively; $2a$ is the channel width; and τ_{HL} is the high-level lifetime in the N base.

In high-voltage JFETs, the channel width $2a$ is much shorter than the diffusion length for free carriers. In these devices, the carrier concentration becomes homogeneous in the channel and

$$n^* = \bar{n} = n^{**}$$

(4.110)

Furthermore, recombination in the channel region becomes negligible compared with recombination in the end regions. The free-carrier concentration in the N-base region is then given by

$$\bar{n} = n_i \left\{ \frac{I_G/A_G}{\left[\dfrac{qD_{np}\bar{n}_p}{L_{np}\tanh(W_p/L_{np})} \right] + \left[\dfrac{qD_{pn}\bar{p}_n}{L_{pn}\tanh(W_n/L_{pn})} \right]} \right\}^{1/2}$$

(4.111)

where I_G is the gate current and A_G is the gate junction area. The on-resistance is determined by the drift of these free carriers in the channel and drift regions because the background doping can be neglected:

$$R_{on} = \frac{W}{A_s q(\mu_n + \mu_p)\bar{n}}$$

(4.112)

$$= \frac{\left\{ \left[\dfrac{qD_{np}\bar{n}_p}{L_{np}\tanh(W_p/L_{np})} \right] + \left[\dfrac{qD_{pn}\bar{p}_n}{L_{pn}\tanh(W_n/L_{pn})} \right] \right\}^{1/2}}{A_s q(\mu_n + \mu_p)n_i(I_G/A_G)^{1/2}}$$

(4.113)

where the device dimensions W and A_s must include spreading resistance effects as described earlier for unipolar operation. The expression derived here indicates that the on-resistance will decrease inversely as the square root of the gate current. This has been observed experimentally in high-voltage power JFETs [18].

If a typical N-base modulation level of 10^{16} free carriers/cm^3 is assumed, the specific on-resistance can be calculated as a function of the breakdown voltage of the JFET. The calculated specific on-resistance is shown in Fig. 4.31 for the case of 100-, 200-, 500-, and 1000-V devices. For comparison, the specific on-resistance of corresponding unipolar FETs is also shown in Fig. 4.31. A drastic reduction in the specific on-resistance is observed especially at the higher voltages. For the bipolar mode of operation, the increase in the on-resistance with decreasing background doping level arises from the slight increase in drift region width required to support the voltage. In contrast, the on-resistance for the unipolar mode of operation increases very rapidly with decreasing doping level because the resistance is dependent on the doping level and the depletion layer width. In addition to the advantage of the much lower on-resistance, the

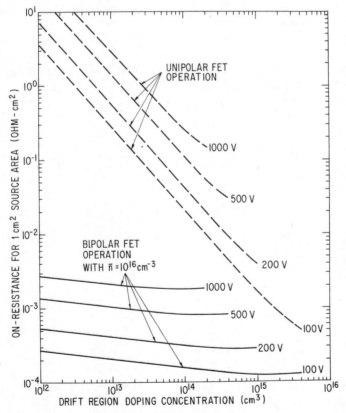

Fig. 4.31. Comparison of the specific on-resistance of high-voltage JFETs operated in the unipolar and bipolar modes.

bipolar mode of operation allows a reduction in channel doping level, which results in a higher forward blocking gain.

4.4.2. Output Characteristic Analysis

Analysis of the on-resistance is based on the assumption that the injected carriers due to gate current flow are uniformly distributed through the drift region. Such a uniform modulation of the channel and drift regions occurs only at very low drain voltages. As the drain voltage increases, the conductivity modulation becomes confined to a region near the source and an unmodulated region forms near the drain that supports the applied drain potential. The solution of the output characteristics must account for the nonuniform distribution of the injected charge [20].

To analyze the output characteristics, the region between source and drain must be considered to consist of three sections. These sections are illustrated in Fig. 4.32. In section I of that Figure there is conductivity modulation because of the presence of a high concentration of minority carriers injected from the gate. In section II there are no minority carriers and current flow occurs by the drift of electrons at a concentration equal to the background doping level N_D. The change in carrier mobility with electric field strength must be accounted for in this region because of the high electric fields at larger drain voltages. When the electric field exceeds a critical value $\mathscr{E}_S$, it is assumed that the carriers attain a saturated drift velocity v_S. In section III the current is space-charge limited and the carriers are transported at the saturated drift velocity.

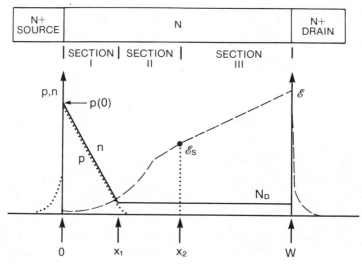

Fig. 4.32. Free-carrier and electric field distribution in the bipolar mode JFET at high drain voltages.

For a constant gate current, as the drain voltage increases, the width of these sections changes and determines the overall output characteristics. At low drain voltage, section I extends all the way from source to drain. The carrier profile and electric field distribution for this case are shown in Fig. 4.33b with the corresponding point in the output characteristic indicated in Fig. 4.33a. Note that the carrier concentration is now much higher than the background doping level throughout the drift region. As the drain voltage increases to point B on the characteristics, the hole concentration at the source increases and an unmodulated region forms near the drain corresponding to section II described above.

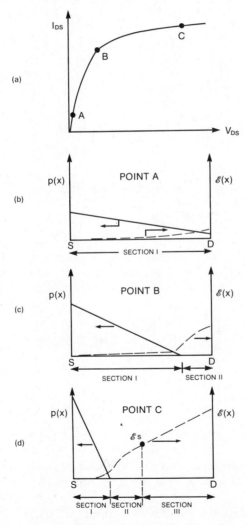

Fig. 4.33. Free-carrier and electric field distribution in the bipolar mode at three typical points on the device characteristics.

The drain voltage is as yet not sufficient to cause saturation in the velocity of the free carriers. With further increase in drain voltage to point C on the characteristics, the carriers attain saturated drift velocity in a region close to the drain as indicated by section III in Fig. 4.33d.

On the basis of the piecewise analysis of the JFET in the bipolar mode of operation, expressions have been derived for its output characteristics [20]. For the low drain voltage region, when the modulation extends throughout the drift region:

$$V_{DS} = \frac{2kT}{q} \ln \frac{\left[\left(\frac{laW}{2\alpha\tau}\right)^2 + \left(\frac{I_G}{q\alpha}\right) + \left(\frac{laW^2 I_D}{4q\alpha\tau D_n A_s}\right)\right]^{1/2} - \left(\frac{laW}{2\alpha\tau}\right) + N_D}{\left[\left(\frac{laW}{2\alpha\tau}\right)^2 + \left(\frac{I_G}{q\alpha}\right) + \left(\frac{laW^2 I_D}{4q\alpha\tau D_n A_s}\right)\right]^{1/2} - \left(\frac{laW}{2\alpha\tau}\right) - \left(\frac{WI_D}{2qD_n A_s}\right) + N_D}$$

$$\tag{4.114}$$

This portion of the $I-V$ characteristics extends until the drain voltage reaches $V_{\text{sat},M}$ where

$$V_{\text{sat},M} = \frac{2kT}{q} \ln\left(1 + \frac{I_D W}{2qD_n A_s N_D}\right) \tag{4.115}$$

When the drain voltage is increased beyond $V_{\text{sat},M}$, an ohmic voltage drop appears across section II in addition to the voltage drop in section I. Then

$$I_D = \frac{2qD_n A_s p(0)}{W} + \frac{\sigma A_s}{W}(V_{DS} - V_{\text{sat},M}) \tag{4.116}$$

The carrier density $p(0)$ is determined by the gate current I_G. When the relationship between the gate current and the carrier concentration as described by Eq. (4.109) is used, the following relationship describing the output characteristics is obtained:

$$I_G = \frac{q}{4}\left(\frac{qN_D}{kT}\right)^2\left(\alpha + \frac{qD_n laA_s}{\tau I_D}\right)\left(\frac{WI_D}{\sigma A_s} - V_{DS} + V_{\text{sat},M}\right)^2 \tag{4.117}$$

where

$$\alpha = \frac{A_s D_{ps}}{W_s N_{DS}} + \frac{A_G D_{nG}}{L_{nG} N_{AG}} \tag{4.118}$$

The expression for α given here is based on the assumption that the thickness of the source region W_s is much smaller than the diffusion length for minority carriers in the source and that the thickness of the gate region is much greater than the diffusion length for minority carriers in it.

From this expression for the active region of the device characteristics, an expression for the current gain can be derived [20]:

$$h_{fe} = \frac{4qD_n^2 A_s^2}{W^2}\left(2\alpha I_D + \frac{qD_n laA_s}{\tau}\right)^{-1} \tag{4.119}$$

In typical devices, the current gain is found to range from over 100 at low currents to less than 10 at high currents [21]. The current gain is controlled by recombination in the source, gate, and drift regions. The relatively low current gain observed in devices results in the need to supply substantial gate current to operate the power JFET in its bipolar mode. This disadvantage and the normally-on characteristics are problems during device application. Both of these issues can be resolved by gating the power JFET in an emitter switching configuration.

4.4.3. Gating Circuits

The problems of normally-on operation of the power JFETs can be overcome by using a low-voltage normally-off device, such as a power MOSFET, in series with the device as shown in Fig. 4.34. When the drain voltage is applied to this circuit with gate short-circuited to the source, the voltage initially appears across the low-voltage power MOSFET (transistor T1) because the JFET (transistor T2) is a normally-on device. As the drain potential D_1 rises, the gate source junction of the JFET becomes increasingly reverse biased and ultimately pinches off the channel. The high-voltage JFET now enters its blocking mode and further applied voltage is supported by the JFET.

In the circuit shown in Fig. 4.34, the second (high-voltage) power MOSFET (transistor T3) is used to provide the gate drive current for the JFET. When a positive gate voltage is applied, both transistors T1 and T3 are turned on. The

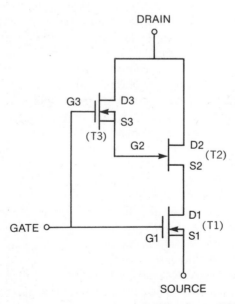

Fig. 4.34. Gating circuit for bipolar operation of high-voltage JFETs with normally-off characteristics.

current flowing through transistor T3 provides the gate current for the JFET required for bipolar operation.

Analysis of this gating scheme for bipolar operation of the JFET indicates that the current density (including the areas of all three transistors) is an order of magnitude higher than for an equivalent power MOSFET. Switching tests performed using 500-V devices show excellent gate controlled turn-off wave-forms [22] with fall times of about 100 nsec. Since this circuit provides a high input impedance interface to control circuitry, it is a promising approach to power switching.

4.5. DEVICE STRUCTURES AND TECHNOLOGY

High-voltage JFETs have been developed by using two basic structures—the buried-gate and surface-gate structures. These structures are illustrated in Fig. 4.1. Historically, the first high-voltage JFETs were fabricated by use of buried gate technology. In France these devices were called *Gridistors* [23], and in Japan they were referred to as *static-induction transistors* (SITs) [24].

A fundamental difference between the buried- and surface-gate structures is the relative location of the source and gate regions. In surface-gate devices, the source and gate regions must be interdigitated, with each source region completely surrounded by the gate junction to pinch off drain–source current flow. Because of the small gate junction spacing necessary to obtain high blocking gain, the interdigitation of gate and source regions is a difficult technological challenge. The formation of electrical short circuits between gate and source has limited the growth in the size and, hence, the current handling capability of surface gate devices. They are particularly suited for high-frequency applications because a very low series gate resistance can be achieved by contacting the gate at all points along the surface with low-resistance metallization. In contrast, the buried gate structure has an inherent gate series resistance because gate current must flow along the relatively high resistance diffused gate fingers. Although this resistance can be decreased by bringing the buried grids to the surface often and placing metal contacts at these points, the active (source) area is reduced by the increasing area consumed by the gate contacts. The buried gate devices are most suitable for achieving high current, large area, devices that are intended for operation at lower frequencies than surface gate devices. In these devices, the gate junctions can be placed very close together to achieve high blocking gains.

If the power JFET is to be operated in the bipolar mode, the surface-gate structure is more suitable than the buried-gate structure. During bipolar operation, a substantial gate current must flow through the gate regions during the on state. The high resistance of the diffused buried gate regions can cause substantial voltage drop along the fingers. This debiases the gate junction and leads to inhomogeneous current distribution with most of the current confined to regions close to the gate contact. The debiasing is negligible for surface gate-devices as a result of the very low resistance of the gate contact metallization.

Because of the rapid increase in the on-resistance with increasing breakdown voltage for power JFETs, the development of these transistors has been constrained to operating voltages below 1000 V. The width of the drift region of these devices is less than 100 μm. Since it is impractical to process wafers with thicknesses of less than 10 mils (250 μm), it becomes essential to fabricate these power junction gate FETs by using epitaxial layers grown on heavily doped N-type substrates. These substrates are usually doped with antimony to a concentration of $2-5 \times 10^{18}$ atoms/cm^3. Antimony is chosen as the dopant because it has a lower diffusion coefficient than phosphorus, and also because of its low vapor pressure at epitaxial growth temperature. This minimizes the width of the transition region between the epitaxial layer and the substrate. The misfit of the antimony atom in the silicon crystal degrades the substrate dislocation density at doping levels above 5×10^{18}/cm^3. This limits the substrate resistivity to greater than 0.01 $\Omega \cdot$cm. When a lower resistivity is desired, arsenic-doped substrates with resistivities of 0.001 $\Omega \cdot$cm must be used. The autodoping problem is significantly greater in this case.

The epitaxial layer growth conditions are also of great importance in determining the quality of the deposits. High-voltage devices require epitaxial layers with low doping concentrations and relatively large thicknesses when compared with other applications. Extreme care is required to maintain a low background level in the reactor. Much higher growth rates than those used in integrated-circuit fabrication are employed to keep the deposition time reasonable. Despite these high growth rates, it is important to obtain a low density of hillocks (protrusions on the epitaxial layer surface) and stacking faults (crystallographic disorientation) because high localized electric fields occur at these sites as a result of the segregation of impurities during device processing. The hillocks also create serious difficulties during photolithography when patterning the device structure. It has been found that a lower hillock and stacking fault density can be achieved by increasing the epitaxial growth temperature and keeping the growth rate below 2 μm/min [25].

4.5.1. Buried-Gate Technology

Power JFETs have been fabricated by using two different buried-grid technologies: (1) the epitaxial buried-grid process and (2) the ion-implanted buried-grid process. In both approaches, the gate regions consists of stripes of P^+ regions located below the upper surface of the device. Contact to the buried-gate fingers is then made by either diffusing a contact region from the upper surface or by etching down to the grids.

The epitaxial buried grid process sequence is illustrated in Fig. 4.35. The first step consists of the diffusion of a P-type impurity such as boron in selective areas by using a suitable mask such as silicon dioxide, followed by the epitaxial growth of N-type silicon over the diffused areas to form the buried grids. The epitaxial growth is usually performed by using vapor-phase epitaxy [24]. A major problem with this process is the occurrence of autodoping as a result of

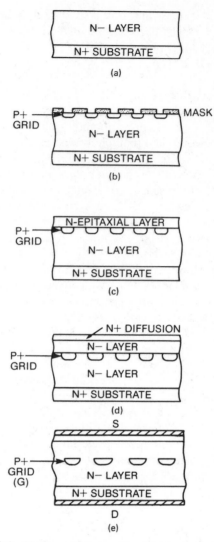

Fig. 4.35. Epitaxial buried-grid process sequence: (*a*) epitaxial growth; (*b*) selective grid diffusion; (*c*) epitaxial growth; (*d*) source diffusion; (*e*) metallization.

the transport of boron from the diffused region into the vapor phase. Since the boron is reincorporated in the epitaxial layer grown over the grids, a *P*-type connecting layer forms between the grids, preventing the field-effect operation. The autodoping has been overcome by the addition of *N*-type dopants to the input gas during epitaxial growth to compensate for the presence of boron, but the process is complicated by the need to achieve low epitaxial layer doping concentrations in the presence of high autodoping boron levels. These difficulties can be circumvented by growing the epitaxial layers from the liquid phase. This has been successfully accomplished by using tin as the solvent. The complete suppression of autodoping and the ability to fabricate heavily boron-doped grids that are spaced close to each other has been achieved by this technique [26]. By means of the epitaxial buried grid process, devices with forward blocking voltages of 1500 V have become commercially available. These devices are capable of switching 100 A in under 500 nsec and have an on-resistance of 0.5 Ω [27]. Typical blocking gains range from 20 to 40.

The ion-implanted buried grid process sequence is illustrated in Fig. 4.36. In this process, the first step consists of the deep ion implantation of a *P*-type impurity such as boron into the *N*-type epitaxial layer in selective areas. By

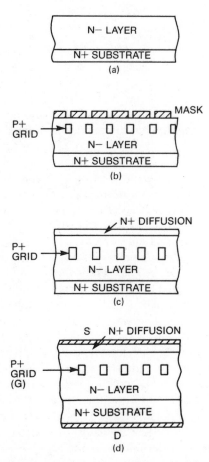

Fig. 4.36. Ion-implanted buried-grid fabrication sequence: (*a*) epitaxial growth; (*b*) ion implantation; (*c*) source diffusion; (*d*) metallization.

adjusting the implantation energy between 600 and 900 keV for boron implantations, a P-type grid with a height of 0.8 μm can be created at 1 μm below the surface [28]. Because of the high implant energy involved, conventional masking techniques with the use of photoresist are inadequate and it is necessary to use a metal mask such as gold that is patterned on the wafer surface prior to the implantation. This metal film must be removed after the implant because of the subsequent high-temperature processing. To remove the implant damage and activate the impurities, it is necessary to perform an anneal at 900°C for 20 mins. Boron is the most suitable P-type impurity for this process because it is the lightest P-type dopant and its diffusion rate in silicon is the slowest. It should be noted that so far the devices made using this process have had low breakdown voltages. Although this technology can potentially be used to produce power JFETs, the fabrication of high-voltage devices has not yet been demonstrated.

4.5.2. Surface-Gate Technology

Power JFETs with gate fingers continuously accessible to metallization at the upper surface are necessary to achieve high speed switching and for bipolar operation. The simplest approach for achieving this is by using planar diffusion technology to fabricate the gate region. The process sequence for the fabrication of planar diffused gates is illustrated in Fig. 4.37. The process uses conventional planar diffusion technology with silicon dioxide as a mask to form the gate and source regions. The planar diffusion of the gate regions has been accomplished by using boron predeposition from conventional sources such as boron nitride or diborane [29, 30], by using boron doped polysilicon as the source [31], and by self-aligned ion implantation of boron [32]. During the fabrication of the planar diffused gate structure, it is necessary to perform three critical alignment steps: (1) to locate the source diffusion between the gate regions, (2) to open contact windows to the gate and source regions, (3) to define the gate and source metallization. Misalignment during any of these steps can lead to the formation of electrical short circuits between gate and source. This is a serious drawback of this process that has limited the size and current handling capability of these devices. Furthermore, when planar diffusion technology is used to form the gate region, the gate junction boundary has a cylindrical shape, which creates an open-channel structure with a small channel aspect ratio. As a result, the blocking gains of these devices are low, typically in the range of 2 to 5. This creates a serious problem for achieving device operation at voltages above 200 V. Moreover, for a given photolithographic delineation capability that determines the size of the gate window, the lateral diffusion increases the area occupied by the gate. This reduces the area available on the chip for the conduction of the source-drain current, which in turn increases the on-resistance. In addition, the lateral diffusion increases the area of the gate region, which causes the input gate capacitance to increase and the frequency response to decrease. Despite these limitations, devices have been made by using planar

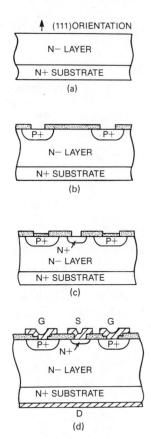

Fig. 4.37. Planar diffused surface gate fabrication sequence: (*a*) epitaxial growth; (*b*) gate diffusion; (*c*) source diffusion; (*d*) contact window–metal.

diffusion with blocking capabilities of up to 150 V with cutoff frequencies approaching 1 GHz [33].

To overcome the low blocking gain limitations of the planar diffused process, a vertically walled gate region must be created. A comparison of the pinch-off in the channel for the vertically walled and planar diffused gate devices is illustrated in Fig. 4.38. In the vertically walled gate structure, the channel potential barrier extends over a longer distance along the channel as shown on the right-hand side of the figure. It is more difficult for the drain potential to penetrate the channel and produce a lowering of the barrier height in this case. Higher blocking gains can be expected in these devices than achievable by planar diffusion. In addition, a high channel aspect ratio can be achieved without consuming a large surface area because the gate extends vertically down from the window defined at the top surface. This also reduces the gate junction area, resulting in a low input capacitance, which is desirable for maximizing the frequency response.

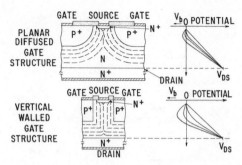

Fig. 4.38. Comparison of the depletion layer spreading for a vertically walled gate structure with a cylindrically walled gate structure.

The first process developed to fabricate vertically walled gate regions was based on etching vertically walled grooves and refilling them with boron-doped silicon [10]. The epitaxial-refill process sequence is shown in Fig. 4.39. In this process (110)-oriented wafers are used because the (111) crystal planes lie perpendicular to this orientation. Deep grooves with vertical sidewalls are then formed in the gate areas by using preferential etching with silicon dioxide as a

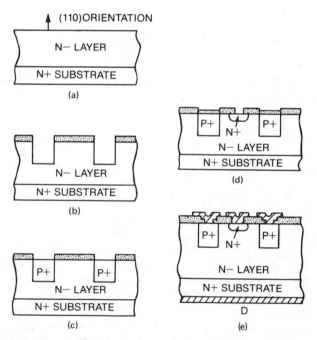

Fig. 4.39. Epitaxial-refill surface-gate fabrication sequence: (*a*) epitaxial growth; (*b*) groove etching; (*c*) epitaxial refill; (*d*) source diffusion; (*e*) contact–metallization.

mask. A typical etch for this process consists of a mixture of potassium hydroxide and isopropanol [34]. With this etch, grooves can be formed to depths of up to 40 μm with less than 1 μm of undercutting. This is followed by the selective epitaxial deposition of *P*-type silicon in the grooves. This has been accomplished by both vapor-phase epitaxy [10] and liquid-phase epitaxy [35]. In the case of vapor-phase epitaxial refill it is important to reduce the polycrystalline silicon deposits on the silicon dioxide masking layer by the addition of hydrogen chloride gas during the epitaxial growth. Nevertheless, some polycrystalline deposits are inevitably observed on the oxide, which creates serious difficulties during subsequent device processing. In contrast, no deposits are observed during epitaxial refill from the liquid phase because the melt does not wet silicon dioxide and no nucleation occurs on the oxide surface. After epitaxial refill, the source region is formed by conventional planar diffusion. By forming vertically walled gate regions, the channel aspect ratio is greatly improved without increasing the area occupied by the gate region. This allows achievement of high blocking gains (on the order of 20 to 40) in these devices while retaining a large device active area and a low gate input capacitance. As a consequence of the high blocking gains achievable with this technology, the development of surface gate devices capable of blocking up to 400 V has been achieved [10].

Although the epitaxial-refill process overcomes the most serious performance limitation of the planar diffused structure (i.e., the poor blocking gain), the problems of fabricating the interdigitated gate–source regions are not addressed by this technology. In fact, it requires the same three critical mask alignment steps discussed earlier for the planar diffused devices. These problems, as well as the demanding epitaxial refill step, can be circumvented by using the recessed-gate structure [14] shown in Fig. 4.40. This structure retains the advantages of

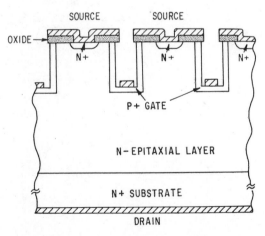

Fig. 4.40. Recessed-gate high-voltage JFET structure.

a vertically walled gate junction. However, instead of refilling the gate region, the metal contact to the gate regions are formed by using a self-aligned metallization process that relies on an oxide overhang at the edges of the grooves. The process sequence for the recessed-gate device structure is illustrated in Fig. 4.41. The process also begins with the growth of an epitaxial layer on (110)-oriented, N^+ substrates. An oxide layer is first grown and patterned to open windows at both the source and gate regions. The simultaneous patterning of the gate and source regions by use of a single mask is an important feature because it provides self-alignment of these regions. The perfect alignment of the source region between the gate regions maximizes the breakdown voltage between gate and source. The next step Fig. 4.41c is the deposition of a silicon nitride layer, followed by its delineation over the gate regions by using photolithography. Note that considerable misalignment can be tolerated at this step. A phosphorus diffusion is now performed to create the source N^+ regions, and an oxide is grown in the source windows. In addition to acting as a diffusion mask, the nitride layer prevents oxide growth in the gate windows. The nitride

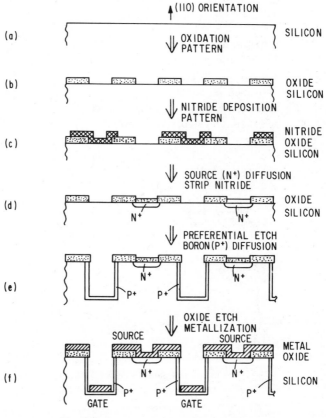

Fig. 4.41. Recessed-gate fabrication sequence.

is now selectively etched away as shown at point d (Fig. 4.41d) in the process. Vertically walled grooves can now be etched by using preferential etching with potassium hydroxide-based solutions. The etch composition is chosen so as to create a small overhang at the edges of the masking oxide. The overhang is required to obtain a automatic source–gate metal definition. A shallow boron diffusion is performed after the groove is etched to create the gate junction. The thin oxide in the source diffusion windows is now washed out, without the need for a masking step in order to open the source contact windows. Any oxide grown in the groove during boron diffusion is also removed by this step. Evaporation of aluminum normal to the wafer surface results in an automatic separation of the source and gate metallization without the need for a masking step. Thus the recessed gate JFET process shown in Fig. 4.41 allows vertically walled gate junction formation without any of the three critical alignment steps used in the previous surface gate processes. Very high device yields have been achieved with this process.

It should be noted that the device design requires the recessed-gate metal in the gate fingers to extend into a wide recessed-gate pad to which external electrical contact can be made during device packaging. With the use of this structure, high blocking gains (>10) have been achieved. These devices have the highest forward blocking capability (600 V) among surface gate structures. The low-resistance gate contact made possible by the recessed-gate structure results in a small signal, unity power gain, cutoff frequency of over 0.5 GHz. At high current levels, forced gate turn-off times in the range of 25 nsec have been observed. The recessed gate devices are also well suited for bipolar operation [21].

A modification of the recessed gate process, developed for silicon devices, has been applied for the fabrication of Schottky barrier gate FETs in gallium arsenide. Because of differences in the lattice planes containing arsenic and gallium atoms, the gate structure takes on a unique shape (Fig. 4.42) called the

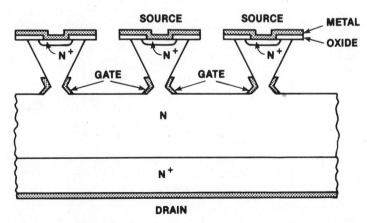

Fig. 4.42. Trapezoidal groove gate structure developed for high-voltage gallium arsenide power MESFETs.

trapezoidal gate [36]. The necked-down shape of the groove aids the pinch-off process in the channel while providing a large area for source contact at the upper surface. A key difference between this structure and that for the vertically walled silicon device is the need to rotate the gallium arsenide wafer with respect to the evaporation source during deposition of the gate metal to achieve coverage extending part of the way along the sidewalls of the grooves. Because of the much greater processing difficulties for gallium arsenide, devices with blocking voltages of less than 100 V have been achieved with recessed gates. In comparison, vertical channel gallium arsenide power JFETs with forward blocking capability of 150 V have been fabricated with the buried grid structure [37].

4.6 TRENDS

In the future, the development of vertical-channel power JFETs is expected to progress in several directions. The push to increase the power ratings is aimed at the application of these devices in audio amplifiers. These devices can be fabricated with buried gate structures. There is also a need for power devices that can be operated at microwave frequencies directly from the AC line. High-voltage Schottky gate FETs fabricated from gallium arsenide should be well suited for this application. The development of commercially available devices is constrained by the difficult technological challenges that must be overcome to create a reliable manufacturing process. In the meantime, the performance of silicon devices is beginning to approach frequencies that are useful for creating microwave power sources. Thus, an improvement in current handling capability and an increasing frequency response can be expected for these devices.

REFERENCES

1. W. Shockley, "A unipolar field effect transistor," *Proc. IRE*, **40**, 1365–1376 (1952).

2. L. J. Sevin, *Field Effect Transistors*, McGraw-Hill, New York, 1965.

3. A. B. Grebene and S. K. Ghandhi, "General theory for pinched operation of the junction-gate FET," *Solid State Electron.*, **12**, 573–589 (1969).

4. D. P. Kennedy and R. R. Obrien, "Computer aided two-dimensional analysis of the junction field-effect transistor," *IBM J. Res. Devel.*, **14**, 95–116 (1970).

5. J. A. Guerst, "Theory of insulated-gate field effect transistors near and beyond pinch-off," *Solid State Electron.*, **9**, 129–142 (1966).

6. G. F. Newmark and E. S. Rittner, "Transition from pentode- to triode-like characteristics in field effect transistors, *Solid State Electron.*, **10**, 299–304 (1967).

7. B. J. Baliga, "Semiconductors for high voltage, vertical channel, field effect transistors," *J. Appl. Phys.*, **53**, 1759–1764 (1982).

8. K. Yamaguchi and H. Kodera, "Optimum design of triode-like JFET's by two-dimensional computer simulation," *IEEE Trans. Electron Devices*, **ED-24**, 1061–1069 (1977).

9. P. Plotka and B. Wilamowski, "Interpretation of exponential type drain characteristics of the static induction transistor," *Solid State Electron.*, **23**, 693–694 (1980).

10. B. J. Baliga, "A power junction gate field-effect transistor structure with high blocking gain," *IEEE Trans. Electron Devices*, **ED-27**, 368–373 (1980).

11. K. Sakai, Y. Komatsu, and H. Kobayashi, "Complementary power FETs with vertical structures," *IEEE Power Electronics Specialists Conference Record*, pp. 214–221 (1974).

12. X. C. Kun, "Calculation of amplification factor of static induction transistor," *IEE Proc.*, **131**, 87–93 (1984).

13. M. S. Adler and B. J. Baliga, "A simple method for predicting the forward blocking gain of gridded field effect devices with rectangular grids," *Solid State Electron.*, **23**, 735–740 (1980).

14. B. J. Baliga, "High voltage, junction gate field effect transistor with recessed gates," *IEEE Trans. Electron Devices*, **ED-29,** 1560–1570 (1982).

15. J. R. Hauser, "Small signal properties of field effect devices," *IEEE Trans. Electron Devices*, **ED-12**, 605–618 (1965).

16. Y. Yukimoto, Y. Kajiwara, G. Nakamura, and M. Aiga, "1 GHz, 20 W static induction transistor," *Jpn. J. Appl. Phys.* **17**, 241–244 (1978).

17. T. Shino, H. Kamo, K. Aoki, and S. Okano, "2 GHz, high power silicon SITs," *Jpn. J. Appl. Phys.* **19**, 283–287 (1980).

18. B. J. Baliga, "Bipolar operation of power junction field effect transistors," *Electron. Lett.*, **16**, 300–301 (1980).

19. A. Herlet, "The forward characteristics of silicon power rectifiers at high current densities," *Solid State Electron.*, **11**, 717–742 (1968).

20. S. Bellone, A. Caruso, P. Spirito, and G. Vitale, "A quasi-one-dimensional analysis of vertical JFET devices operated in the bipolar mode," *Solid State Electron.*, **26**, 403–413 (1983).

21. B. J. Baliga, "High temperature characteristics of bipolar mode power JFET operation," *IEEE Electron Device Lett.*, **EDL-4**, 143–145 (1983).

22. B. J. Baliga, "The MAJIC–FET: A high speed power switch with low on-resistance," *IEEE Electron Device Lett.*, **EDL-3**, 189–191 (1982).

23. S. Teszner and R. Gicquel, "Gridistor—a new field effect device," *Proc. IEEE*, **52**, 1502–1513 (1964).

24. J. Nishizawa, T. Terasaki, and J. Shibata, "Field effect transistor versus analog transistor (static induction transistor," *IEEE Trans. Electron Devices*, **ED-22**, 185–197 (1975).

25. B. J. Baliga, "Defect control during silicon epitaxial growth using dichlorosilane," *J. Electrochem. Soc.*, **129**, 1078–1084 (1982).

26. B. J. Baliga, "Buried-grid fabrication by using silicon liquid phase epitaxy," *Appl. Phys. Lett.*, **34**, 789–790 (1979).

27. F. Goodenough, "Fast static induction transistors control up to 150 kW," *Electron. Design*, 190 (January 26, 1984).

28. D. P. Lecrosnier and G. P. Pelous, "Ion-implanted FET for power applications," *IEEE Trans. Electron Devices*, **ED-21**, 113–118 (1974).

29. J. L. Morenza and D. Esteve, "Entirely diffused vertical channel JFET: Theory and experiment," *Solid State Electron.*, **21**, 739–746 (1978).

30. M. Kotani, Y. Higaki, M. Kato, and Y. Yukimoti, "Characteristics of high-power and high breakdown voltage static induction transistor with high maximum frequency of oscillation," *IEEE Trans. Electron Devices*, **ED-29**, 194–198 (1982).

31. O. Ozawa, H. Iwasaki, and K. Muramoto, "A vertical channel JFET fabricated using silicon planar technology," *IEEE J. Solid State Circuits*, **SC-11**, 511–517 (1976).

32. O. Ozawa and H. Iwasaki, " A vertical FET with self-aligned, ion-implanted source and gate regions," *IEEE Trans. Electron Devices*, **ED-25**, 56–57 (1978).

33. J. I. Nishizawa and K. Yamamoto, "High-frequency, high-power static induction transistor," *IEEE Trans. Electron Devices*, **ED-25**, 314–322 (1978).

34. D. L. Kendall, "Vertical etching of silicon at very high aspect ratios," *Annu. Rev. Materials Sci.* **9**, 373–403 (1979).

35. B. J. Baliga, "Refilling silicon grooves by liquid phase epitaxy," *J. Electrochem. Soc.*, **129**, 2819–2823 (1982).

36. P. M. Campbell, W. Garwacki, A. Sears, and B. J. Baliga, "Trapezoidal Groove Schottky Gate Vertical Channel GaAs FET," *IEEE International Electron Devices Meeting Digest*, Abstract 7.3, pp. 186–189 (1984).

37. P. M. Campbell, R. S. Ehle, P. V. Gray, and B. J. Baliga, "150 volt vertical channel GaAs FET," *IEEE International Electron Devices Meeting Digest*, Abstract 10.4, pp. 258–260 (1982).

PROBLEMS

4.1. Consider a JFET with channel doping of $10^{14}/cm^3$ and gate junction separation of 10 μm. Calculate the pinch-off voltage.

4.2. What would the pinch-off voltage be if the channel doping were increased to $10^{15}/cm^3$ for the preceding device structure?

4.3. How much must the gate junction separation be reduced for the channel doping of $10^{15}/cm^3$ if the same pinch-off voltage as in Problem 4.1 is to be achieved?

4.4. Calculate the channel resistance per square centimeter for a JFET with channel doping of $10^{15}/cm^3$ and junction separation of 10 μm when a vertically walled gate structure with gate depth of 10 μm is used. Assume the gate region width is 5 μm.

4.5. Calculate the breakdown voltage and ideal specific on-resistance for a JFET fabricated using a drift region doping of $10^{15}/cm^3$. What is the drift region thickness needed to support this voltage?

4.6. A JFET is fabricated by using a vertically walled gate structure as described in Problem 4.4. It also contains a drift region as calculated in Problem 4.5. Determine the specific on-resistance for the device.

4.7. Calculate the percentage increase in the on-resistance due to the channel resistance and spreading resistance for the device in Problem 4.6 over the ideal device analyzed in Problem 4.5.

4.8. What is the blocking gain of the JFET described in Problem 4.6?

4.9. What is the blocking gain for the device described in Problem 4.8 if the gate structure were to be altered to buried grids with a radius of 2.5 μm and a pitch of 15 μm?

4.10. Calculate the transit-time limited frequency response of JFETs with breakdown voltages of 100, 300, and 600 V. Assume that a channel length of 10 μm is used for all cases.

4.11. What is the specific on-resistance of the JFET analyzed in Problem 4.5 if it is operated in the bipolar mode with an injected carrier density of $10^{16}/\text{cm}^3$? Compare this to the unipolar case.

5 POWER FIELD-CONTROLLED DIODES

The power field-controlled diode (FCD) is a three-terminal device that evolved from the JFET. The device consists of a $P-i-N$ rectifier structure in which a gate junction has been introduced to control the current flow between the anode and cathode regions. The gate controlling mechanism is similar to that for JFETs, but these devices differ from JFETs because the forward conduction occurs with very strong conductivity modulation of the drift region. Since these devices exhibit a rectifierlike on-state characteristic with both forward and reverse blocking capability, they have also been called *field-controlled thyristors* and *static-induction thyristors*. Another term used to describe these devices is the bipolar Gridistor, because of the gridlike gate junction used in the device structure.

Because of the extremely high injected minority carrier level into the drift region, FCDs operate at very high current densities with a low forward voltage drop. This makes it possible to design high-voltage (> 1000-V) devices that are capable of handling large currents. Unfortunately, the large minority carrier density injected into the drift regions limits the switching speed of the devices to below 1 MHz. Furthermore, these devices conduct current in the absence of a gate bias. This normally-on characteristic can be a problem when it is used in power circuits. Therefore, special gating schemes must be used to circumvent this problem. Barring this issue, FCDs have been shown to offer several unique features such as very high (dV/dt) capability, high radiation tolerance, and good high-temperature characteristics.

5.1. BASIC STRUCTURES AND CHARACTERISTICS

For the reasons cited in the case of the power JFET, power FCDs are made with a vertical current conduction path through the chip. The cathode and gate regions are located on the upper surface, whereas the anode is placed at the bottom of the wafer. This arrangement allows achievement of high forward and reverse blocking capability as well as providing these devices with high current handling capability.

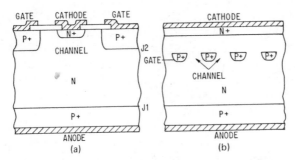

Fig. 5.1. Field-controlled diode structures: (*a*) surface-gate structure; (*b*) buried-gate structure.

Many structures for FCDs have been explored. They can be generally classified as buried-gate structures or surface-gate structures as illustrated in Fig. 5.1. In making this distinction, the surface-gate structure has been defined as one in which the gate region is accessible at all points along the length of the gate finger to allow making a low-resistance contact to the gate. By necessity, the buried-gate regions must also be brought to the surface at some locations to make an ohmic contact to them with the gate metal. These portions represent inactive regions in this device structure, and the device must be designed to minimize the area occupied by the contact region.

In both types of device structure, the devices are operated by grounding the cathode and applying bias voltages to the anode and gate regions with respect to the cathode. In the absence of a gate bias, the device behaves essentially like a *P–i–N* rectifier. When a positive voltage is applied to the anode, the device operates in its forward conduction mode with junction J1 forward-biased. Minority carriers are injected from junction J1 into the *N*-drift region. As the forward-bias voltage increases, the injected carrier density increases until it exceeds the background doping level of the drift region, and it can be many orders of magnitude larger than the background doping level. Since these injected carriers are present throughout the region between the anode and the cathode, the resistance of the drift region is drastically reduced by the resulting conductivity modulation. A high forward current density can be achieved with a low forward voltage drop in this mode of operation.

When a negative voltage is applied to the anode, the anode junction (J1) becomes reverse biased and supports the applied voltage without allowing significant current flow (except for the leakage current of the reverse-biased junction) until its avalanche breakdown voltage is reached. This provides these devices with their reverse blocking capability.

These characteristics are similar to those of a *P–i–N* rectifier. However, unlike the *P–i–N* rectifier, the FCD is capable of providing forward blocking capability equal to its reverse blocking capability. This is achieved by using the gate junction. If a negative bias is applied to the gate with respect to the source, the gate junction becomes reverse biased and a depletion layer extends from the metallurgical junction into the channel region. When the gate bias is sufficiently high, the depletion layer of adjacent gate regions merge in the channel and pinch off the current flow between the anode and the cathode. This operation is similar to that described for the junction gate FET. The gate voltage establishes a potential barrier in the channel that impedes the transport of electrons from the cathode to the anode. As the anode voltage increases, this potential barrier is lowered until current flow commences. The resulting device characteristics are triodelike in appearance, as illustrated in Fig. 5.2. Since larger gate bias voltages increase the channel potential barrier height, anode current flow is shifted to higher anode voltages until the avalanche breakdown voltage of the upper junction (J2) is reached.

In comparison with power JFETs, the power FCD characteristics are similar in the forward blocking region of operation. The characteristics of these devices strongly differ in the forward conduction and reverse blocking regimes of operation. Field-controlled diodes do not display the pentodelike forward conduction characteristics of power JFETs at high current levels because of the increasing conductivity modulation of the drift region with increasing anode current. Field-controlled diodes are also capable of blocking current flow in the reverse direction, which makes them applicable to AC circuits.

In the preceding description of device operation, it was assumed that the gate bias is present prior to the application of the anode voltage. The device is then being biased into a static forward blocking condition. This mode of operation would occur in AC circuits where the devices are turned on during the conducting half of the cycle and allowed to commutate to the off state at the end of the conduction half of the cycle. The current and voltage waveforms for this case are provided in Fig. 5.3 for reference. Note that the negative gate volt-

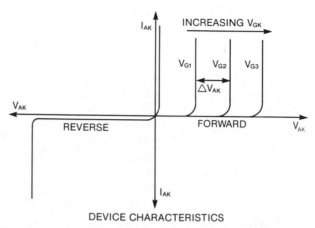

Fig. 5.2. Output characteristics of the field-controlled diode illustrating the triodelike forward blocking region.

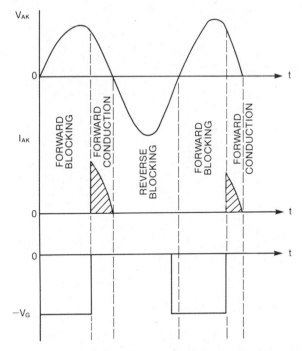

Fig. 5.3. Typical anode current–voltage and gating waveforms for operation of a FCD in an AC phase-control circuit.

age required to hold the device in its forward blocking mode is supplied prior to the application of the positive anode voltage to the device.

Field-controlled diodes are also capable of forcibly turning off the anode current during current conduction. The current and voltage waveforms applicable to this mode of operation are provided in Fig. 5.4 for the case of a resistive load. Here the reverse gate voltage is removed at time t_1 and the device is switched into its forward conducting mode. At time t_2 in the conduction half of the cycle, the reverse gate bias is reapplied. The FCD must forcibly interrupt the anode current at this point in time. Since the device is in its on-state prior to time t_2, the N-drift region is flooded with minority carriers. To switch the device into a blocking state, these carriers must be removed. The reverse gate bias applied at time t_2 serves to remove minority carriers (holes) from the channel region. Only after these carriers have been removed can a gate depletion layer be established to create a channel potential barrier that is capable of holding off the anode voltage. A substantial gate current flow is typically required in power FCDs to remove the charge in the channel and switch the device into its forward blocking mode. It should be noted that the ability to provide forced gate turn-off capability is an important feature of these devices. It not only

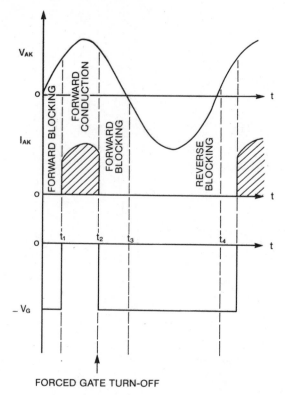

FORCED GATE TURN-OFF

Fig. 5.4. Typical anode current–voltage and gating waveforms for forced gate turn-off operation of FCDs in an AC circuit.

makes their use in AC circuits more versatile, but allows their application to DC circuits as well.

Because of the presence of the anode junction J1 in the current flow path between the anode and cathode, the forward conduction characteristics of the power FCD are similar to those of a diode. These characteristics contain a knee as shown in Fig. 5.5. This knee occurs at about 0.7 V at room temperature. Very little anode current flow is observed below the knee. Above the knee the characteristics can be considered to enter a resistive region at the higher current levels. Note that the typical forward conduction operating point occurs well above the knee. Because of the offset in the characteristics along the voltage axis, it is not possible to define a fixed on-resistance for these devices because this parameter is dependent on the location of the operating point along the I–V curve. Instead, power FCDs are characterized in terms of the forward voltage drop V_F at a known current density (I_F/A_a), where A_a is the anode area. Occasionally, an on-resistance defined by the straight line (V_F/I_F) is used for comparison with other power devices. It is preferable to avoid this approach since the power dissipation due to current conduction is not determined by this

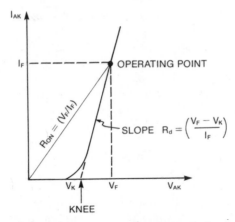

Fig. 5.5. Diodelike forward conduction characteristics of a FCD.

on-resistance but is instead dependent on the knee voltage and the dynamic resistance $R_d = [(V_F - V_K)/I_F]$:

$$P_D = V_K I_F + R_d I_F^2 \tag{5.1}$$

but

$$P_D \neq R_{on} I_F^2 \tag{5.2}$$

As in the case of power JFETs, the forward blocking characteristics have triodelike characteristics. Higher forward anode voltages can be supported with increasing gate bias. It is appropriate to define a blocking gain for the FCD in a manner similar to that used for the power JFET:

$$g_B = \left. \frac{\Delta V_{AK}}{\Delta V_{GK}} \right|_{I_{AK}} \tag{5.3}$$

The blocking gain of the power FCD is controlled by the channel design. In general, higher blocking gains have been achieved in FCDs than in power JFETs because the forward drop in the FCD is not as sensitive to the channel length and doping concentration.

During the description of forced gate turn-off operation of the FCD as illustrated in Fig. 5.4, it was pointed out that substantial gate current must be provided during turn-off to achieve the transition of the device from its forward conduction to its forward blocking condition. Although very little gate current is required to maintain the FCD in its steady-state forward blocking condition, the gate circuit must be capable of delivering the peak gate current during turn-off. The peak gate current is a function of the anode current at the time when the reverse gate voltage is applied, because the anode current determines the minority carrier density in the channel. The ratio of these currents is a valuable measure of device performance. It is called the *turn-off current gain*:

$$G_T = \frac{I_{AK}}{I_{GK(peak)}} \tag{5.4}$$

For a given anode current, if the peak gate current is increased by either increasing the gate reverse voltage or decreasing the gate circuit impedance, the minority carriers in the channel are removed at a faster rate, resulting in a higher turn-off speed.

5.2. DEVICE ANALYSIS

The analysis of both the static and dynamic characteristics of FCDs differs from those of the JFET. The differences arise primarily from the existence of minority carrier injection in this device. During device analysis, it is necessary to include both drift and diffusion current flow. The strong concentration gradients created by minority carrier injection produce a diffusion current component that is comparable to the drift current component. The analysis of minority carrier transport also requires the inclusion of carrier recombination statistics. Because of the high injection levels prevalent in these devices during current flow, the recombination rate is not only determined by the usual Shockley–Read–Hall statistics, but is affected by Auger processes as well.

5.2.1. Static Characteristics

Since current conduction in FCDs occurs with the injection of a high concentration of minority carriers into the drift region, the forward characteristics can be analyzed by using P–i–N rectifier theory. The presence of the gate junction has been theoretically shown to affect the current flow, but its impact is small and can be neglected.

The maximum forward and reverse blocking capabilities of the FCD are determined by the gate and anode junctions, respectively. In both cases, the maximum voltage supported across the FCD structure is determined by the open-base PNP transistor formed between the gate and the anode (see Fig. 5.1). For the forward blocking mode, it is also necessary to analyze the channel potential barrier because current flow between anode and cathode can occur by injection of electrons over this barrier.

5.2.1.1. Reverse Blocking. During reverse blocking, the negative voltage applied to the anode of the FCD reverse-biases the anode junction (junction J1 in Fig. 5.1). As the voltage supported across this junction increases, the depletion layer extends from the anode toward the gate junction (J2). The current gain of the open-base P–N–P transistor formed between the anode and gate increases with increasing reverse voltage because the undepleted base width decreases. The breakdown voltage is determined by the voltage at which the product of the multiplication factor and the current gain α becomes equal to unity. The analysis of open-base transistor breakdown is discussed in Section 3.7. From that analysis it can be concluded that the highest breakdown voltage is obtained when the minority carrier diffusion length is small. For any desired reverse blocking capability, an optimum drift region width and doping level exists, as illustrated in Fig. 3.48.

The drift region doping and thickness not only determine the reverse blocking capability, but also affect its other characteristics. The design of the drift region must take into account its effect on forward drop and blocking gain. As the drift region thickness increases, the reverse blocking capability will improve, but it has the adverse effect of increasing the forward voltage drop and decreasing the turn-off speed. The drift region doping level affects not only the reverse blocking capability, but also the blocking gain. The blocking gain is affected by the channel doping level. For devices with symmetrical blocking capability, a homogeneously doped N-base region is used with equal drift and channel region doping concentrations. Although a reduction in channel doping level favors achievements of high blocking gains, the attendant reduction in drift region doping promotes punch-through breakdown at low reverse-bias voltages. In designing field controlled doides, it is customary to select the drift region doping and thickness on the basis of optimizing the open-base transistor breakdown voltage. The channel geometry is then designed to obtain the required forward blocking gain.

5.2.1.2. Forward Blocking. In the absence of a gate bias, the FCD will conduct current between the anode and cathode terminals; that is, it has normally-on characteristics. When a negative bias is applied to the gate that is sufficient to completely deplete the channel, the device operates in its forward blocking mode. In this mode of operation, a potential barrier is formed in the channel that prevents current flow between the anode and the cathode. This operation is identical to that discussed for the junction gate FET in Chapter 4, and triode-like characteristics are observed for the field controlled diode at low anode current levels.

In the triode regime, the anode current is determined by carrier injection across the channel potential barrier:

$$I_A = I_0 \, e^{-(qV_B/kT)} \tag{5.5}$$

The variation of the barrier height with anode and gate voltage can be derived empirically [1]:

$$V_B = \alpha(V_G + V_{bi}) - \beta V_A^n \tag{5.6}$$

where α, β, and n are constants for a given device and V_{bi} is the built-in potential of the gate junction. Substituting this expression in Eq. (5.5) yields

$$I_A = I_0 \exp \frac{q}{kT} [\beta V_A^n - \alpha(V_G + V_{bi})] \tag{5.7}$$

The exponential increase in anode current with decreasing gate bias voltage, described by Eq. (5.7), has been observed in all FCDs. A typical example is

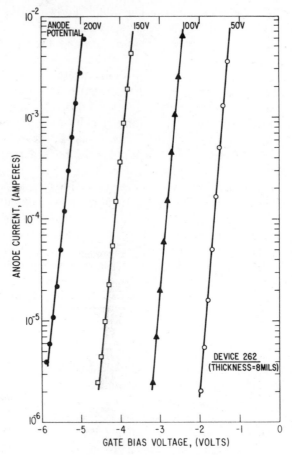

Fig. 5.6. Exponential increase of anode current with decreasing gate voltage. (After Ref. 2, reprinted with permission from the IEEE © 1982 IEEE.)

shown in Fig. 5.6 for the asymmetric device structure [2]. Note that all the curves have the same slope, which justifies treating α as a constant.

In the asymmetric FCD, the anode current is also found to vary exponentially with anode voltage. A typical example is provided in Fig. 5.7, where the anode current is plotted as a function of anode voltage with gate bias as a parameter. Once again all the curves have equal slope, which justifies treating β as a constant. It is worth noting that although n is found to be equal to unity for the asymmetric FCD structure, this does not apply in general to all structures. For vertically walled, symmetrical FCD devices, the best fit to the characteristics is obtained for a value of $n = 0.2$ [1].

Equation (5.7) predicts a monotonic increase in anode current with increasing anode voltage in the forward blocking regime. During early device development, this was regarded as an advantage of the FCD over conventional thyristors.

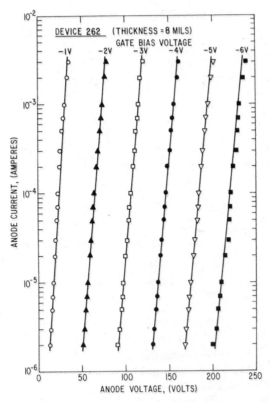

Fig. 5.7. Exponential increase in anode current with increasing anode voltage. (After Ref. 2, reprinted with permission from the IEEE © 1982 IEEE.)

In contrast to the conventional thyristor, the absence of a regenerative four-layer structure within the FCD was assumed to make breakover in this device impossible. More recently, the existence of a breakover characteristic for FCDs was observed experimentally [2]. The breakover in FCD characteristics has been shown to arise from a gate debiasing phenomenon quite different from the regenerative phenomenon responsible for the breakover in the characteristics of conventional thyristors.

In the triode regime, as the anode voltage increases, the channel potential barrier is reduced, resulting in increasing anode current flow. It has been found that this increase in anode current is accompanied by a proportional increase in the gate current [2]. A typical example of the variation in gate current with anode current is shown in Fig. 5.8. At low anode currents, the gate current is independent of the anode current because it is determined by space-charge generation in the gate depletion layer. At higher anode currents relevant to the breakover region, the gate current increases linearly with anode current:

$$I_G = \gamma I_A \tag{5.8}$$

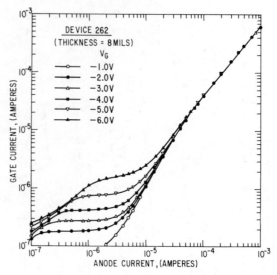

Fig. 5.8. Measured increase in gate current with increasing anode current during operation of the FCD in its forward blocking mode. (After Ref. 2, reprinted with permission from the IEEE © 1982 IEEE.)

where for a given device γ is a constant. This behavior in the FCD arises because the anode current consists of both electron and hole components. Although the electrons flow from the cathode, the holes arriving from the anode junction are collected by the gate junction because of the strong electric field attracting the holes to this region. This hole current is determined by the injected electron concentration across the potential barrier.

The hole current flow into the gate region represents a negative current flowing into the gate supply. This gate current will produce a potential drop in any series gate resistance formed internally or externally and lead to debiasing of the gate junction. The gate voltage is then given by

$$V_G = V_{GB} - \gamma I_A R_G \tag{5.9}$$

where V_{GB} is the gate supply voltage and R_G is the total (external plus internal) series gate resistance The debiasing of the gate results in an increase in the anode current. This can be analytically demonstrated by substituting Eq. (5.9) in Eq. (5.7) to obtain the expression:

$$I_A = I_0 \exp \frac{q}{kT} \left[\beta V_A^n - \alpha(V_{GB} - \gamma I_A R_G) \right] \tag{5.10}$$

$$= I_0 \exp\left(\frac{q\beta}{kT} V_A^n\right) \exp\left[-\left(\frac{q\alpha}{kT} V_{GB}\right) \right] \exp\left(\frac{q\alpha\gamma}{kT} I_A R_G\right) \tag{5.11}$$

The gate current flow establishes a positive feedback between the anode current flow and the lowering of the channel potential barrier by the gate debiasing. This leads to continual reduction in barrier height with increasing anode current. Ultimately, the device can no longer hold off the anode voltage and it breaks over into the on-state. A closed-form analytical expression describing the I–V curves for the breakover characteristics can be obtained from Eq. (5.11):

$$V_A = \frac{kT}{q}\frac{1}{\beta}\ln\left(\frac{I_A}{I_0}\right) + \frac{\alpha}{\beta}V_{GB} - \frac{\alpha\gamma}{\beta}I_A R_G \tag{5.12}$$

The maximum anode voltage that can be supported by the gate voltage V_{GB} in the presence of a gate series resistance R_G is called the *breakover voltage*. At the breakover voltage, the rate of change of the anode voltage with anode current becomes equal to zero. This point can be obtained by differentiating Eq. (5.12) with respect to I_A and setting it equal to zero:

$$\frac{dV_A}{dI_A} = \frac{kT}{q}\frac{1}{\beta}\frac{1}{I_{A,bo}} - \frac{\alpha\gamma}{\beta}R_G = 0 \tag{5.13}$$

where $I_{A,bo}$ is the current at the breakover point. From Eq. (5.13), an expression for the anode breakover current is obtained:

$$I_{A,bo} = \frac{kT}{q}\frac{1}{(\alpha\gamma)}\frac{1}{R_G} \tag{5.14}$$

Using this expression in Eq. (5.12), an expression for the anode breakover voltage can be derived:

$$V_{A,bo} = \frac{kT}{q}\frac{1}{\beta}\ln\left[\frac{kT}{q}\frac{1}{(\alpha\gamma)}\frac{1}{R_G I_0}\right] - 1 + \frac{\alpha}{\beta}V_{GB} \tag{5.15}$$

If the device parameters α, β, n, and I_0 are known, the anode breakover voltage can be calculated as a function of the gate bias voltage and the gate resistance. As the gate series resistance increases, the preceding expressions indicate an inversely proportional reduction in the breakover current and a relatively gradual (logarithmic) decrease in the anode breakover voltage. A comparison of the calculated and measured breakover characteristics for a typical asymmetric FCD is provided in Fig. 5.9. These characteristics were calculated by using the following values for the device parameters: $n = 1$, $\alpha = 0.23$, $\beta = 0.0066$, $I_0 = 0.15$, and $\gamma = 0.4$. Note that the blocking gain G is related to the parameters α and β by

$$G = \frac{dV_A}{dV_G} = \frac{\alpha}{\beta} \tag{5.16}$$

For the device whose characteristics are shown in Fig. 5.9, $G = 38$. Extremely good agreement between the theoretically calculated and experimental data has been found not only for the case of $R_G = 1000\,\Omega$ shown in Fig. 5.9, but over a very wide range of gate resistances.

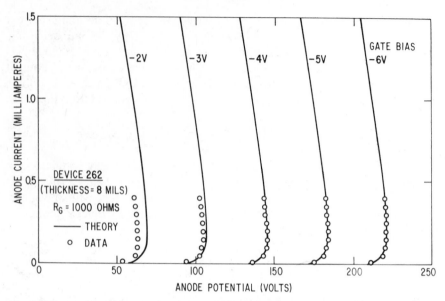

Fig. 5.9. Comparison of measured and theoretically calculated curves for the breakover region of an asymmetric FCD. (After Ref. 2, reprinted with permission from the IEEE © 1982 IEEE.)

The experimentally observed decrease in anode breakover current with increasing gate resistance is compared with the theoretically calculated values obtained from Eq. (5.14) in Fig. 5.10. The anode breakover current decreases inversely with increasing gate resistance as predicted by the analysis. Similarly, the decrease in anode breakover voltage predicted by the analysis has been confirmed experimentally as illustrated in Fig. 5.11.

The presence of breakover in the output characteristics of FCDs is important because it can lead to an inadvertent switching of the devices from the forward blocking state into the forward conduction state leading to catastrophic failure. The preceding analysis indicates that to avoid breakover, it is necessary to keep the gate current and gate series resistance small. A low gate current flow can be maintained by applying gate bias voltage well in excess of the minimum value required to hold off the highest anode voltage at which the devices are being operated.

5.2.1.3. Forward Conduction. An important feature of FCDs is their ability to operate at high current densities with low forward voltage drops. This reduces the power dissipation in the on-state and allows these devices to handle large anode currents even when the device size is small. To analyze this behavior, it is sufficient to treat the FCD as a $P-i-N$ rectifier existing between the anode and cathode terminals. Although the gate junction does affect minority carrier transport near the channel, its effect on the forward conduction characteristics

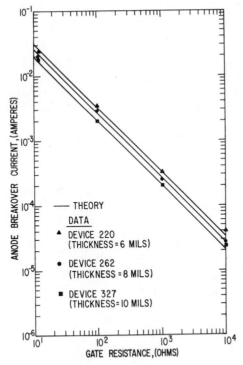

Fig. 5.10. Effect of gate series resistance on the anode breakover current. (After Ref. 2, reprinted with permission from the IEEE © 1982 IEEE.)

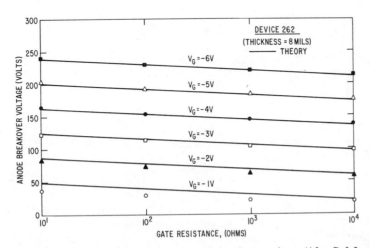

Fig. 5.11. Impact of gate series resistance on the anode breakover voltage. (After Ref. 2, reprinted with permission from the IEEE © 1982 IEEE.)

are small and can be neglected to obtain a simplified analytical solution to the current conduction characteristics.

In the case of the $P-i-N$ rectifier, the injection of minority carriers plays a major role in determining the forward conduction characteristic. Here, the nature of the current–voltage characteristic depends on the injection level. At very low injection levels, the current flow is dominated by recombination occurring within the space-charge layer of the $P-N$ junction due to the presence of deep levels. This current is given by [3]

$$J_{SC} = -\frac{q n_i W}{2\tau_{SC}} \left[e^{(q V_A/2kT)} - 1 \right] \tag{5.17}$$

where W is the width of the depletion layer, τ_{SC} is the space-charge generation lifetime, V_A is the applied voltage, k is Boltzmann's constant, and T is the absolute temperature.

As the applied voltage and forward current increases, the current becomes limited by diffusion of the minority carriers that are injected from the anode into the N-base region. The minority carrier distribution in the N-base region is given by

$$p_N = p_N(0)\, e^{-(x/L_p)} \tag{5.18}$$

where L_p is the minority carrier diffusion length and $P_N(0)$ is the minority carrier density at the edge of the junction given by

$$p_N(0) = \bar{p}_N\, e^{(q V_A/kT)} \tag{5.19}$$

where $\bar{p}_N$ is the hole concentration at thermal equilibrium in the N-base region. For a uniformly doped N-base region, current flow at low injection levels occurs exclusively by diffusion:

$$J_p = q D_p \frac{dp_N}{dx}\bigg|_{x=0} \tag{5.20}$$

where D_p is the diffusion coefficient for holes. Using the exponential minority carrier distribution described by Eq. (5.18) in Eq. (5.20), it can be shown that:

$$J = \frac{q D_p \bar{p}_N}{L_p} \left[e^{(q V_A/kT)} - 1 \right] \tag{5.21}$$

The current flow expressions under space-charge generation and at low injection levels differ in the rate of variation of the current with voltage. If the current is plotted on a semilogarithmic scale, the slope of the $I-V$ characteristics changes when the current transport shifts from the generation current dominated regime at very low forward voltage drops to the diffusion-limited current transport regime under low-level injection conditions.

The current flow mechanisms described above occur at low currents well below the current ratings of power FCDs. As the current increases, the injection level also increases and ultimately exceeds the relatively low background doping of the *N*-base region as illustrated in Fig. 5.12. This condition is called *high-level injection*. When the injected hole density becomes much greater than the background doping, charge neutrality in the *N* base requires that the concentration of holes and electrons become equal:

$$n(x) = p(x) \tag{5.22}$$

Under steady-state conditions, the current flow can be accounted for by the recombination of holes and electrons in the *N*-base regions and the anode–cathode end regions. Consider the case where the recombination in the end regions is negligible; that is, the end regions have unity injection efficiency [4]. The current density is then determined by recombination in the *N*-base region:

$$J = \int_{-d}^{+d} qR \, dx \tag{5.23}$$

where *R* is the recombination rate given by

$$R = \frac{n(x)}{\tau_{\text{HL}}} \tag{5.24}$$

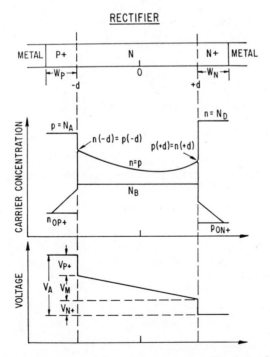

Fig. 5.12. Free-carrier and potential distribution in a *P–i–N* rectifier during forward conduction.

If the high-level lifetime is independent of the carrier density, it follows that

$$J = \frac{2q\bar{n}d}{\tau_{HL}} \tag{5.25}$$

where $\bar{n}$ is the average carrier density and d is half the N-base width.

The continuity equation in the N-base region can be written as

$$\frac{\partial n}{\partial t} = 0 = -\frac{n}{\tau_{HL}} + D_a \frac{\partial^2 n}{\partial x^2} \tag{5.26}$$

where D_a is the ambipolar diffusion coefficient discussed in Chapter 2.

If an ambipolar diffusion length is defined as

$$L_a = \sqrt{D_a \tau_{HL}} \tag{5.27}$$

the preceding equation takes the form

$$\frac{d^2 n}{dx^2} - \frac{n}{L_a^2} = 0 \tag{5.28}$$

The boundary conditions needed to solve this equation are obtained by the current transport occurring at the P^+ and N^+ ends of the diode:

$$J = J_n(+d) = q\mu_n n(+d) \,\mathscr{E}(+d) + qD_n \frac{dn}{dx}\bigg|_{x=+d} \tag{5.29}$$

The electric field is in turn determined by the carrier density:

$$\mathscr{E}(+d) = \frac{kT}{q} \frac{1}{n(+d)} \frac{dn}{dx}\bigg|_{x=+d} \tag{5.30}$$

Substituting Eq. (5.30) in Eq. (5.29), it can be shown that

$$J = 2qD_n \frac{dn}{dx}\bigg|_{x=+d} \tag{5.31}$$

Analysis of the boundary condition at the opposite end of the device gives

$$J = J_p(-d) = -2qD_p \frac{dn}{dx}\bigg|_{x=-d} \tag{5.32}$$

Note that the charge neutrality condition described by Eq. (5.22) has been used in deriving this expression. The solution of Eq. (5.28) with these boundary conditions is given by [5]

$$n = p = \frac{\tau_{HL} J}{2qL_a} \left[\frac{\cosh(x/L_a)}{\sinh(d/L_a)} - \frac{1}{2} \frac{\sinh(x/L_a)}{\cosh(d/L_a)} \right] \tag{5.33}$$

This carrier distribution is illustrated in Fig. 5.12 together with the P–i–N rectifier structure being analyzed. The hole and electron concentrations are the highest at the $P^+ - N(-d)$ and $N - N^+(+d)$ junctions, with the minimum closer to the cathode side as a result of the difference in the mobility of electrons and holes. The extent of the drop in the carrier concentration away from the junctions is dependent on the ambipolar diffusion length. At medium current densities, this diffusion length is controlled by the high-level lifetime.

To determine the voltage drop across the rectifier, it is necessary to first obtain the electric field distribution. The current flow in the base region is related to the electric field by

$$J = J_p + J_n \tag{5.34}$$

with

$$J_p = q\mu_p \left(p\mathscr{E} - \frac{kT}{q} \frac{dp}{dx} \right) \tag{5.35}$$

and

$$J_n = q\mu_n \left(n\mathscr{E} - \frac{kT}{q} \frac{dn}{dx} \right) \tag{5.36}$$

From these equations and the charge neutrality condition defined by Eq. (5.22), it can be shown that

$$\mathscr{E} = \frac{J}{q(\mu_n + \mu_p)n} - \frac{kT}{2q} \frac{1}{n} \frac{dn}{dx} \tag{5.37}$$

In this equation, the first term on the right-hand side accounts for the ohmic drop due to current flow and the second term accounts for the asymmetric concentration gradient produced by the unequal electron and hole mobilities in silicon. The voltage drop across the N-base region (V_l) can be obtained by integrating the electric field distribution, using the carrier distribution provided by Eq. (5.33) [5]:

$$V_l = \frac{kT}{q} \left\{ \frac{3}{2} \frac{\sinh(d/L_a)}{\sqrt{1 - \frac{1}{4}\tanh^2(d/L_a)}} \arctan \sqrt{1 - \frac{1}{4}\tanh^2(d/L_a)} \sinh(d/L_a) \right.$$

$$\left. + \frac{1}{2} \ln \left[\frac{1 + \frac{1}{2}\tanh^2(d/L_a)}{1 - \frac{1}{2}\tanh^2(d/L_a)} \right] \right\} \tag{5.38}$$

It is important to note that the voltage drop in the middle region is independent of the current density because the free-carrier concentration increases in proportion to the current density. A normalized plot of the voltage drop across the middle (N-base) region is provided as a function of the normalized base width in Fig. 5.13. Note the very rapid increase in the ohmic voltage drop with increasing ratio (d/L_a). As the half-base width d becomes longer than the diffusion length, the voltage drop across the middle regions causes an appreciable increase in the forward voltage drop as the result of a reduction in the conductivity modulation in the central portion of the N-base region. Field-controlled diodes are generally designed with half-base widths of less than the

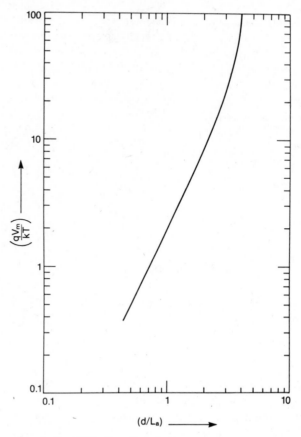

Fig. 5.13. Voltage drop in the middle (i) region of a P–i–N rectifier as a function of the ratio of half the base width to the diffusion length.

diffusion length. In devices that must operate at higher switching frequencies, the lifetime must be reduced to such an extent that the middle-region voltage drop becomes an important contributor to the forward drop.

The voltage drops across the anode (P^+–N) and cathode (N^+–N) junctions also contribute to the forward voltage drop as illustrated in the potential profile shown in the lower portion of Fig. 5.12. The injected minority carrier density at these junctions can be related to the voltage drop across them:

$$p(-d) = p_0 \, e^{qV_{P^+}/kT} \tag{5.39}$$

where p_0 is the minority carrier density in the N-base region at thermal equilibrium and V_{p+} is the voltage drop across the anode junction. Form this expression, the voltage drop across the anode junction is given by

$$V_{P+} = \frac{kT}{q} \ln\left[\frac{p(-d)N_D}{n_i^2}\right] \tag{5.40}$$

Similarly, at the cathode junction

$$n(+d) = n_0\, e^{qV_{N+}/kT} \tag{5.41}$$

resulting in

$$V_{N+} = \frac{kT}{q} \ln\left[\frac{n(+d)}{N_D}\right] \tag{5.42}$$

The total voltage drop across the end regions is obtained from Eqs. (5.40) and (5.42):

$$V_{P+} + V_{N+} = \frac{kT}{q} \ln\left[\frac{n(+d)\, n(-d)}{n_i^2}\right] \tag{5.43}$$

The charge neutrality condition has again been used in deriving this equation. After combination with Eqs. (5.33) and (5.38), the current density of a forward-bias diode at high-level injections in the absence of recombination in the end regions is given by

$$J = \frac{2qD_a n_i}{d} F\left(\frac{d}{L_a}\right) e^{qV_A/2kT} \tag{5.44}$$

where

$$F\left(\frac{d}{L_a}\right) = \frac{[(d/L_a)\tanh(d/L_a)]}{\sqrt{1 - \frac{1}{4}\tanh^4(d/L_a)}}\, e^{-(qV_M/kT)} \tag{5.45}$$

A plot of the function $F(d/L_a)$ is provided in Fig. 5.14. A low forward drop occurs when the function F is large. From Fig. 5.14, it can be seen that the highest value for F occurs at (d/L_a) values close to unity. For a fixed value of d, as required to obtain the desired forward and reverse blocking capability, the forward drop goes through a minimum as the diffusion length L_a increases. The variation in the forward drop with increasing (d/L_a) ratio is shown in Fig. 5.15 for a typical operating current density of 280 A/cm^2 [6]. The lowest forward drop occurs when the (d/L_a) ratio becomes equal of unity. The minimum in the forward drop arises because the voltage drop across the middle region decreases with increasing diffusion length. Higher diffusion lengths imply larger high-level lifetime, which increases the middle region conductivity modulation [see Eq. (5.25)] and reduces the ohmic drop across it. However, the resulting higher injected free-carrier concentration at the anode and cathode junctions must be provided by a higher voltage drop across these junctions as described by Eq. (5.43). At small values of (d/L_a), the junction drops are predominant and the forward drop decreases with increasing (d/L_a) ratio. At large values of (d/L_a), the middle-region ohmic drop becomes predominant and the forward drop increases with increasing (d/L_a) ratio. The minima occurs at (d/L_a) of unity. It should be noted that this discussion applies only when the end region and Auger recombination are neglected.

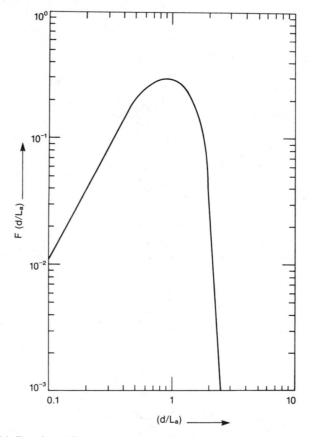

Fig. 5.14. Function $F(d/L_a)$ used to compute the forward drop of a P–i–N rectifier.

From the preceding discussion of the forward conduction characteristics as a function of injection level, it can be concluded that (1) at extremely low current densities, the current should vary as $\exp(qV_A/2kT)$ as a result of space-charge generation current flow; (2) at low current densities where low level injection prevails, the current should vary as $\exp(qV_A/kT)$ as a result of diffusion-limited current flow; and (3) at medium current densities when high-level injection prevails, the current should vary as $\exp(qV_A/2kT)$. This is indeed observed in the field-controlled diodes as illustrated in the forward characteristic shown in Fig. 5.16 [7]. The space-charge generation current flow at very low forward voltage drops has been observed but is not shown in this figure. At very high current levels, the output characteristic is found to deviate from the exponential behavior discussed above. This occurs because of the onset of significant recombination in the end regions as well as a reduction in the difusion length as a result of carrier–carrier scattering.

At high current densities, the injection of minority carriers into the anode and cathode regions becomes significant. Since the minority carrier lifetime de-

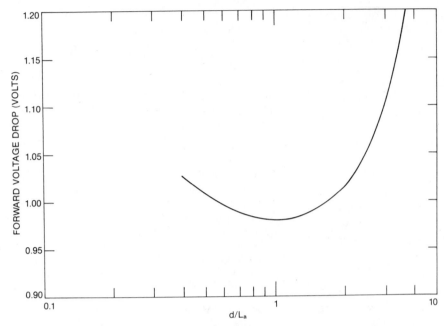

Fig. 5.15. Calculated forward drop of a P–i–N rectifier as a function of the (d/L) ratio. (After Ref. 6, reprinted with permission from the IEEE © 1970 IEEE.)

creases rapidly with increasing doping level [8], recombination in the end regions adds additional components to the forward current:

$$J = J_{P^+} + J_M + J_{N^+} \tag{5.46}$$

The current density due to injected carriers in the end regions can be derived from low level injection theory because the end regions are heavily doped:

$$J_{P^+} = \frac{qD_{nP^+}\bar{n}_P}{L_{nP^+}\tanh(W_P/L_{nP^+})} e^{(qV_{P^+}/kT)} \tag{5.47}$$

$$= J_{P+S}\, e^{(qV_{P^+}/kT)} \tag{5.48}$$

and

$$J_{N^+} = \frac{qD_{pN^+}\bar{p}_N}{L_{pN^+}\tanh(W_N/L_{pN^+})} e^{(qV_{N^+}/kT)} \tag{5.49}$$

$$= J_{N+S}\, e^{(qV_{N^+}/kT)} \tag{5.50}$$

where W_P and W_N are the thickness of the anode and cathode regions, respectively. The parameters J_{P+S} and J_{N+S} are called the *saturation current densities* of heavily doped P^+ and N^+ regions, respectively [5]. The saturation current densities are a measure of the quality of the end regions and depend on the method of preparation. Typical values are in the range of 1–4×10^{-13} A/cm^2.

With quasi-equilibrium at the P^+-N and N^+-P junctions, the injected carrier concentrations on either side of the junction are related by

$$\frac{p_{P^+}(-d)}{p(-d)} = \frac{n(-d)}{n_{P^+}(-d)} \tag{5.51}$$

where p_{P^+} and n_{P^+} are, respectively, hole and electron concentration on the anode side (P^+) of the junction. Because of the low injection level on the heavily doped side:

$$p_{P^+} = \bar{p}_{P^+} \tag{5.52}$$

The injected electron concentration in the P^+ anode region is related to the voltage across the anode junction:

$$n_{P^+}(-d) = \bar{n}_{P^+}\, e^{qV_{P^+}/kT} \tag{5.53}$$

Using these expressions in Eq. (5.51), gives

$$n(-d)p(-d) = \bar{p}_{P^+}\bar{n}_{P^+}\, e^{qV_{P^+}/kT} \tag{5.54}$$

$$= n_{ieP^+}^2\, e^{qV_{P^+}/kT} \tag{5.55}$$

where n_{ieP^+} is the effective intrinsic concentration in the P^+ anode, including the effect of band gap narrowing discussed in Chapter 2. Using the charge neutrality condition, it can be shown that

$$e^{qV_{P^+}/kT} = \left[\frac{n(-d)}{n_{ieP^+}}\right]^2 \tag{5.56}$$

Substitution of this expression into Eq. (5.48) gives

$$J_{P^+} = J_{P+S}\left[\frac{n(-d)}{n_{ieP^+}}\right]^2 \tag{5.57}$$

A corresponding derivation at the cathode end of the rectifier yields

$$J_{N^+} = J_{N+S}\left[\frac{n(+d)}{n_{ieN^+}}\right]^2 \tag{5.58}$$

where n_{ieN^+} is the intrinsic carrier concentration on the cathode side including heavy doping effects. In deriving these equations, it has been assumed that the anode and cathode junctions are abrupt and that these end regions are uniformly doped. An important conclusion that can be made from the last two equations is that, if the recombination in the middle region is negligible and end region recombination becomes dominant, the injected carrier density in the N base will no longer increase linearly with current density but will increase as the square root of the current density. The forward drop will then increase more rapidly with increasing current as observed in the upper portion of Fig. 5.16.

Two additional phenomena that affect the current conduction characteristics are carrier–carrier scattering and Auger recombination. These phenomena were

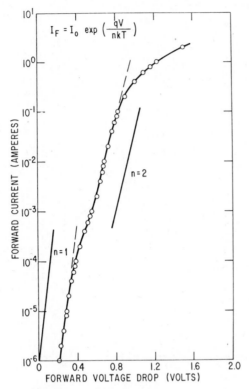

Fig. 5.16. Measured forward conduction characteristic of a typical FCD. (After Ref. 7, reprinted with permission from the IEEE © 1978 IEEE.)

discussed in Chapter 2. Carrier–carrier scattering occurs in the middle region at high current densities because of the simultaneous presence of a high concentration of both holes and electrons. The greater probability of mutual coulombic scattering causes a reduction in the mobility and diffusion length for both carriers. The reduction in diffusion length with increasing current density produces a decrease in the conductivity modulation in the central portion of the middle region, which, in turn, results in a higher ohmic voltage drop. The effect of including carrier–carrier scattering in the calculation of the forward conduction characteristics is shown in Fig. 5.17 [9] by the dashed line. At typical operating current densities of 200–300 A/cm², the forward drop is increased by more than 0.5 V by carrier–carrier scattering. The effect of carrier–carrier scattering is even greater under current surge conditions during which the current density exceeds 1000 A/cm².

At high current densities, the carrier density in the N-base region becomes sufficiently large that Auger recombination begins to affect the carrier statistics. In addition to the Shockley–Read–Hall recombination described by Eq. (5.24),

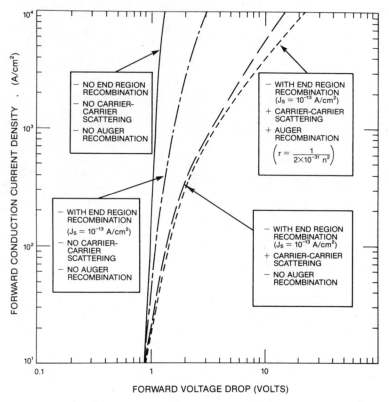

Fig. 5.17. Impact of end region recombination, carrier–carrier scattering, and Auger recombination on the forward conduction characteristics of a *P–i–N* rectifier. (After Ref. 9, reprinted with permission from *Solid State Electronics*.)

the rate of recombination must include the additional Auger recombination process:

$$R = \frac{n(x)}{\tau_{HL}} + C_A[n(x)]^2 \qquad (5.59)$$

where C_A is the Auger recombination coefficient. The inclusion of the Auger recombination term modifies the carrier distribution. Equation (5.26) must now be rewritten in the form

$$D\frac{d^2n(x)}{dx^2} = R = \frac{n(x)}{\tau_{HL}} + C_A[n(x)]^2 \qquad (5.60)$$

A solution for this differential equation has been obtained [10]:

$$n(x) = n_0\left[cn\left(\frac{x-x_0}{L}\bigg|m\right)\right]^{-1} \qquad (5.61)$$

where $cn(u/m)$ is a Jacobian elliptic function of argument u. In this expression, x_0 is the value of x at which $dn/dx = 0$ and $n_0 = n(x_0)$. The parameter m is given by

$$m = \frac{(C_A \tau_{HL} n_0^2/2) + 1}{C_A \tau_{HL} n_0^2 + 1} \tag{5.62}$$

and

$$L = \sqrt{D_a \tau_{HL}(2m - 1)} \tag{5.63}$$

is the modified ambipolar diffusion length.

The effect of including the Auger recombination on the output conduction characteristic is shown in Fig. 5.17 [9] by the dotted lines. To perform these calculations, an Auger coefficient of 2×10^{-31} cm^6/sec was used. The impact of Auger recombination is small at operating current densities of 200–300 cm^2 and becomes significant only at very high current densities.

Auger recombination also impacts current transport in the end regions. These regions are doped to very high concentrations, generally exceeding 10^{19} atoms/cm^3. Because of the high majority carrier density, Auger recombination alters the minority carrier lifetime. The resulting decrease in the minority carrier diffusion lengths (L_{nP+}, L_{pN+}) increases the recombination current flow (J_{P+}, J_{N+}) into the end regions. The effect of Auger recombination in the end regions is included in the output characteristics shown in Fig. 5.17 by factoring it into the end region saturation current J_s.

It was previously shown with the aid of Fig. 5.15 that, for a fixed base width, the forward drop goes through a minimum when the diffusion length L_a is varied. This minimum was found to occur at (d/L_a) of unity when no end region recombination or carrier–carrier scattering is included in the analysis. When these phenomena are included, the forward voltage drop no longer increases rapidly with decreasing diffusion length as shown on the left-hand side of Fig. 5.15 at low (d/L_a) values. Instead, the forward voltage drop becomes nearly independent of the lifetime in the N-base region. This is illustrated in Fig. 5.18, where the forward voltage drop is shown as a function of the high-level lifetime in the N-base region. The curves shown in this figure were obtained by including carrier–carrier scattering [6]. Note that in addition to the flattening out of the curves at higher base region lifetimes when the end region recombination is included, there is also an increase in the minimum achievable forward voltage drop when the saturated end region current density exceeds 10^{-14} A/cm^2.

The processing of the end regions of the FCDs must be tailored to obtain the optimum depth and concentration profile that will minimize the saturation current densities to obtain the lowest achievable forward voltage drop. Although a high doping of the anode and cathode regions is to be recommended to obtain a high injection efficiency, the onset of band gap narrowing and Auger recombination in these regions with increasing doping levels leads to a decrease in the saturation current densities J_{Ps} and J_{Ns}. Calculations of the variation of the saturation current density with increasing doping of the end regions for a step

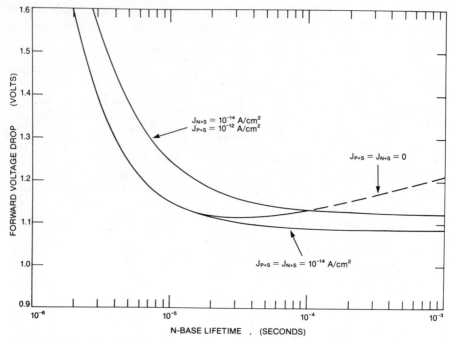

Fig. 5.18. Impact of end region recombination on the variation of the forward voltage drop as a function of the lifetime in the N-base region. (After Ref. 6, reprinted with permission from the IEEE © 1970 IEEE.)

junction indicate that an optimum doping level for the end region exists at which the saturated current density reaches its maximum value [11]. However, the measured saturated current densities [9] in devices fabricated by means of various processing techniques (including alloying and diffusion) indicate only a weak dependence of the saturated current density on the doping concentration in the end region as well as the process used to fabricate these regions.

The discussion of the forward conduction of the P–i–N rectifier presented above can be directly applied to the FCD. The presence of gate regions in the path of the current flow has been indicated to influence the forward conduction characteristics of the device. Two-dimensional numerical modeling [12] of the current flow in these devices indicates that some of the holes that are injected from the anode into the channel region are diverted into the gate at the bottom of the channel as a result of the built-in potential of the gate junction, and then reinjected from the top of the gate. This diversion of the holes into the gate tends to reduce the conductivity modulation of the channel and has, therefore, been predicted to increase the forward voltage drop. This effect can be expected to become more prominent as the gate region depth increases. Measurement of the forward drop as a function of the current density in FCDs [13], fabricated with vertically walled gates extending over a large range of gate depths, indicates no significant impact of the gate depth on current conduction in these devices. The diversion of holes into the gate region apparently has a negligible influence on the output characteristics.

5.2.2. Dynamic Characteristics

The high concentration of minority carriers injected during forward conduction provides FCDs with excellent forward conduction characteristics. These carriers are absent during the forward and reverse blocking modes. The switching of the devices between these states must involve the introduction and removal of the minority carriers. The turn-on process is controlled by the minority carrier transit time. Typical turn-on times are below 1 μsec and do not limit the switching speed because the turn-off times are significantly longer. The turn-off process, achieved by either reversal of anode voltage or application of a gate voltage, involves recombination of the minority carriers with typical lifetimes in excess of 1 μsec. Nevertheless, the turn-on process must be analyzed from the point of view of the maximum (di/dt) that can be tolerated by the devices. In addition to these dynamic switching characteristics, the devices are subjected to a very high rate of change of anode voltage (dV/dt) during switching from the on-state to the forward blocking state. It was initially believed that the FCD would be immune from (dV/dt)-induced turn-on. It has been recently demonstrated that the devices do possess a (dV/dt) limit.

5.2.2.1. Turn-on Analysis.   To switch the field controlled diode from its blocking state to the on-state, the gate potential must be rapidly reduced to the cathode potential. On an elementary basis, the rate at which anode current can build up is determined by the rate of removal of the potential barrier and the formation of the conducting channel between the gate junctions. This process is controlled by the gate junction depletion layer movement in the channel, which is determined by the gate capacitance and the series resistance of the gate drive circuit. A low-impedance gate drive circuit is desirable to allow faster turn-on of FCDs. A further improvement in the turn-on speed can be achieved by driving the gate positive since this ensures the maximization of the undepleted channel width and minimization of the channel current density during turn-on.

In the ideal case, where all portions of the active area of the device are assumed to be identical, the FCD can be expected to turn on uniformity over the entire active area. In contrast, the turn-on of conventional thyristors occurs at the edge of the cathode that is in closest proximity with the gate contact. The on-region then spreads out from the gate area until the entire area under the cathode becomes active. The rate of spreading of the conduction area in conventional thyristors is thus found to limit their (di/dt) capability.

In practice, the assumption that uniform turn-on occurs in the field controlled diode over its entire active area breaks down for several reasons. First, these devices must be fabricated with a highly interdigitated gate and cathode finger structure to achieve high device current ratings. The external connections to the device are made at the gate and cathode pads that interconnect one end of the gate and cathode fingers to their respective contact pads. During turn-on, the gate bias is not instantaneously applied to all areas of the device as a result of the RC charging time constant of the gate finger, which behaves as a transmission line. As a result, areas of the device closest to the gate pads can be expected to turn on before areas at the other extremity of the gate fingers.

Second, the rate at which the channel potential barrier is removed, and subsequently the width of the conducting channel formed under the cathode region, are not equal at all portions of the active area. The channel formation is dependent on the rate of change of the depletion width with applied gate bias. Since the depletion width is dependent on the doping concentration in the channel area, the undepleted channel width is given by

$$W_{\mathrm{CH}} = 2\left[a - \sqrt{\frac{2\epsilon_s}{qN_\mathrm{D}}(V_\mathrm{G} + V_{\mathrm{bi}})} \right] \qquad (5.64)$$

where $2a$ is the width between the gate junctions, N_D is the channel doping concentration, V_G is the applied gate bias, V_{bi} is the junction built-in potential, q is the electronic charge, and ϵ_s is the dielectric constant of the semiconductor. From this equation, it can be inferred that, for a given applied gate bias voltage, a wider channel will form in those sections of the device that have a lower background doping level N_D. The turn-on in these portions of the device will also precede that in other areas; this, in turn, will cause the localization of the cathode current during turn-on, which can lead to destructive failure in FCDs.

Destructive failure in FCDs during turn-on due to the mechanisms mentioned above has been observed experimentally [14]. It was found that the (di/dt) failure was always accompanied by the localized melting of the cathode finger metallization and that (1) the melting of the metal occurs only on the cathode fingers and not on the gate fingers; (2) the melting is localized to the ends of the cathode fingers, where they join the cathode bonding pad; and (3) the melting occurs in a localized area of the device.

These observations can be explained as follows. The turn-on in the FCD occurs at a region close to the gate pad at a point where the silicon substrate has the highest resistivity. The turn-on current is funneled to this area. It is then collected by the cathode fingers located in this area that feed the current to the cathode bond pad. As a result, the maximum current density in the cathode finger metallization occurs where the cathode fingers join the cathode bonding pad. The abnormally high current density in those cathode fingers, which carry the localized turn-on current, causes sufficient heat dissipation to result in melting of the aluminum. Once the local temperature exceeds the melting point of aluminum, it can form a eutectic with silicon. This leads to deep penetration of the aluminum into the silicon at the ends of the diffusion window in the cathode fingers. Since the cathode diffusion depth in these devices is shallow (typically 1 μm), the aluminum spikes through this diffusion. Under reverse-biasing, the gate depletion layer extends under the cathode diffusion.

The aluminum spike formed during the localized turn-on extends into the gate depletion layer and severely degrades its reverse blocking characteristics.

To minimize localized device turn-on, it is necessary to minimize variations in the resistivity of the base material. This can be accomplished by using neutron transmutation-doped silicon that has superior resistivity homogeneity (within $\pm 1\%$) when compared with pulled crystals (usually $\pm 20\%$) as discussed in Chapter 2. Second, the device should be designed with short gate and cathode fingers. The importance of this can be demonstrated with the aid of Fig. 5.19, which illustrates current flow in a single cathode finger. In this figure, $J_A(x)$ is the local anode current density flowing vertically through the FCD structure during turn-on. The cathode current density in the finger (J_F) can be obtained by integration of the vertical current collected by the finger along its entire length L:

$$J_F = \frac{\int_0^L J_A(x)\,dx}{t} \tag{5.65}$$

To achieve a high (di/dt) rating without device failure, it is necessary to reduce J_F. This can be accomplished by using small finger lengths L. Decreasing the cathode finger length also improves the rate of charging of the gate capacitance during turn-on, which will improve the (di/dt) capability. However, it is worth pointing out that as the finger length L is shortened, more device area is taken up by the gate and cathode pads, reducing the average current handling capability of the chip.

The (di/dt) rating could also be increased by increasing the thickness t of the cathode metallization. To define thicker metallization, it becomes necessary to increase the cathode finger width W to allow metal patterning by photolithography. Although such changes in the cathode width W do not influence the (di/dt) capability, an increase in cathode width results in a reduction in the blocking gain of these devices. These practical considerations are expected to

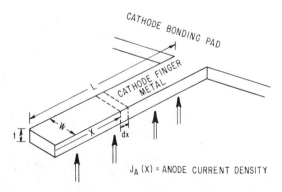

Fig. 5.19. Illustration of current collection in the cathode fingers of a FCD during dynamic turn-on.

limit the (di/dt) rating of the surface-gate FCDs to less than 200 A/μsec unless a large sacrifice in the blocking gain can be tolerated.

It has been shown in the preceding discussion that the (di/dt) limitation arises from the lateral current flow in the cathode finger metallization to the cathode bonding pad. Device failure due to this mechanism can be circumvented by developing improved contacting schemes that allow the cathode current to flow vertically through the device structure. One method for achieving this vertical current flow is illustrated in Fig. 5.20. In this technique the contact to the gate and cathode regions is formed with patterned aluminum metallization as in earlier devices. The gate metal is then selectively covered with a polyimide passivant by using photolithography. A large thick contact plate can now be attached to all the cathode fingers by use of solder reflow. This plate allows the anode current to flow vertically out of the cathode metal without the lateral current flow that has been demonstrated to be responsible for the (di/dt) failure in these devices. The implementation of this cathode contacting technique is expected to result in at least another order of magnitude increase in the (di/dt) ratings of FCDs.

5.2.2.2. Reverse Recovery.   Field-controlled diodes undergo reverse recovery when switched from the forward conducting state into the reverse blocking state. The circuit waveforms for this type of switching are illustrated in Fig.

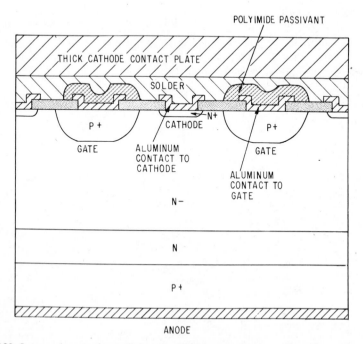

Fig. 5.20. Improved contacting technique for FCDs to prevent nonuniform dynamic turn-on.

5.3. During forward conduction, the N base of the devices contains a high concentration of holes and electrons injected from the anode and cathode regions. These carriers must be removed from the N base before a space charge (depletion) layer can be formed across the anode junction to sustain the reverse blocking voltage. The removal of the carriers occurs by both the flow of reverse current out of the anode prior to the establishment of a junction depletion layer and by the recombination of the free carriers. The reverse recovery process in the FCD is similar to that for a $P–i–N$ rectifier.

Consider the case of a forward-biased $P–i–N$ diode operating at a high injection level with a forward current I_F. At time $t = 0$, if the voltage across the diode is reversed to a fixed value V_R with a resistance R connected in series with the diode, the current in the circuit will initially be limited to $I_R = V_R/R$. This external current is supported by the diffusion of holes from the middle N-base region into the P^+ region and the diffusion of electrons from the middle N-base region into the N^+ region. Thus, to maintain current continuity, the currents flowing out of the end regions must be equal:

$$J_R = \frac{V_R}{AR} = qD_p \frac{dp}{dx}\bigg|_{x=-d} = -qD_n \frac{dn}{dx}\bigg|_{x=+d} \tag{5.66}$$

where A is the diode area and d is half the width of the diode. Since the diffusion coefficient for electrons (D_n) is larger than that for holes (D_p), it follows that

$$\frac{dp}{dx}\bigg|_{x=-d} > -\frac{dn}{dx}\bigg|_{x=+d} \tag{5.67}$$

Thus the charge removal from the anode side of the junction proceeds faster than from the cathode side.

The current J_R will remain constant only until the concentration of holes reaches the background level in the N base. Beyond this time, a space-charge region will begin to develop at the anode junction and a part of the reverse-bias voltage V_R will be supported by the diode. The current flowing in the circuit $[(V_R - V_D)/R]$ will, now, decrease with time. This produces a change in the carrier distribution within the N-base region as illustrated in Fig. 5.21 at various points during the turn-off transient [15]. The time at which the reverse current begins to decrease (t_L) can be derived with the aid of Fig. 5.21b [30], which shows a linear approximation for the charge distribution at time t_L, on the basis of which

$$J_R = \frac{2qD_p \, n(-d)}{\delta} \tag{5.68}$$

From charge conversation:

$$J_R t_L = Q_L + Q_R \tag{5.69}$$

$$= Q_L \left(1 + \frac{1}{b}\right) \tag{5.70}$$

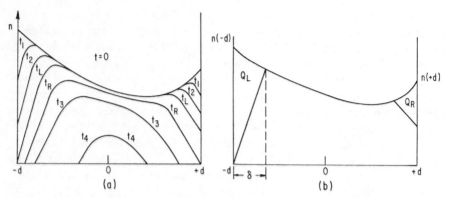

Fig. 5.21. Carrier distribution in a P–i–N rectifier during reverse recovery. (After Ref. 15, reprinted with permission from the IEEE © 1967 IEEE.)

where b is the ratio of the charge removed from the anode side to the charge removed from the cathode side. This ratio is determined by the ratio of the diffusion coefficients for electrons and holes. At low injection levels $b = 3$, but at high injection levels the ratio b decreases as a result of carrier–carrier scattering and tends to unity (see Chapter 2). Combination of Eq. (5.68) and (5.69) gives

$$t_{\rm L} = q^2 D_p \left(1 + \frac{1}{b}\right) \left[\frac{n(-d)}{J_{\rm R}}\right]^2 \tag{5.71}$$

Similarly, the time $t_{\rm R}$ at which a space-charge region begins to form at the cathode junction can be derived under the assumption that the current is still approximately constant:

$$t_{\rm R} = q^2 D_n (1 + b) \left[\frac{n(+d)}{J_{\rm R}}\right]^2 \tag{5.72}$$

Beyond the time $t_{\rm R}$, the voltage across the diode builds up by the formation of these space-charge regions on both sides of the device. The space-charge regions then sweep toward each other as the carriers are continuously swept out of the N-base region. A detailed analysis of this process indicates that the voltage across the diode will be built up as the $\frac{3}{2}$ power of time [15].

5.2.2.3. Forced Gate Turn-off.

If a large reverse gate voltage is applied to the field-controlled diode during forward conduction, the holes injected into the channel from the anode can be pulled out of the gate. If the gate current is sufficiently large, in addition to the removal of the holes in the channel, a space-charge (depletion) layer can be established at the gate junction. This depletion layer will extend into the channel and create the channel potential barrier, which will then prevent anode current flow. Typical waveforms for this type of circuit application are illustrated in Fig. 5.4. The forced gate turn-off capability of FCDs makes them useful for AC and DC circuits.

An abrupt termination of the anode current at the time of application of the reverse gate voltage (t_2 in Fig. 5.4) is highly desirable to minimize power loss during switching from the on-state to the off-state. The idealized current waveform shown in Fig. 5.4 is not observed in actual FCDs. During the forced gate turn-off process, all the minority carriers stored in the n-base region are not removed by the gate current prior to the establishment of the channel potential barrier. The minority carriers left in the n-base region, after the gate junction supports the applied voltage, are moved by recombination. An illustration of typically observed waveforms for the anode and gate currents and voltages, in the case of a resistive load, is provided in Fig. 5.22. The important difference between these waveforms and the idealized waveforms shown in Fig. 5.4 are the tail in the anode current turn-off waveform and the finite gate current observed after application of the gate turn-off voltage. The gradual decay of

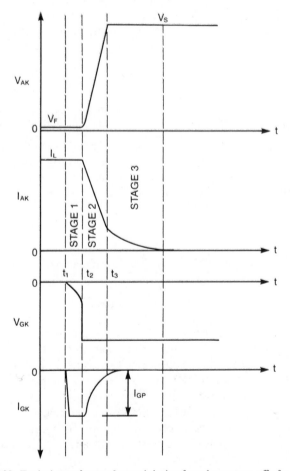

Fig. 5.22. Typical waveforms observed during forced gate turn-off of a FCD.

the anode current leads to power dissipation during turn-off because the anode voltage is large during this time. A faster decay of the anode current after turn-off is desirable especially at higher switching frequencies.

For this case of a gate circuit consisting of a voltage source with a series resistance, it can be seen that a substantial gate current flow occurs immediately after the application of the reverse gate bias at time t_1. During this gate current flow, the holes in the channel are being removed prior to establishing the gate depletion layer, and the gate voltage remains low. At time t_2, the gate voltage rises to its applied value. This is simultaneously accompanied by the sharp fall in the anode current and the rise in the anode voltage. Thus the gate turn-off occurs in three stages. In the first stage, there is a substantial gate current flow but the anode current remains essentially constant. The time taken for the first stage is dependent on the magnitude of the peak gate current. During this storage time, the charge in the channel region can be removed more rapidly if the peak gate current is larger. Although larger peak gate current favors faster turn-off, it requires greater input gate power. An important measure of the effectiveness of the forced gate turn-off process is the turn-off current gain, defined by

$$G_T = \frac{I_A}{I_{GP}} \qquad (5.73)$$

For a given turn-off time, a larger turn-off current gain indicates a superior device design. Unfortunately, typical turn-off gains range from 1 to 5 for high-speed turn-off of FCDs. The design of the gate circuit must include the capability to deliver these large pulse currents to ensure forced gate turn-off.

In general, the gate circuit can be regarded as a voltage source V_G with a fixed series impedance R_G. The peak gate current is then given by

$$I_{GP} = \frac{V_G}{R_G} \qquad (5.74)$$

The turn-off time is related to the peak gate current by

$$t_{off} \propto \frac{Q_s}{I_{GP}} \propto \frac{Q_s R_G}{V_G} \qquad (5.75)$$

where Q_s is the stored charge. The inversely proportional reduction in turn-off time with increasing gate bias voltage has been observed experimentally for a variety of FCDs. Some typical examples are provided in Fig. 5.23 for vertically walled devices [13]. Note that the turn-off time is smaller for the same gate bias voltage when the groove depth increases because the gate junction extends further into the N-base region where the stored charge exists.

Other measurements of the forced gate turn-off time indicate an increase in gate turn-off time with increasing anode current and voltage. When the anode current being turned off increases, the stored charge Q_s also increases. If the device is operating in the high-level injection (exponential) regime where the

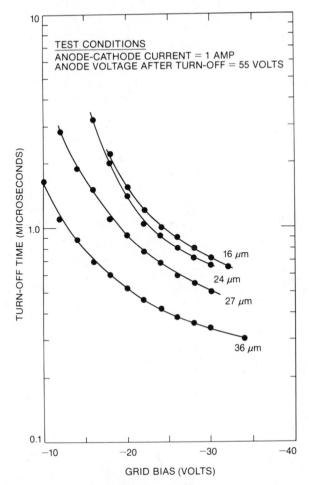

Fig. 5.23. Reduction of turn-off time with increasing gate bias voltage, with groove depth as a parameter. (After Ref. 13, reprinted with permission from *Solid State Electronics.*)

carrier density in the N base is proportional to the current density:

$$t_{off} \propto Q_s \propto \bar{n} \propto J_F \tag{5.76}$$

The turn-off time will then increase linearly with increasing anode current. This has been observed in the vertically walled FCDs as illustrated in Fig. 5.24a [7].

The increase in turn-off time with increasing anode forward blocking voltage is shown in Fig. 5.24b. The turn-off time becomes larger at higher anode blocking voltages for two reasons: (1) the minimum gate voltage required to establish the channel potential barrier increases as indicated by the arrows in Fig. 5.24b, and (2) a larger depletion layer must be established across the gate junction to support a larger anode voltage. More stored charge must be removed to accomplish this which slows down the turn-off process.

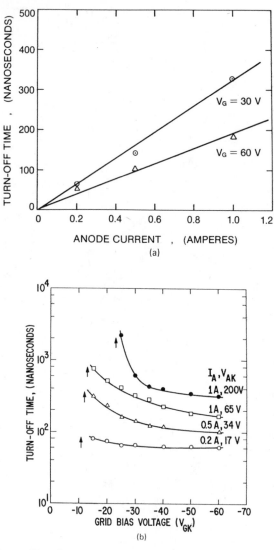

Fig. 5.24. Dependence of forced gate turn-off time on the anode current and voltage. (After Ref. 7, reprinted with permission from the IEEE © 1978 IEEE.)

5.2.2.4. (dV/dt) Capability. During forced gate turn-off, the anode voltage rises very rapidly from the forward voltage drop V_F to the supply voltage V_s. The increase in anode voltage is illustrated in Fig. 5.22 at a constant rate. In conventional thyristors, the application of a rapidly rising anode voltage (dV/dt) has been observed to cause switching of the device from the forward blocking mode to the on-state at well below the forward blocking capability. The existence of breakover in the forward blocking characteristics for FCDs indicates

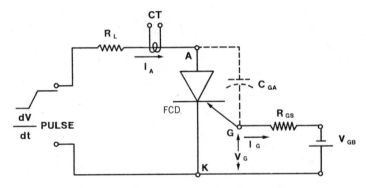

Fig. 5.25. Field-controlled diode circuit with high (dV/dt) applied across anode–cathode terminals.

that these devices may also be susceptible to turn-on under high (dV/dt) conditions.

Consider the FCD operating in a circuit shown in Fig. 5.25. In this circuit, the device is biased into the forward blocking mode by the gate supply consisting of a voltage source V_{GS} with series gate resistance R_{GS}. Note that the gate to anode capacitance of the field controlled diode is shown by the dashed lines. This is the gate junction capacitance, as illustrated in the cross section of the device in Fig. 5.26. During the application of a high (dV/dt) to the anode, a capacitive gate current flows into the gate circuit by means of the gate capacitance C_{GA}. The capacitive gate current is given by

$$I_G = C_{GA} \left[\frac{dV}{dt} \right] \tag{5.77}$$

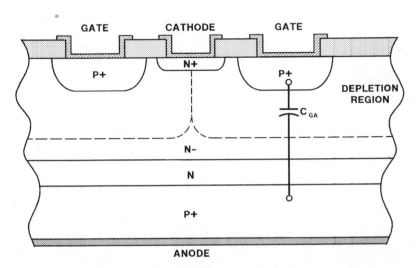

Fig. 5.26. Cross section of a FCD illustrating the anode-gate capacitance

The gate current flow debiases the gate as a result of the voltage drop across the gate resistance. In the analysis of the breakover in FCDs, it was shown that the breakover current is given by

$$I_{A,bo} = \frac{kT}{q} \frac{1}{\alpha\gamma} \frac{1}{R_{GS}} \tag{5.78}$$

At this current, the maximum blocking voltage is observed with $(dV_A/dI_A) = 0$. However, it has been observed [16] that the switching from the forward blocking mode to the on-state does not occur until the anode current reaches approximately five times this value. On the basis of this observation

$$\frac{dV}{dt} = 5 \frac{kT}{q} \frac{1}{\alpha\gamma} \frac{1}{R_{GS}C_{GA}} \tag{5.79}$$

This expression is valid for small gate resistances. The (dV/dt) capability decreases inversely with increasing gate series resistance. For large gate resistances, the gate current does not remain constant during the application of the anode voltage ramp because of the large RC time constant represented by the equivalent circuit, which consists of the gate capacitance C_{GA} in series with the gate resistance R_{GS}. Instead

$$I_G(t) = \left[\frac{dV}{dt}\right] C_{GA} [1 - e^{-(t/C_{GA}R_{GS})}] \tag{5.80}$$

Substituting this equation into the breakover current expression:

$$\frac{dV}{dt} = 5 \frac{kT}{q} \frac{1}{\alpha\gamma} \frac{1}{R_{GS}C_{GA}} [1 - e^{-(t/C_{GA}R_{GS})}]^{-1} \tag{5.81}$$

This general expression reduces to the simple expression given by Eq. (5.79) when R_{GS} is small, and the exponential term becomes negligible in Eq. (5.81). For very large gate series resistances

$$(C_{GA}R_{GS}) \gg t \tag{5.82}$$

and

$$e^{-(t/C_{GA}R_{GS})} \simeq 1 - \frac{t}{C_{GA}R_{GS}} \tag{5.83}$$

Then

$$\frac{dV}{dt} = 5 \frac{kT}{q} \frac{1}{\alpha\gamma} \frac{1}{t} \tag{5.84}$$

The (dV/dt) capability becomes independent of the series gate resistance at very high gate series resistances. The experimentally observed decrease in the (dV/dt) capability with increasing gate series resistance is compared with the preceding theoretically derived expressions in Fig. 5.27 [16]. The theoretical curve was generated by using parameters of the asymmetric FCDs obtained by using the barrier-controlled current flow analysis described in Section 5.2.1.2.

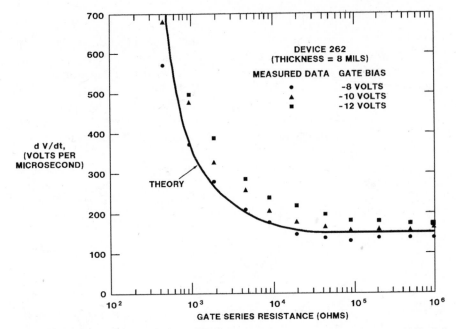

Fig. 5.27. Decrease of (dV/dt) capability of FCDs with increasing gate series resistance. (After Ref. 16, reprinted with permission from the IEEE © 1983 IEEE.)

The preceding analysis indicates that it is important to design the gate drive circuitry to obtain a low series gate resistance if a high (dV/dt) capability is desired. By extrapolating the measured results to zero external series gate resistance, it is possible to obtain the intrinsic (dV/dt) capability of the device. For the practical surface gate devices, the internal series gate resistance is less than 10 Ω. Consequently, their intrinsic (dV/dt) capability is estimated to be over 50,000 V/μsec. This value is substantially higher than that achievable in the conventional thyristor (typically 2000 V/μsec), even when designed with high cathode short-circuiting densities, because its gate sensitivity is coupled to its (dV/dt) capability. The FCD is, therefore, an attractive device for applications where high rate of change of voltage (dV/dt) may be encountered.

5.3. HIGH-TEMPERATURE PERFORMANCE

Compared with other three-terminal gate turn-off devices, FCDs offer exceptional high-temperature characteristics [17]. It has been found that they retain their forward blocking capability at high temperatures and maintain a low forward voltage drop during current conduction. These features make them attractive for applications where high ambient temperatures may be encountered. The effect of increasing temperature on the device characteristics is discussed in this section.

5.3.1. Reverse Blocking Capability

In the reverse blocking mode of operation, the anode junction of the field-controlled diode is reverse-biased. The breakdown voltage of the device is then determined by the characteristics of the open-base transistor formed between the anode and the gate regions. This breakdown voltage limitation is identical to that of a conventional thyristor in its reverse blocking mode. High blocking voltage capability can be maintained even at elevated temperatures if the base width of the device is designed to prevent punch-through breakdown of the open-base transistor.

As long as the current gain of the parasitic anode–gate bipolar transistor is small, the anode leakage current is determined by the generation of carriers in the depletion layer (space-charge generation current) and the generation of carriers in the neutral base region (diffusion current):

$$I_L = I_{LG} + I_{LD} \tag{5.85}$$

$$= \frac{q W_D A n_i}{\tau_{SC}} + \frac{q A D_p n_i^2}{L_p N_D} \tag{5.86}$$

where q is the electronic charge, n_i is the intrinsic carrier concentration, W_D is the width of the depletion layer, τ_{SC} is the space charge generation lifetime, D_p is the diffusion coefficient for holes, L_p is the diffusion length for holes, N_D is the base region doping concentration, and A is the active area of the device. Although the space-charge generation current is the dominant component of the leakage current at room temperature in the case of a wide band gap semiconductor such as silicon, the diffusion current becomes significant at higher device operating temperatures. The measured increase in the leakage current of the reverse biased anode junction is shown in Fig. 5.28 as a function of the reverse-bias voltage at various ambient temperatures [17]. When the anode voltage is below 200 V, the leakage current increases as the square root of the anode voltage at below 150°C, as expected from the proportionate increase in the junction depletion layer width W_D with applied anode voltage.

Below 150°C and when the anode voltage is increased above 200 V, a more rapid increase in the leakage current is observed as a result of the increase in the base transport factor of the open-base transistor formed between the anode and the gate:

$$I_A = \frac{I_L}{1 - \alpha} \tag{5.87}$$

where α is the current gain of the open-base transistor. This behavior is observed over the entire temperature range over which the device leakage current measurements were undertaken.

When the temperature is raised above 150°C, the diffusion current component of Eq. (5.86) begins to contribute to the leakage current. This component is dominant at the lower anode voltages where the junction depletion layer

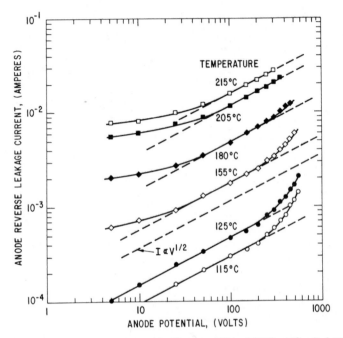

Fig. 5.28. Impact of temperature on reverse blocking capability of FCDs. (After Ref. 17, reprinted with permission from the IEEE © 1981 IEEE.)

width—and, thus, the space charge generation current—is small. These two components of the leakage current have a different dependence on the temperature. In the case of both the space-charge generation current and the diffusion current, their temperature dependence arises primarily from the rapid increase in the intrinsic carrier concentration n_i with temperature (see Chapter 2). The diffusion current increases more rapidly with temperature since it is proportional to the square of the intrinsic carrier concentration, whereas the generation current increases linearly with increasing intrinsic carrier concentration.

5.3.2. Forward Blocking Capability

The forward blocking characteristics of the field-controlled diode are obtained by reverse-biasing the gate and forming a potential barrier in the channel between the grids. The anode–cathode current in the forward blocking mode of operation is consequently determined by the injection of carriers over this potential barrier as well as the leakage current arising from the space charge generation and diffusion current components discussed in the previous section:

$$I_A = I_L + I_0 \exp\left[\frac{q}{kT}(\beta V_A - \alpha V_G)\right] \tag{5.88}$$

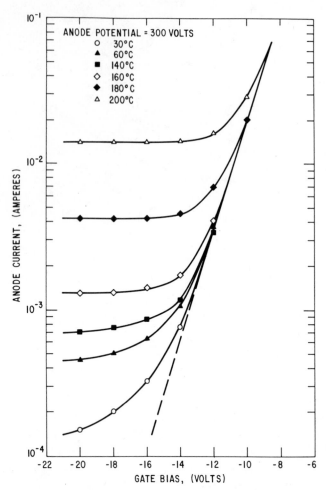

Fig. 5.29. Effect of increasing ambient temperature on the gate-controlled anode current flow. (After Ref. 17, reprinted with permission from the IEEE © 1981 IEEE.)

A typical example of the variation of the anode current with gate bias voltage under forward blocking conditions is shown in Fig. 5.29 at various ambient temperatures for a fixed anode potential of 300 V. It can be seen that the data asymptotically approach an exponential dependence of the anode current on the gate bias voltage. This exponential increase in the anode current with decreasing gate bias is the expected behavior under barrier-controlled current flow. At the larger gate bias voltages, the measured anode current is found to be independent of the gate voltage. This component of the anode current arises from the generation of carriers in the depletion layer of the reverse-biased gate junction and is consequently not influenced by the height of the channel potential barrier. Its magnitude is strongly dependent on temperature. In the forward blocking mode of operation, therefore, the anode current is given by the sum of the leakage current and the barrier-controlled current.

The variation of the anode current with increasing anode voltage is shown in Fig. 5.30 for a fixed gate bias voltage of 16 V at various ambient temperatures. At the larger anode voltages, the anode current shows an exponential increase with anode voltage. As the ambient temperature is increased, a slower rate of increase in the anode current with anode voltage is observed. This behavior of the anode current is consistent with Eq. (5.88), which indicates that the exponent factor of the anode voltage should decrease inversely with increasing temperature. These results demonstrate that even at high ambient temperatures, the forward blocking characteristics of the FCD are controlled by the channel potential barrier height. The maximum forward blocking voltage in the barrier-controlled current flow regime can be derived from Eq. (5.88) by neglecting the

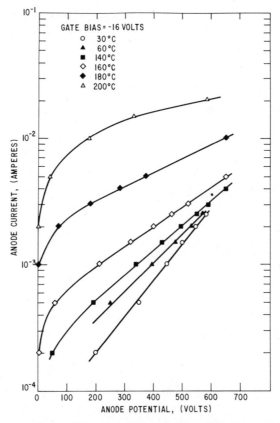

Fig. 5.30. Effect of increasing ambient temperature on the anode current–voltage characteristics in the forward blocking mode. (After Ref. 17, reprinted with permission from the IEEE © 1981 IEEE.)

leakage current term I_L:

$$V_A = \frac{1}{\beta}\left[\alpha V_G + \frac{kT}{q}\ln\left(\frac{I_A}{I_0}\right)\right] \tag{5.89}$$

Here I_A is the anode leakage current at which the blocking voltage is measured. This equation predicts an increase in anode blocking voltage capability with increasing temperature for any fixed gate bias voltage. This has been observed in FCDs as illustrated in Fig. 5.31 [17] at lower temperatures and higher anode leakage currents. At higher temperatures, the leakage component I_L becomes large leading to an apparent sharp reduction in forward blocking capability. If the anode current at which the blocking voltage is measured is increased, however, the devices are found to continue to exhibit a higher anode forward blocking capability. The data shown in Fig. 5.31 indicate that these devices can be operated even at 200°C.

5.3.3. Forward Conduction Characteristics

The forward conduction characteristic of the FCD has been shown to be similar to that of a $P–i–N$ rectifier. In the high-level injection regime the current increases exponentially with anode voltage as described by Eq. (5.44)

$$J_F = \frac{2qD_a n_i}{d} F\left(\frac{d}{L_a}\right) e^{qV_A/2kT} \tag{5.90}$$

From this expression, it can be concluded that for a fixed forward drop, the current will increase with increasing temperature. This has been observed in FCDs as shown in Fig. 5.32. Equation (5.90) holds true only in the high-level

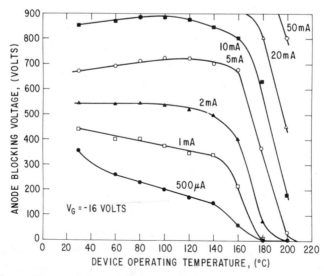

Fig. 5.31. Measured increase in forward blocking capability with increasing temperature for FCDs. (After Ref. 17, reprinted with permission from the IEEE © 1981 IEEE.)

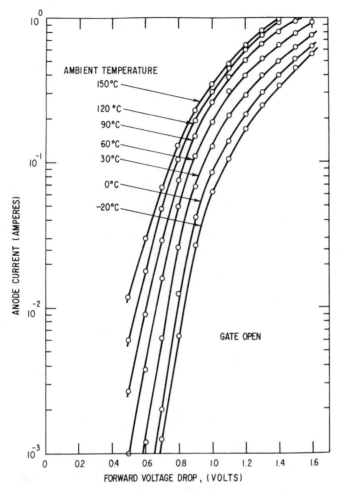

Fig. 5.32. Forward conduction characteristics of an FCD with temperature as a parameter. (After Ref. 17, reprinted with permission from the IEEE © 1981 IEEE.)

injection (exponential) regime. At higher current levels where the forward drop exceeds 2 V, the curves bend over and cross each other. The crossover occurs as a result of the reduction in ambipolar mobility at higher temperatures. The crossover of the characteristics with increasing temperature is important for ensuring stable operation with uniform current distribution within the device.

5.3.4 Gate Turn-off Time

The forced gate turn-off time of FCDs has been measured as a function of temperature [17]. It has been found that the turn-off time increases with increasing temperature. Some examples of the increase in turn-off time with temperature are shown in Fig. 5.33. For typical deep-level impurities, as temperature

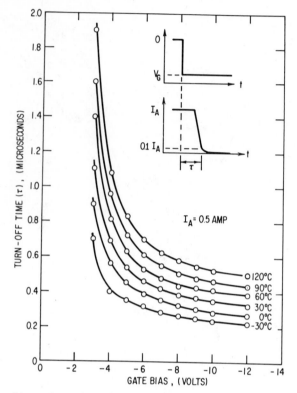

Fig. 5.33. Effect of increasing ambient temperature on forced gate turn-off time. (After Ref. 17, reprinted with permission from the IEEE © 1981 IEEE.)

increases, the minority carrier recombination lifetime increases. Since the decay in the anode current is prolonged by a slower recombination rate, the gate turn-off time increases in FCDs with increasing temperature. The slower turn-off process results in a higher power dissipation during switching at elevated temperatures.

5.4. FREQUENCY RESPONSE

Because of the high concentration of minority carriers injected into the N-base region during forward conduction, the switching speed of FCDs is lower than that for power JFETs. Typical forced gate turn-off times are on the order of 1 μsec. Such slow speeds would limit the application of these devices to low-frequency circuits. Their switching speed can be improved by the introduction of recombination centers into the N-base region, to enhance the removal of the stored charge during the turn-off process. Although the diffusion of impurities such as gold or platinum can be used to increase the switching speed, electron irradiation provides much greater control over the lifetime [18, 31]. Electron irradiation on FCDs has been demonstrated to provide an excellent method to optimize their characteristics [19].

5.4.1. Gate Turn-off Time

The introduction of recombination centers into the N-base region by the electron irradiation reduces the decay time for the anode current during forced gate turn-off. The variation of the turn-off and fall times are shown in Fig. 5.34 as a function of the radiation dose given in megarads. One megarad corresponds to a fluence of about $10^{14}/cm^2$. The turn-off time includes both the storage time and the fall time. Both the storage time and the fall time decrease with increasing radiation dose.

The minority carrier lifetime is related to the radiation dose ϕ by means of the radiation damage coefficient K:

$$\frac{1}{\tau_f} = \frac{1}{\tau_i} + K\phi \tag{5.91}$$

where τ_i and τ_f are the lifetimes before and after irradiation. If the initial lifetime is large, the lifetime after irradiation decreases inversely with radiation dose.

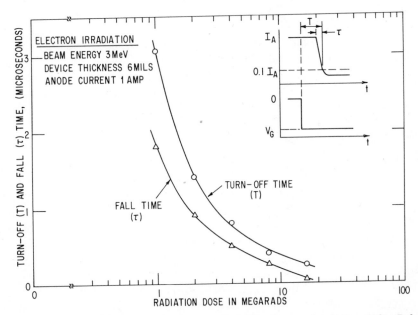

Fig. 5.34. Reduction in FCD forced gate turn-off time by electron irradiation. (After Ref. 19, reprinted with permission from the IEEE © 1982 IEEE.)

For a fixed forward conduction current density, the concentration of the stored charge can be obtained as a function of the lifetime from Eq. (5.25):

$$\bar{n} = \frac{J_F \tau_{HL}}{2qd} = \frac{J_F}{2qdK\phi} \tag{5.92}$$

From this equation, it can be seen that the stored charge, and hence the storage time $(T - \tau)$, should decrease with increasing electron irradiation. A lower lifetime reduces not only the decay time of the anode current, but also the storage time. It is noteworthy that the electron irradiation process has been used to tailor the turn-off time from 20 μsec down to as low as 200 nsec.

5.4.2. Forward Conduction Characteristics

Electron irradiation reduces the minority carrier lifetime. The reduction in the lifetime is accompanied by a corresponding decrease in diffusion length:

$$L = \sqrt{D\tau} = \sqrt{D/K\phi} \tag{5.93}$$

During the discussion of the forward conduction in a P–i–N rectifier, it was shown that the forward voltage drop is a function of the diffusion length. When half the base width of a P–i–N rectifier becomes greater than the diffusion length, the forward voltage drop increases rapidly, as discussed earlier with the aid of Fig. 5.18. Consequently, the forward voltage drop of field controlled diodes can be expected to increase following electron irradiation.

The impact of electron irradiation on the forward voltage drop of field controlled diodes is illustrated in Fig. 5.35 for three wafer thicknesses. As long as the radiation dose is less than 1 Mrad, the forward voltage drop remains relatively unchanged. At higher doses, the forward drop begins to increase rapidly. The dose at which the rapid increase occurs is a function of the N-base width. The rapid rise in forward voltage drop occurs when the diffusion length becomes smaller than half the N-base width. Consequently, higher radiation doses can be tolerated by devices with narrower N-base regions.

5.4.3. Trade-off Curve

Because of the undesirable increase in forward voltage drop with increasing radiation dose, it is necessary to compromise between reducing the power dissipation during turn-off and achieving a low power loss during current conduction. A useful method for achieving this is to plot the forward voltage drop at the operating current against the gate turn-off time, thus creating a trade-off curve. Typical examples of trade-off curves are shown in Fig. 5.36. Since the power dissipation during current conduction becomes excessive if the forward drop exceeds 3 V, it can be seen that the turn-off time will be limited to about 0.5 μsec for the 6-mil-thick device and about 1 μsec for the 8–10-mil-thick devices. The thicker devices are needed for achieving higher forward blocking capability. It can be concluded that as the forward blocking capability increases, the maximum switching speed will be reduced.

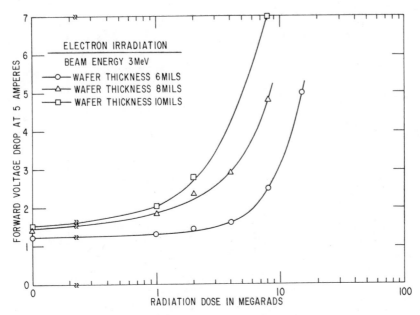

Fig. 5.35. Increase in forward voltage drop of FCDs after electron irradiation. (After Ref. 19, reprinted with permission from the IEEE © 1982 IEEE.)

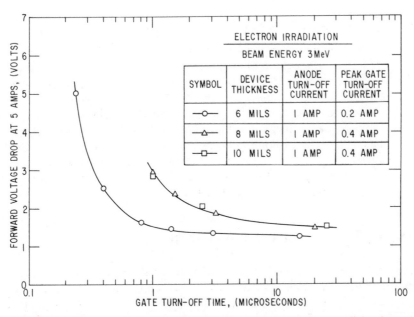

Fig. 5.36. Trade-off curves between forward voltage drop and forced gate turn-off time for typical FCDs. (After Ref. 19, reprinted with permission from the IEEE © 1982 IEEE.)

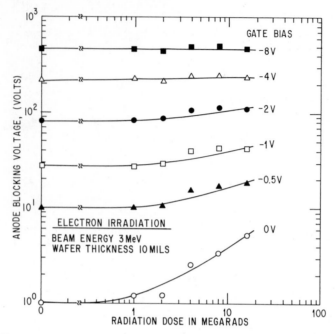

Fig. 5.37. Slight increase in forward blocking capability observed in FCDs after electron irradiation. (After Ref. 19, reprinted with permission from the IEEE © 1982 IEEE.)

5.4.4. Forward Blocking Capability

Electron irradiation has been found to have minimal impact on the forward blocking capability. In addition to reducing the lifetime, the deep levels produced by electron irradiation result in carrier removal due to compensation. Compensation increases the gate depletion layer width for any given gate bias voltage. The increase in the gate depletion width results in lowering of the channel pinch-off voltage and an increase in the channel potential barrier height, especially at the smaller gate–cathode voltages. Both these effects combine to increase the forward blocking capability at low gate bias voltages as illustrated in Fig. 5.37. However, the impact of electron irradiation on the forward blocking characteristics at high gate voltages is negligible.

5.5. NEUTRON RADIATION TOLERANCE

After neutron radiation at high fluences, the forward blocking capability of conventional thyristors deteriorates as a result of the increase in leakage current, enhanced by the internal regenerative action of the $P-N-P-N$ structure. These devices have also been found to be difficult to trigger into the on-state following neutron radiation because of the severe reduction in the current gains of the internal transistors that control the regenerative turn-on. In the case of bipolar transistors, lifetime reduction by the neutron radiation damage causes a severe decrease in their current gain. Power MOSFETs also exhibit poor neutron radiation tolerance because it introduces charge in the gate oxide, resulting in large threshold shifts. In comparison, FCDs continue to exhibit good forward blocking capability following neutron radiation because no regenerative $P-N-P-N$ structure exists in these devices. Since these devices exhibit a forward current conduction characteristic in the on-state similar to that of a $P-i-N$ rectifier, their neutron radiation tolerance is similar to that of diodes, and considerably superior to that of other three-terminal power device structures [20].

5.5.1. Forward Conduction

The impact of neutron radiation on two types of FCD structures has been measured [20]. In both cases, the forward voltage drop remains relatively unaffected until a critical radiation dose is reached. It then rises rapidly with further radiation as shown in Fig. 5.38. The neutron fluence at which the rapid increase in forward drop occurs is a function of the N-base width. As in the case of electron

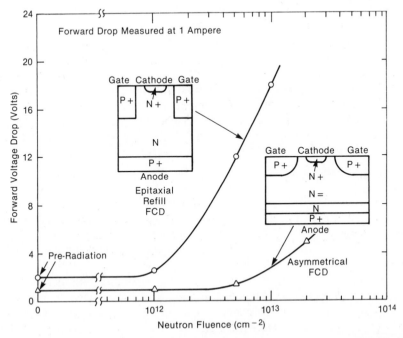

Fig. 5.38. Effect of neutron radiation on the forward voltage drop of two types of FCD. (After Ref. 20, reprinted with permission from the IEEE © 1983 IEEE.)

irradiation, the neutron irradiation creates minority carrier recombination centers. For high fluences ϕ, the minority carrier lifetime decreases inversely with fluence:

$$\tau = \frac{K}{\phi} \tag{5.94}$$

where the damage coefficient K is on the order of 10^7 neutron sec/cm^2 at high injection levels (independent of the N-base resistivity) [21]. The neutron radiation produces a corresponding reduction in the diffusion length:

$$L = \sqrt{D\tau} = \sqrt{KD/\phi} \tag{5.95}$$

With this equation, the diffusion length can be calculated as a function of the neutron fluence. The diffusion length is found to become less than half the N-base width (100 μm) for the epitaxial refill devices at a fluence of 10^{12} neutrons/cm^2. Similarly, for the asymmetric devices, the diffusion length is found to become less than half the N-base width (75 μm) at a fluence of 4×10^{12} neutrons/cm^2. The forward voltage drop of the FCD should increase rapidly beyond these fluences, as can be seen in Fig. 5.38. An important conclusion that can be drawn from these results is that the radiation tolerance is a strong function of the N-base width.

The fluence at which a rapid increase in forward drop will occur is inversely proportional to the square of the N-base width. For symmetrical FCDs, the N-base width increases approximately as the square root of the breakdown voltage. The critical neutron fluence beyond which the forward drop will increase rapidly for these devices will decrease as $(V_B)^{0.25}$. In contrast, for the asymmetric device structure, the N-base width increases approximately linearly with increasing forward breakdown voltage. The critical neutron fluence beyond which the forward drop will increase rapidly for these devices will decrease as $(V_B)^{0.5}$. For a given forward blocking capability, the asymmetric structure will have greater neutron radiation tolerance because its N-base width is smaller than for the corresponding symmetrical device structure.

5.5.2. Forward Blocking

The effect of the neutron radiation on the forward blocking capability has been measured for both FCD device structures. It was found that devices of both types continued to exhibit good forward blocking capability up to the highest neutron fluence examined. Typical results of measured anode forward blocking voltage as a function of neutron fluence are shown in Fig. 5.39. It can be seen that there is an increase in the blocking voltage capability with increasing neutron fluence. This unusual behavior compared with other power devices results from an enhancement of the channel pinch-off in FCDs following radiation.

Neutron irradiation results in carrier removal with a measured carrier removal rate of about 1.5–3/(neutron · cm) irrespective of the original resistivity [22]. This carrier removal results in a doubling of the N-base resistivity for the

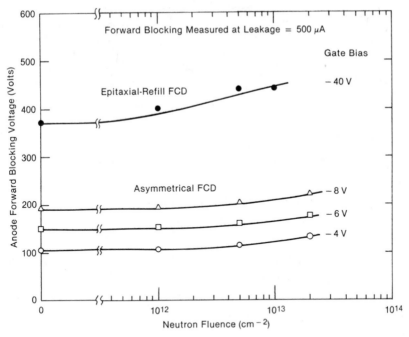

Fig. 5.39. Measured increase in forward blocking capability of FCDs after neutron radiation. (After Ref. 20, reprinted with permission from the IEEE © 1983 IEEE.)

epitaxial-refill devices at a fluence of 10^{13} neutrons/cm^2, which accounts for the observed increase in their anode blocking capability from 370 to 450 V. An even greater increase in resistivity occurs for the very lightly doped N base in the case of the asymmetric devices. In this case, however, the channel pinch-off occurs at very low gate voltages even prior to neutron radiation. Consequently, a further increase in N-base resistivity due to carrier removal has a much smaller impact on the forward blocking capability than that in the epitaxial-refill devices. It should be noted that although carrier removal effects are observed for both types of device, there is no evidence of type conversion up to the highest neutron fluences that have been examined.

5.6. GATING TECHNIQUES

The FCD is fundamentally a normally-on device. It will conduct current in the absence of a gate bias voltage. To prevent forward conduction, it is necessary to supply a negative gate bias with respect to the cathode for n-channel devices. The basic gating circuit for the device consists of a voltage source V_G with a switching device S1 in series to control the gating waveform as shown in Fig. 5.40. When switch S1 is open, the FCD will be in its on-state. When switch S1

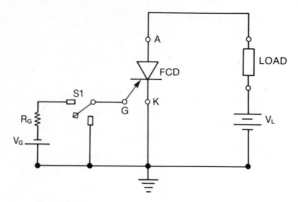

Fig. 5.40. Basic gating circuit for FCDs.

is closed, the gate bias V_G is applied and the device will block forward current flow. To ensure forward blocking, it is necessary to satisfy the condition

$$V_G > \frac{V_L}{G_B} \tag{5.96}$$

where G_B is the DC forward blocking gain. During the discussion of the forward breakover voltage and (dV/dt) capability, it was shown that the series gate resistance R_G must be small to maximize these parameters. In addition, the circuit requires operation of the devices under forced gate turn-off. The speed at which the turn-off can be accomplished is proportional to the peak gate current given by

$$I_{GP} = \frac{V_G}{R_G} \tag{5.97}$$

The gate series resistance, including that of the switching device S1, must be designed to provide adequate gate current for achieving the desired switching speed.

The gating technique described above operates the FCD as a normally-on device. In many power circuits, normally-on performance is not acceptable, and this has curtailed the application of FCDs. To overcome this problem, two approaches have been taken: (1) the FCD structure is designed for strong pinch-off at zero gate bias to create a normally-off device, and (2) the gating circuit is altered to create an effectively normally-off power switch.

5.6.1. Normally-off Structures

If the FCD structure is designed in such a manner that the depletion layer width due to the built-in junction potential is larger than half the channel width, a potential barrier will be created in the channel without the application of an external gate bias. If the barrier height created by the built-in junction potential is sufficiently large, it can be used to produce a substantial forward blocking capability. These devices can be switched to the on-state by the application of a positive gate bias (for *n*-channel devices) that reduces the barrier and allows current conduction.

The depletion layer width resulting from the junction built-in potential is given by

$$W_{bi} = \sqrt{\frac{2\epsilon_s}{qN_D}V_{bi}} \tag{5.98}$$

The built-in potential is given by

$$V_{bi} = \frac{kT}{q}\ln\left(\frac{N_A N_D}{n_i^2}\right) \tag{5.99}$$

where N_A and N_D are the doping concentration on the opposite sides of an abrupt gate junction. To achieve normally-off performance, it is necessary to satisfy the condition that half the channel width *a* be smaller than W_{bi}. Two methods for achieving the preceding condition are the reduction of the channel width and channel doping. The normally-off region of operation for the FCD corresponds to the purely triodelike regime of operation for JFETs. To design a normally-off FCD structure, the channel width and doping concentration must be chosen to lie in the lower right-hand side of Fig. 5.41, below the solid line representing the zero-bias depletion width. Along this line, the depletion

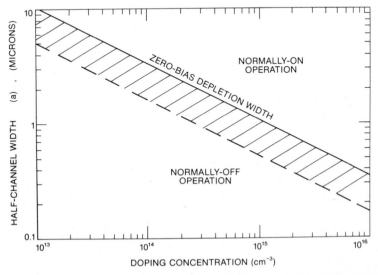

Fig. 5.41. Region of channel widths and doping concentrations required to obtain normally-off operation for FCDs.

layers will barely punch through in the channel and no significant channel potential barrier will be created. To establish a potential barrier sufficiently large to support substantial anode voltage, it is necessary in practice to design the normally-off device structure for operation in the region below the dashed line.

Normally-off FCDs have been fabricated with both surface- and buried-gate structures. To fabricate the surface-gate devices, it is necessary to utilize high-resolution [very large scale integration (VLSI)] processing. A very narrow channel width ($2a$) of 1 μm has been achieved by using LOCOS processing borrowed from integrated-circuit technology [23]. The channel doping used for these devices was $1 \times 10^{14}/\text{cm}^3$. These device parameters allow operation well within the normally-off region of operation below the dashed line in Fig. 5.42. The devices have been found to be capable of blocking up to 500 V. For a 7×10-mm chip containing 9000 channels, the devices were found to conduct 25 A at a forward voltage drop of 1.5 V.

To fabricate the normally-off buried-gate devices, conventional lithography was used with 5-μm resolution technology. The normally-off behavior was accomplished by reducing the N-base doping to about $1 \times 10^{12}/\text{cm}^3$ and fabricating a vertically walled buried grid by using an epitaxial-refill process [24]. The combination of the very low channel doping level and the vertically walled gate junction produces a strong channel potential barrier at zero gate bias. Devices capable of blocking 500 V have been fabricated with forward current handling capability of 10 A. It should be noted that the barrier height and hence

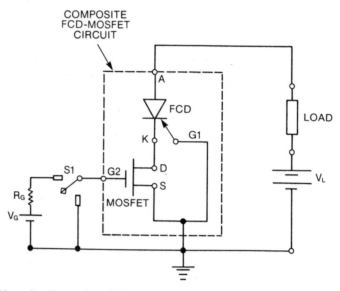

Fig. 5.42. Normally-off operation of FCDs achieved by gating with a low-voltage power MOSFET connected in series. (After Ref. 25, reprinted with permission from *Solid State Electronics*.)

the blocking capability deteriorates with increasing temperature, rendering these devices of limited value.

5.6.2 Normally-off Gating

In the previous section, methods were described for altering the field-controlled diode structure to obtain normally-off behavior. An alternative approach is to develop a gating technique that utilizes a normally-on FCD and a separate normally-off three-terminal device to create a composite device that exhibits normally-off characteristics. The most favorable circuit for achieving normally-off operation is shown in Fig. 5.42 [25]. In this circuit, a low-voltage, normally-off MOSFET is connected in series with the FCD. Note that in addition to the connection of the cathode of the FCD to the drain of the MOSFET, the gate of the FCD is connected to the source of the MOSFET. These connections place the MOSFET in parallel with the gate–cathode junction [i.e., V_{DS} (power MOSFET) $= V_{GK}$ (field-controlled diode)].

In the forward blocking mode, the MOSFET gate G2 is short-circuited to its source and it is maintained in the off-state. As the load voltage V_L increases, the MOSFET initially supports the voltage because the FCD is a normally-on device. As the drain voltage increases, it produces a reverse bias across the gate–cathode junction. When the load voltage reaches the pinch-off voltage, the gate depletion layers punch-through in the FCD channel and form a potential barrier. Beyond the punch-through voltage, most of the load voltage is supported by the FCD. An example of the variation of the MOSFET drain voltage with increasing load voltage is shown in Fig. 5.43. Note that although the applied load voltage exceeds 500 V, the MOSFET drain voltage remains below 50 V. This is a very important feature of this gating method because it allows the use of a low-breakdown-voltage MOSFET for gating the field controlled diode. Such low-voltage MOSFETs can be designed with very low on-resistance. Since the load current flows through the MOSFET, it is essential to use a low-breakdown-voltage MOSFET in this circuit to avoid a prohibitively high composite forward drop across the power devices.

Another feature of this gating technique is that forced gate turn-off is achieved essentially by discharging the input gate capacitance of the MOSFET. Typical peak gate currents required to discharge the input capacitance are below 100 mA. When the MOSFET turns off, the drain voltage rises and the anode current is diverted to the gate. The turn-off occurs in the conditions that prevail in the circuit shown in Fig. 5.40 with a gate turn-off current equal to the anode current, that is, under unity current gain G_T . The large gate current flow removes the stored charge rapidly, resulting in fast turn-off. It is worth emphasizing that, although the turn-off is occurring under unity current gain conditions, the actual gate current supplied by the gating circuit is several orders of magnitude smaller than the load current. Thus the gating technique shown in Fig. 5.42 allows high-speed forced gate turn-off at a very high current gain. A typical variation of the turn-off time with increasing anode current is shown in Fig. 5.44. Note

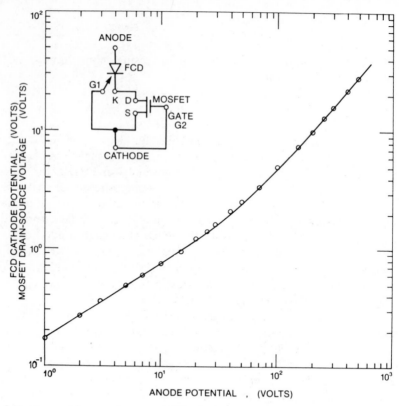

Fig. 5.43. Measured increase in power MOSFET drain voltage with increasing supply voltage. (After Ref. 25, reprinted with permission from *Solid State Electronics*.)

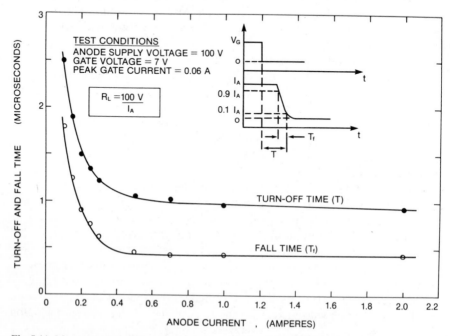

Fig. 5.44. Measured turn-off time for the gating technique using a power MOSFET in series with the FCD. (After Ref. 25, reprinted with permission from *Solid State Electronics*.)

that the turn-off time decreases rapidly at low anode currents and then remains essentially independent of the anode current. Turn-off times of less than 1 μsec have been achieved. During the testing of these devices it was found that the input gate current was 60 mA irrespective of the size of the anode current. Since the composite circuit could be operated at load currents of up to 10 A, turn-off current gains of over 150 can be achieved by using this gating method.

5.7. DEVICE STRUCTURES AND TECHNOLOGY

Because of the similarity between the forward blocking characteristics of the FCDs and power JFETs, the gate structure of FCDs can be fabricated by using any of the processes discussed for the JFETs in Chapter 4. A key difference between these devices is that the channel doping of the FCDs can be made much lower than for JFETs because the channel resistance is modulated by the injected carriers during forward conduction. The use of a lower channel doping level has allowed the fabrication of devices with very high blocking gain, including the normally-off devices described earlier. A precaution that must be taken with regard to lowering the channel doping is that the forward blocking capability can deteriorate as a result of punch-through breakdown of the $P-N-P$ transistor formed between the gate and anode regions.

5.7.1. Symmetrical Structure

The best performance for a symmetrical FCD structure has been obtained by using the vertically walled, epitaxial-refill gate technology described earlier with the aid of Fig. 4.39. By using an N-base region with a doping level of $5 \times 10^{13}/cm^3$, very high blocking gains have been achieved for devices capable of blocking up to 1000 V in both directions. A plot of the increase in the blocking gain observed with increasing gate depth is shown in Fig. 5.45. Differential blocking gains of over 300 have been achieved with this structure, making operation of these devices possible with gate drive voltages of below 20 V [13]. At the same time, these devices can turn off in less than 500 nsec, as described earlier with the aid of Fig. 5.23. The major disadvantage of this structure is the technological difficulty in producing a good epitaxial refill.

For the symmetrical FCD, it is easier to achieve a high forward blocking gain by using the buried-grid structure. In fact, the first FCDs were fabricated by means of this technology [26, 27]. Buried-grid technology can be used to fabricate larger-area devices than achievable with the use of surface-gate technology because of the greater probability of short circuits between gate and cathode in surface-gate structures. Buried-gate FCDs with current handling capability of over several hundred amperes have been achieved with blocking voltages of over 1000 V. The most significant problem with the fabrication of the buried-grid devices is the autodoping phenomenon, which can produce a short-circuiting layer between the grids. This problem has been overcome by

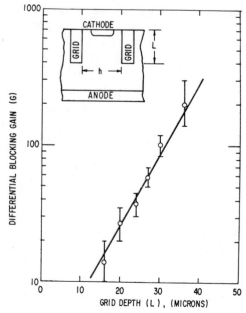

Fig. 5.45. Increase in differential blocking gain obtained by increasing the channel aspect ratio of vertically walled FCDs. (After Ref. 13, reprinted with permission from *Solid State Electronics.*)

using silicon liquid-phase epitaxy to fabricate these devices [28], but this technology has yet to be commercialized.

5.7.2. Asymmetric Structure

If no reverse blocking capability is required as in the case of DC circuits, it is possible to create a device with excellent forward blocking characteristics and high blocking gain with the use of conventional planar processing. This structure, called the *asymmetric field-controlled diode (FCD) structure* [29], is contrasted with the conventional structure in Fig. 5.46. In the new device structure, the homogeneously doped N base of the conventional device has been replaced by a two-layer structure consisting of a very lightly doped portion near the gate and a more heavily doped portion near the anode. This can be most easily achieved by starting with very-high-resistivity ($>2000\ \Omega\cdot\text{cm}$) silicon and doing a phosphorus diffusion on the anode side with a low surface concentration as indicated by the doping profiles shown in the middle of Fig. 5.46. This change in doping concentration in the N-base produces a very significant change in the electric field profile in the device during both the forward and reverse blocking mode as shown on the right-hand side of the figure. Because of the very low doping concentration near the gate, the electric field remains high

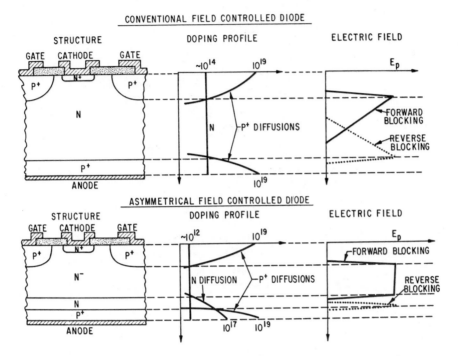

Fig. 5.46. Comparison of structure and electric field distribution for the asymmetric FCD structure with the symmetrical FCD structure. (After Ref. 29, reprinted with permission from the IEEE © 1980 IEEE.)

throughout the N-base region. For the same total water thickness, the asymmetric FCD diode structure can support nearly twice the voltage that can be blocked by the conventional FCD under forward blocking conditions. This greatly improves the trade-off between the forward voltage drop and the forward blocking voltage. However, the presence of the heavily doped N-region near the anode, which is necessary to prevent punch-through breakdown, results in a low reverse blocking capability.

The very low doping of the N-base (typically $1 \times 10^{12}/\text{cm}^3$) near the gate results in a very high blocking gain in these devices despite the open-channel structure obtained by the planar diffusion process used to fabricate these devices. This is due to the increased gate depletion layer spreading even at low gate voltage. The blocking characteristics of typical devices fabricated by using wafers with thickness ranging from 5 to 10 mils are shown in Fig. 5.47. Devices capable of blocking over 900 V with gate bias voltages of less than 15 V have been achieved with this structure [29]. From Fig. 5.47, it can be seen that the blocking gain increases with increasing wafer thickness. This is a unique feature of the asymmetric device structure, which allows the fabrication of high-voltage devices without substantial increase in the gate control voltage. The DC blocking gains observed here (>65) are the highest as yet observed in FCDs.

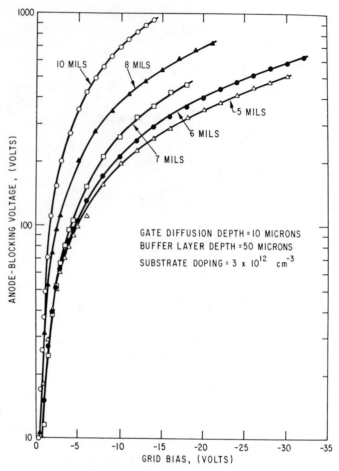

Fig. 5.47. Forward blocking capability of asymmetric FCDs fabricated with various *N*-base thicknesses. (After Ref. 29, reprinted with permission from the IEEE © 1980 IEEE.)

In addition to the high blocking gain, these devices exhibit excellent forward current conduction characteristics as shown in Fig. 5.48. In obtaining these characteristics, a forward gate bias of 0.8 V has been used. This gate bias is required to ensure conduction over the entire channel width. In the absence of this forward gate bias, the built-in diffusion potential of the gate junction can be sufficient to cause substantial gate depletion layer extension into the channel as a result of the very low doping level of the *N* base.

The gate turn-off speed of this device structure can be expected to be higher than that of the conventional device structure because even at low gate voltages, the gate depletion layer will extend through the entire *N* base. This will assist the rapid removal of most of the stored charge in the *N* base and allow rapid switching. The removal of a larger amount of stored charge requires a lower

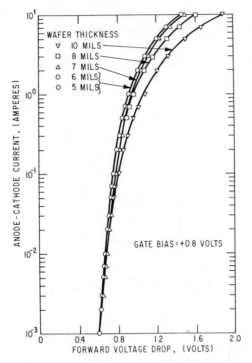

Fig. 5.48. Forward conduction characteristics of asymmetric FCDs fabricated with various *N*-base widths. (After Ref. 29, reprinted with permission from the IEEE © 1980 IEEE.)

gate drive impedance and larger gate currents during turn-off. These drawbacks can be overcome by using a MOSFET to gate these devices as described in Section 5.6.2.

5.8. TRENDS

Field-controlled diodes have been shown to exhibit excellent forward conduction characteristics even when designed for high-voltage operation. The major draw-back that has curtailed their application in power switching circuits is their normally-on characteristics. If this feature can be utilized, these devices offer many unique advantages. They exhibit very high (dV/dt) capability and can be operated at high ambient temperatures without significant degradation in either the forward conduction or forward blocking capability. Their reverse blocking capability makes them applicable to AC circuits as well. Furthermore, FCDs have been found to exhibit good neutron radiation tolerance.

By using the power MOSFET gating technique, the FCD can be operated as a normally-off device. Although this may be expected to make it suitable for general power switching circuits, this has not transpired because of the recent development of another device called the *insulated-gate transistor* (IGT). In fact, the IGT was conceived during an attempt to integrate the power MOSFET and FCD structure into a single device structure. Any future developments in FCDs are expected to be focused on raising their voltage ratings above 2000 V, where they can compete with gate turn-off thyristors.

REFERENCES

1. B. J. Baliga, "Barrier-controlled current conduction in field controlled thyristors," *Solid State Electron.*, **24**, 617–620 (1981).

2. B. J. Baliga, "Breakover phenomena in field controlled thyristors," *IEEE Trans. Electron Devices*, **ED-29**, 1579–1587 (1982).

3. C. T. Sah, R. N. Noyce, and W. Shockley, "Carrier generation and recombination in *P–N* junctions and *P–N* junction characteristics," *Proc. IRE*, **45**, 1228–1243 (1957).

4. R. N. Hall, "Power rectifiers and thyristors," *Proc. IRE*, **40**, 1512–1518 (1952).

5. A. Herlet, "The forward characteristics of silicon power rectifiers at high current densities," *Solid State Electron.*, **11**, 717–742 (1968).

6. S. C. Choo, "Effect of carrier lifetime on the forward characteristics of high power devices," *IEEE Trans. Electron Devices*, **ED-17**, 647–652 (1970).

7. B. W. Wessels and B. J. Baliga, "Vertical channel field controlled thyristors with high gain and fast switching speeds," *IEEE Trans. Electron Devices*, **ED-25**, 1261–1265 (1978).

8. B. J. Baliga and M. S. Adler, "Measurement of carrier lifetime profiles in diffused layers of semiconductors," *IEEE Trans. Electron Devices*, **ED-25**, 472—477 (1978).

9. J. Burtscher, F. Dannhauser, and J. Krausse, "Die rekombination in thyristoren und gleichrichtern aus silizium," *Solid State Electron.*, **18**, 35–63 (1975).

10. N. G. Nilsson, "The influence of Auger recombination on the forward characteristics of semiconductor power rectifiers at high current densities," *Solid State Electron.*, **16**, 681–688 (1973).

11. A. Muñoz-Yague and P. Leturq, "High-level behavior of power rectifiers," *IEEE Trans. Electron Devices*, **ED-25**, 42–49 (1978).

12. M. S. Adler, "Factors determining forward voltage drop in the field-terminated diode (FTD)," *IEEE Trans. Electron Devices*, **ED-25**, 529–536 (1978).

13. B. J. Baliga, "Grid depth dependence of the characteristics of vertical channel field controlled thyristors," *Solid State Electron.*, **22**, 237–239 (1979).

14. B. J. Baliga, "The *di/dt* capability of field controlled thyristors," *Solid State Electron.*, **25**, 583–588 (1982).

15. H. Benda and E. Spenke, "Reverse recovery processes in silicon power rectifiers," *Proc. IEEE*, **55**, 1331–1354 (1967).

16. B. J. Baliga, "The *dV/dt* capability of field controlled thyristors," *IEEE Trans. Electron Devices*, **ED-30**, 612–616 (1983).

17. B. J. Baliga, "Temperature dependence of field controlled thyristor characteristics," *IEEE Trans. Electron Devices*, **ED-28**, 257–264 (1981).

18. B. J. Baliga and E. Sun, "Comparison of gold, platinum, and electron irradiation for controlling lifetime in power rectifiers," *IEEE Trans. Electron Devices*, **ED-24**, 685–688 (1977).

19. B. J. Baliga, "Electron irradiation of field controlled thyristors," *IEEE Trans. Electron Devices*, **ED-29**, 805–811 (1982).

20. B. J. Baliga, "Neutron radiation tolerance of field controlled thyristors," *IEEE Trans. Electron Devices*, **ED-30**, 1832–1834 (1983).

21. B. L. Gregory, "Minority carrier recombination in neutron irradiation silicon," *IEEE Trans. Nucl. Sci.*, **NS-16**, 53–62 (1960).

22. R. F. Bass, "Influence of impurities on carrier removal and annealing in neutron-irradiated silicon," *IEEE Trans. Nucl. Sci.*, **NS-14**, 78–81 (1967).

23. Y. Nakamura, T. Tadano, S. Sugiyama, I. Igarashi, T. Ohmi, and J. Nishizawa, "Normally-off type high speed SI-thyristor," *IEEE International Electron Devices Meeting Digest*, Abstract 18.1, pp. 480–483 (1982).

24. B. J. Baliga, "A novel buried grid device fabrication technology," *IEEE Electron Device Lett.*, **EDL-1**, 250–252 (1980).

25. B. J. Baliga, "High gain power switching using field controlled thyristors," *Solid State Electron.*, **25**, 345–353 (1982).

26. R. Baradon and P. Laurenceau, "Power bipolar Gridistor," *Electron. Lett.*, **12**, 486–487 (1976).

27. J. I Nishizawa, T. Terasaki, and J. Shibata, "Field effect transistor versus analog transistor (static induction transistor)," *IEEE Trans. Electron Devices*, **ED-22**, 185–197 (1975).

28. B. J. Baliga, "Buried grid field-controlled thyristors fabricated using silicon liquid phase epitaxy," *IEEE Trans. Electron Devices*, **ED-27**, 2141–2145 (1980).

29. B. J. Baliga, "The asymmetrical field controlled thyristor," *IEEE Trans. Electron Devices*, **ED-27**, 1262–1268 (1980).

30. S. K. Ghandhi, *Semiconductor Power Devices*, Wiley, New York, 1977.

31. P. Rai-Choudhury, J. Bartko, and J. E. Johnson, "Electron irradiation induced recombination centers in silicon minority carrier lifetime control", *IEEE Trans. Electron Devices*, **ED-23**, 814–818 (1976).

PROBLEMS

5.1. Consider a FCD with a symmetrical blocking voltage capability of 600 V. Determine the doping level and width of the N-base region, given that the low-level minority carrier lifetime is 0.5 μsec.

5.2. At what minority carrier lifetime will the minimum forward voltage drop occur for the device described in Problem 5.1?

5.3. Calculate the anode breakover voltage for a FCD at gate bias voltages of 10, 20, and 30 V. Assume the following device parameters: $n = 1$, $\gamma = 0.5$, $I_0 = 0.2$, $\alpha = 0.5$, and $G_0 = 20$. The gate circuit has a resistance of 1000 Ω.

5.4. What is the anode breakover current for the device described in Problem 5.3?

5.5. Determine the impact of changing the gate circuit resistance to 10 Ω on the anode breakover voltage and current.

5.6. For the device discussed in Problem 5.1, calculate the current density at which high-level injection begins to occur. Assume that the high-level lifetime is equal to the ideal value determined in Problem 5.2.

5.7. Compare the (dV/dt) capability of a field-controlled diode fabricated using a buried-grid structure with an internal resistance of 1000 Ω to a surface-grid structure with internal resistance of 10 Ω when driven by use of a gate circuit with series resistance of 1000 Ω. Assume all other device parameters to be identical.

5.8. Perform the same comparison as in Problem 5.7, assuming that the gate circuit series resistance is reduced to 10 Ω.

5.9. Estimate the electron irradiation dose to decrease the turn-off time of a FCD from a preradiation value of 15 μsec to a postradiation value of 1 μsec.

6 POWER METAL-OXIDE–SEMICONDUCTOR FIELD-EFFECT TRANSISTORS

Power metal-oxide–semiconductor field-effect transistors (MOSFETs) are the most commercially advanced devices discussed in this book. These devices have evolved from MOS integrated-circuit technology. Prior to the development of power MOSFETs, the only device available for high-speed, medium-power applications was the power bipolar transistor. The power bipolar transistor was first developed in the early 1950s, and its technology has matured to a high degree, allowing the fabrication of devices with current handling capability of several hundred amperes and blocking voltages of 600 V. Despite the attractive power ratings achieved for bipolar transistors, there exist several fundamental drawbacks in their operating characteristics. First, the bipolar transistor is a current-controlled device. A large base drive current, typically one-fifth to one-tenth of the collector current, is required to maintain them in the on-state. Even larger reverse base drive currents are necessary for obtaining high-speed turn-off. These characteristics make the base drive circuitry complex and expensive. The bipolar transistor is also vulnerable to a second breakdown failure mode under the simultaneous application of a high current and voltage to the device as commonly required in inductive power circuits. Furthermore, it is difficult to parallel these devices. The forward voltage drop in bipolar transistors decreases with increasing temperature. This promotes diversion of the current to a single device unless emitter ballasting schemes are utilized.

The power MOSFET was developed to solve the performance limitations experienced with power bipolar transistors. In this device, the control signal is applied to a metal gate electrode that is separated from the semiconductor surface by an intervening insulator (typically silicon dioxide). The control signal required is essentially a bias voltage with no significant steady-state gate current flow in either the on-state or the off-state. Even during the switching of the devices between these states, the gate current is small at typical operating frequencies (< 100 kHz) because it only serves to charge and discharge the input gate capacitance. The high input impedance is a primary feature of the power MOSFET that greatly simplifies the gate drive circuitry and reduces the cost of the power electronics.

The power MOSFET is a unipolar device. Current conduction occurs through transport of majority carriers in the drift region without the presence of minority carrier injection required for bipolar transistor operation. No delays are observed as a result of storage or recombination of minority carriers in power MOSFETs during turn-off. Their inherent switching speed is orders of magnitude faster than that for bipolar transistors. This feature is particularly attractive in circuits operating at high frequencies where switching power losses are dominant.

Power MOSFETs have also been found to display an excellent safe operating area; that is, they can withstand the simultaneous application of high current and voltage (for a short duration) without undergoing destructive failure due to second breakdown. These devices can also be easily paralleled because the forward voltage drop of power MOSFETs increases with increasing temperature. This feature promotes an even current distribution between paralleled devices.

These characteristics of power MOSFETs make them important candidates for many applications. They are being used in audio/radiofrequency circuits and in high-frequency inverters such as those used in switch mode power supplies. Other applications for these devices are for lamp ballasts and motor control circuits.

6.1. BASIC STRUCTURES AND OPERATION

The fundamental operation of the power MOSFET relies on the formation of a conductive layer at the surface of the semiconductor. The modulation of the charge at the surface of a semiconductor by using the bias on a metal plate with an intervening insulating layer was first demonstrated in 1948 [1]. Through application of this principle, the first surface FET was fabricated in silicon with thermally grown silicon dioxide as the insulator [2]. Since then, the insulated-gate field-effect transistor (IGFET) has been applied to the fabrication of a large variety of integrated circuits. Until recently, the development of discrete devices has followed the basic concept of the lateral channel structure used in these earlier applications. Such devices have the drain, gate, and source terminals on the same surface of the silicon wafer. Although this feature makes them well suited for integration, it is not optimum for achieving a high power rating. As mentioned in Chapter 4, the vertical channel structure with source and drain on opposite surfaces of the wafer is more suitable for a power device because more area is available for the source region and because the electric field crowding at the gate is reduced.

Two discrete vertical-channel power MOSFET structures have evolved. A cross section of these structures is provided in Fig. 6.1. The DMOS structure is fabricated by using planar diffusion technology with a refractory gate such as polysilicon. In these devices, the P-base region and the N^+-source region are diffused through a common window defined by the edge of the polysilicon

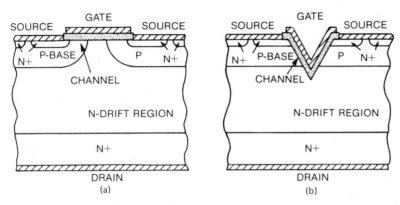

Fig. 6.1. Cross-sectional view of the basic power MOSFET cell structures: (a) DMOS structure; (b) VMOS structure.

gate. The *P*-base region is driven in deeper than the N^+ source. The difference in the lateral diffusion between the *P*-base and N^+-source regions defines the surface channel region.

The vertical-channel VMOS power FET structure requires a different fabrication process. In this case, an unpatterned *P*-base diffusion is performed followed by the N^+-source region diffusion. A V-shaped groove is then formed extending through these diffusions. The gate electrode is next placed so as to overlap the N^+ source and extend into the groove beyond the bottom of the *P*-base region. The channel region for this structure is formed along the walls of the V groove.

In both of these structures, the *P*–*N* junction between the P-base region and the *N*-drift region provides the forward blocking capability. Note that the *P*-base region is connected to the source metal by a break in the N^+-source diffusion. This is important to establishing a fixed potential to the *P*-base region during device operation. If the gate electrode is externally short-circuited to the source, the surface of the *P*-base region under the gate (i.e., the channel region), remains unmodulated at a carrier concentration determined by the doping level. When a positive drain voltage is now applied, it reverse-biases the *P*-base/*N*-drift region junction. This junction supports the drain voltage by the extension of a depletion layer on both sides. Because of the higher doping level of the *P*-base region, the depletion layer extends primarily into the *N*-drift region. Its doping concentration and width must be chosen in accordance with the criteria established in Chapter 3 for avalanche breakdown of *P*–*N* junctions. Higher drain blocking voltage capability requires a lower drift region doping and a greater width.

It is important to connect the gate electrode to the source to establish its potential at the lowest point during the forward blocking state. If the gate is left floating, its potential can rise through capacitive coupling to the drain potential. This induces modulation of the channel region, which can produce an undesirable current flow at drain voltages well below the avalanche break-down limit. Thus power MOSFETs will not support large drain voltages unless the gate is grounded (i.e., connected to the source during forward blocking).

To carry current from drain to source in the power MOSFET, it is essential to form a conductive path extending between the N^+-source regions and the N-drift region. This can be accomplished by applying a positive gate bias to the gate electrode. The gate bias modulates the conductivity of the channel region by the strong electric field created normal to the semiconductor surface through the oxide layer. For a typical gate oxide thickness of 1000 Å and gate drive voltage of 10 V, an electric field of 10^6 V/cm is created in the oxide. The gate-induced electric field attracts electrons to the surface of the P-base region under the gate. This field strength is sufficient to create a surface electron concentration that overcomes the P-base doping. The resulting surface electron layer in the channel provides a conductive path between the N^+-source regions and the drift region. The application of a positive drain voltage now results in current flow between drain and source through the N-drift region and the channel. This current flow is controlled by the resistance of these regions. Note that the current flow occurs solely by transport of majority carriers (electrons for n-channel devices) along a resistive path comprising the channel and drift regions. No minority carrier transport is involved for the power MOSFET during current conduction in the on-state.

To switch the power MOSFET into the off-state, the gate bias voltage must be reduced to zero by externally short-circuiting the gate electrode to the source electrode. When the gate voltage is removed, the electrons are no longer attracted to the channel and the conductive path from drain to source is broken. The power MOSFET then switches rapidly from the on-state to the off-state without any delays arising from minority carrier storage and recombination as experienced in bipolar devices. The turn-off time is controlled by the rate of removal of the charge on the gate electrode because this charge determines the conductivity of the channel. Turn-off times of under 100 nsec can be achieved with a moderate gate drive current flow arising from the discharge of the input gate capacitance of the device.

An examination of the power MOSFET structures illustrated in Fig. 6.1 reveals the existence of a parasitic $N^+-P-N-N^+$ vertical structure. This parasitic bipolar transistor must be kept inactive during all modes of operation of the power MOSFET. To accomplish this, the P-base region is short-circuited to the N^+-emitter region by the source metallization as shown in the cross section of the devices. The short circuit between the P-base and the N^+ emitter can be provided within each cell of the device structure as illustrated in Fig. 6.1, or occasionally at selected locations. The latter approach considerably facilitates the fabrication of the cells because it eliminates an alignment step necessary to form the short circuit within the cells. However, the resistance of the P-base region between the short circuits can become large, and any lateral current flow in the P base, as the result of capacitive currents arising at high applied (dV/dt) to the drain, can lead to forward-biasing of the N^+-P junction at locations remote from the short circuits. Forward-biasing of the N^+-P junction activates the parasitic bipolar transistor and leads to the initiation of minority carrier transport. This can not only slow down the switching of the power MOSFET, but can lead to second breakdown. Because of the high rate of change of voltage observed in the high-frequency applications for which the power MOSFET is particularly well suited, it is common practice to form the short circuit in every cell and minimize the length of the N^+-source region from the edge of the channel to the short circuit.

6.2. BASIC DEVICE CHARACTERISTICS

The power MOSFET is capable of blocking voltage in only one quadrant. For n-channel structures, the devices are operated with a positive voltage applied to the drain. When the gate electrode is short-circuited to the source, the device can support a large drain voltage across the P-base–N-drift region junction. The forward blocking capability is shown in Fig. 6.2 by the lowest trace. Although a finite leakage current is illustrated in the figure, it is very small except at high operating temperatures. The maximum forward blocking voltage is determined by the avalanche breakdown voltage of the P-base–N-drift layer junction. The breakdown voltage is dependent not only on the device termination, but is affected by the internal cell structure of the devices.

When a positive gate bias is applied, the channel becomes conductive. At low drain voltages, the current flow is essentially resistive, with the on-resistance determined by the sum of the channel and drift region resistances. The channel resistance decreases with increasing gate bias, whereas the drift region resistance remains constant. The total on-resistance decreases with increasing gate bias until it approaches a constant value. At large gate bias voltages, the channel resistance becomes smaller than the drift region resistance, and the device on-resistance becomes independent of gate bias. The on-resistance is an important power MOSFET parameter. It is a measure of the current handling capability

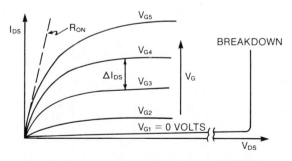

Fig. 6.2. Output characteristics of a power MOSFET.

of the device because it determines the power dissipation during current conduction. The on-resistance is defined as the slope of the output characteristic in the linear region at low drain voltages.

At high drain voltages, the resistance of the power MOSFET increases. Ultimately, the current saturates at high drain voltages as shown on the right-hand side of Fig. 6.2. The current saturation in power MOSFETs can be used to provide a current-limiting function in power circuits as long as the power dissipation in the devices is kept within reasonable bounds.

An important device parameter for power MOSFETs is the transconductance. It is defined by

$$g_m = \frac{\Delta I_{DS}}{\Delta V_{GS}}\bigg|_{V_{DS}} \qquad (6.1)$$

where $\Delta V_{GS} = (V_{G4} - V_{G3})$ is illustrated in Fig. 6.2. A large transconductance is desirable to obtain a high current handling capability with low gate drive voltage and for achieving high frequency response. In the saturated current region of operation, the output characteristics are controlled by the gate-induced channel characteristics. The transconductance is, therefore, determined by the design of the channel and gate structure.

Although the power MOSFET was originally intended for operation in only one quadrant as illustrated in Fig. 6.2 for n-channel devices, it has recently been operated with negative drain voltages as a synchronous rectifier [6]. When a negative drain voltage is applied to the devices illustrated in Fig. 6.1 with their gates connected to the source, the P-base–N-drift layer junction becomes forward-biased. When the drain voltage exceeds a knee voltage V_N of approximately 0.7 V at room temperature, the P–N junction will inject minority carriers and begin to conduct current as illustrated in Fig. 6.3.

If a positive gate voltage is applied to the device to create a channel, an alternate path for current flow between drain and source is created. This current flow occurs by means of majority carriers and is limited by the resistance of

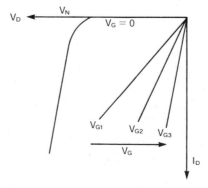

Fig. 6.3. Power MOSFET (n-channel) operated with negative drain voltage.

the channel and drift region. If these resistances are small, the power MOSFET characteristics will resemble those shown in Fig. 6.3. In this figure, the device is shown to carry current with a voltage drop well below the knee point V_N. Under these operating conditions, the $P-N$ junction will not inject any significant concentration of minority carriers. Consequently, the power MOSFET will exhibit a very low forward drop, significantly lower than for a $P-N$ junction rectifier, and retain its high switching speed. For these reasons, it has been suggested that the power MOSFET be used as a replacement for diodes in low-voltage (<5-V), switch-mode power supplies. To accomplish this function, it is essential to provide a gate signal to the power MOSFET in synchronism with the supply voltage to maintain it in the low forward drop on-state, as well as the blocking state, at the appropriate times. The device is, therefore, called a *synchronous rectifier*.

For use as a synchronous rectifier, the power MOSFET must exhibit an extremely low on-resistance if it is to be capable of handling significant drain currents. For example, a device capable of switching 50 A with a forward drop of less than 0.25 V must have an on-resistance of less than 5 mΩ. To achieve this, very-large-area devices are required with breakdown voltages of below 50 V.

6.3. DEVICE ANALYSIS

To understand current flow in the power MOSFET, it is essential to analyze the physics of the MOS structure. Metal-oxide–semiconductor physics controls the characteristics of the channel region, which governs the output characteristics of the devices. The current transport in the channel must then be coupled to the drift region to obtain the total on-resistance of the device. In analyzing the drift region resistance, it is necessary to consider the effect of the pinch-off action arising from the adjacent P-base regions at higher drain voltages.

In many ways, the physics of the operation of power MOSFETs is simpler than that for other power devices because of the absence of minority carrier injection. However, it is necessary to understand the interaction between the cell geometry and the device characteristics before undertaking an accurate device design. In this section, the blocking characteristics are first treated followed by analysis of the on-state characteristics.

6.3.1. Forward Blocking Capability

In the forward blocking mode, the gate electrode of the power MOSFET is externally short-circuited to the source. Under these conditions, no channel forms under the gate at the surface of the P-base region. When a positive drain voltage is applied, the P-base–N-drift layer junction becomes reverse-biased and supports the drain voltage. The breakdown voltage of this junction is not given by the simple parallel-plane analysis described in Chapter 3 for several reasons. First, the power MOSFET structure consists of an N^+-P-N transistor where the P-base region is short-circuited at selected points to the N^+ emitter by the source metallization. Despite the short-circuiting of the N^+ emitter and P base at some locations, this structure will conduct current as soon as the depletion layer in the P-base punches through to the N^+ emitter because the N^+ emitter then becomes forward-biased and injects electrons into the P-base region. The punch-through breakdown condition is determined by the doping profiles used for device fabrication. The second factor that can alter the breakdown voltage is the presence of junction curvature in the DMOS cell. Its impact on the breakdown voltage is dependent on the cell spacing. The third factor is the presence of sharp corners in the VMOS cell, producing high local electric fields and leading to premature breakdown. These factors are discussed in the following sections.

6.3.1.1. Doping Profile. In the forward blocking mode, the depletion layer of the P-base–N-drift layer junction extends on both sides. The P-base doping must be maintained at a relatively low concentration in power MOSFETs to permit inversion of the surface under the gate during the operation of the power MOSFET in its on-state. A typical diffusion profile for the device is shown in Fig. 6.4. The solid lines indicate the dopant distributions, whereas the dashed lines show the resulting carrier concentration profiles, which differ from the dopant profiles as a result of compensation effects. The surface concentration N_{SP} of the P-base diffusion and the N^+-emitter depth combine to determine the peak doping N_{AP} in the P-base indicated by the arrow in Fig. 6.4. The peak P-base doping is an important parameter for the doping profile because it

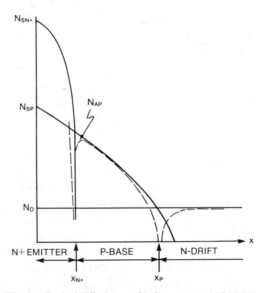

Fig. 6.4. Typical diffusion profile for a power MOSFET.

controls the threshold voltage of the power MOSFET, that is, the minimum gate voltage required to induce a surface channel. For a typical threshold voltage of 2–3 V for an n-channel power MOSFET, the peak base concentration N_{AP} is about $1 \times 10^{17}/\text{cm}^3$.

The channel length is another important design parameter in power MOSFETs because it has a strong influence on the on-resistance and the transconductance. The channel length is determined by the difference between the depths of the P-base and N^+-emitter diffusions [i.e., $(x_P - x_{N^+})$]. (It should be noted that, although a vertical doping profile is measured to characterize devices, the lateral profile under the gate determines the channel properties.) The need to maintain a low peak base concentration and a narrow base width to achieve good on-state characteristics can adversely affect the forward blocking capability. Despite the short-circuiting of the N^+ emitter to the P-base by the source metal, a parasitic N^+–P–N bipolar transistor exists in the power MOSFET. When the P-base–N-drift layer junction is reverse-biased, the depletion layer in the P-base can extend to the N^+-emitter–P-base junction and cause premature punch-through breakdown. It is important to design the P-base diffusion profile so that sufficient charge is resident in it to prevent punch-through of the depletion layer to the N^+ emitter.

The depletion width extension on the diffused side of a P–N junction has been calculated at breakdown using numerical techniques [3]. In Fig. 6.5, the

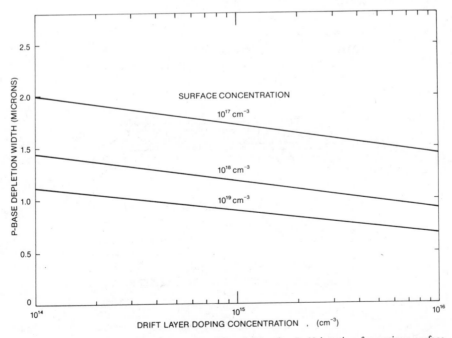

Fig. 6.5. Depletion width extension on the diffused side of a P–N junction for various surface concentrations.

change in the depletion width on the diffused side of the junction is provided as a function of the background doping level and surface concentration of the diffusion. As the drift layer concentration decreases, the junction profile becomes more graded, causing an increase in the depletion width on the diffused side. It should be noted that when the surface concentration is below $10^{18}/cm^3$, the depletion width on the diffused side can extend well over 1 μm, especially for low drift layer concentrations. Care must be taken during device design to prevent punch-through breakdown in these cases. From the data in Fig. 6.5, it can also be concluded that the fabrication of devices with channel lengths of less than 1 μm is difficult to achieve unless the device is required to support only low voltages, that is, for devices with high drift layer concentrations.

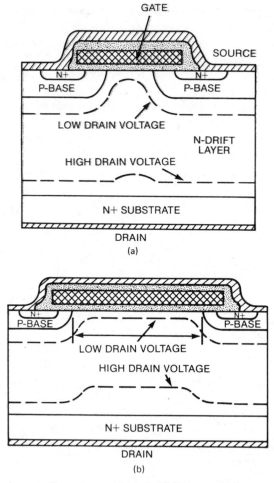

Fig. 6.6. Depletion layer profiles at low and high drain voltages for DMOS devices with (a) small and (b) large cell spacing.

6.3.1.2. Cell Spacing.   To achieve a large channel width for good on-state characteristics, power MOSFETs are fabricated with a repetitive pattern of small
cells. In the DMOS cell structure shown in Fig. 6.1a, the P-base–N-drift layer
is brought to the surface between the individual cells. Each cell contains a
planar junction edge. During forward blocking the depletion layer extends out
from the junction as well as from the gate overlap region between the cells.
The depletion layer profile at low and high drain voltages is shown for two
examples of a very small cell spacing and a large cell spacing in Fig. 6.6. In the
case of the DMOS design with small cell spacing, the depletion layer curvature
becomes small at high drain voltages. These devices will break down at the
edge termination. If the cell spacing is large, however, the curvature of the
depletion layer at the edges of the P-base regions is significant, resulting in
lower breakdown voltage, as was discussed for the case of cylindrical or spherical junctions in Chapter 3.

Numerical analysis of the effect of cell spacing on the breakdown voltage
has been performed for the DMOS structure [4]. As an example, the calculated
breakdown voltage is shown as a function of the gate width in Fig. 6.7 for
several drift region doping levels. It can be seen that the breakdown voltage
increases when the cell spacing decreases below 10 μm. Note that the rise in
the breakdown voltage occurs at larger cell spacing for the case of lower drift
layer doping concentrations. This is consistent with the larger depletion layer

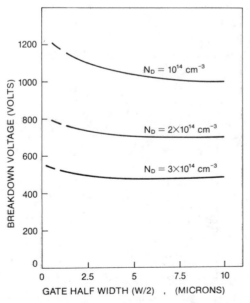

Fig. 6.7. Impact of DMOS cell spacing on the breakdown voltage. (After Ref. 4, reprinted with
permission from the IEEE © 1984 IEEE.)

spreading at lower doping levels, which tends to alleviate the junction curvature. In general, a 15% increase in the breakdown voltage can be obtained by bringing the cells close together. However, the cells cannot be placed arbitrarily close together because of an increase in the on-resistance, as discussed later in this chapter.

6.3.1.3. V-Groove Corner.

In the case of the VMOS structure shown in Fig. 6.1*b*, the depletion layer from the *P*-base–*N*-drift layer junction expands into the drift region from either side of the V-groove. The depletion layer in the *N*-drift layer extends well beyond the tip of the V groove, as illustrated in Fig. 6.8. A higher electric field is created at the vicinity of the sharp tip of the V groove. The higher local electric fields at the tip of the groove cause avalanche breakdown along the paths indicated in Fig. 6.8 prior to avalanche breakdown in the parallel-plane portion of the junction. To alleviate the increase in electric field at the sharp pointed end of the V groove, a truncated V-groove structure has been developed as illustrated in Fig. 6.9. The truncation of the groove is achieved by using a wider window during etching of the groove and stopping the etch before a sharp point is created. This creates a wide drain overlap region and eliminates the sharply pointed end of the groove. Nevertheless, a discontinuity in the semiconductor surface still remains at point *A*. The impact of this discontinuity on the breakdown voltage has been analyzed by numerical techniques [5]. An example of the reduction in the breakdown voltage due to the groove is provided in Fig. 6.10. As the drain overlap distance x (i.e., the distance over which the gate lies over the drift region) shortens the structure approaches the V-groove case and the breakdown voltage decreases, as shown by the solid line for the VMOS case. In comparison, the breakdown voltage of the DMOS structure (solid line) increases slightly as the drain overlap decreases. These results apply when the gate electrode completely extends between the adjacent cells over the entire drain overlap region.

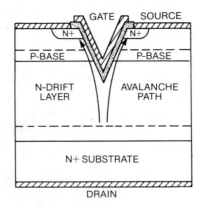

Fig. 6.8. Enhanced avalanche path created in a V-groove power MOSFET structure due to electric field crowding at the sharp point at the end of the groove.

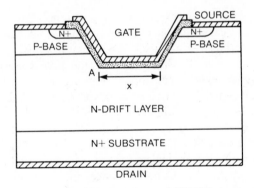

Fig. 6.9. Truncated V-groove power MOSFET structure.

When no overlap of the gate electrode is provided, that is, if the gate extends only to the edge of the junction between the *P*-base and *N*-drift regions, the breakdown voltage is found to rapidly decrease with increasing spacing between the cells for both the DMOS and VMOS structures as shown by the dashed lines in Fig. 6.10. For this reason, it is advisable to let the gate electrode extend completely between the cells during power MOSFET design. An exception to

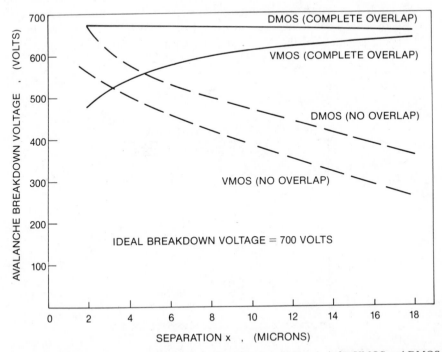

Fig. 6.10. Avalanche breakdown voltages calculated by numerical analysis for VMOS and DMOS power MOSFET structures. (After Ref. 5, reprinted with permission from the IEEE © 1979 IEEE.)

this rule occurs when designing for maximizing the frequency response. The overlap of the gate electrode over the drain region adds significantly to the input capacitance, thus degrading the frequency response. By minimizing the overlap of the gate over the drain and keeping the separation between cells small, it is possible to achieve a reasonable compromise between achieving high breakdown voltage and high frequency response [6]. Another approach to achieving this goal is to use a thicker oxide in the region between the cells.

These results of numerical analysis of the breakdown of VMOS structures indicate that these devices will exhibit a lower avalanche breakdown voltage when compared with DMOS devices for the same epitaxial layer doping level and thickness. This has serious consequences on the minimum on-resistance achievable by use of the VMOS structure. Because of higher specific on-resistance for a given avalanche breakdown voltage for the VMOS structure, most manufacturers have converted to the DMOS structure for commercial fabrication of power MOSFETs.

6.3.2. Forward Conduction Characteristics

Current flow in a power MOSFET during forward conduction is achieved by the application of a positive gate bias voltage for n-channel devices to create a conductive path across the P-base region underneath the gate. The current flow is limited by the total resistance between source and drain. This resistance consists of several components as illustrated in the DMOS cross section in Fig. 6.11. The resistances of the N^+ emitter (R_{N+}) and substrate (R_S) regions are

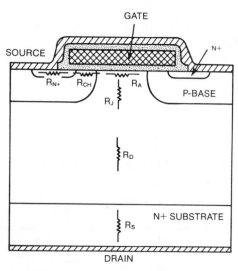

Fig. 6.11. Cross section of DMOS power transistor with internal resistances of the structure indicated.

generally negligible for high-voltage power MOSFETs that have high drift region resistance. They become quite important when the drift and channel resistance become small as in the case of low (<100 V) breakdown voltage devices. The channel (R_{ch}) and accumulation layer (R_A) resistances are determined by the conductivity of the thin surface layer induced by the gate bias. These resistances are a function of the charge in the surface layer and the electron mobility near the surface. The analysis of the charge at the surface is treated in this section. The electron mobility in surface inversion and accumulation layers was treated in Chapter 2.

In addition to these resistances, the drift layer contributes two more components to the total on-resistance. The portion of the drift region that comes to the upper surface between the cells contributes a resistance R_J that is enhanced at higher drain voltages due to the pinch-off action of depletion layers extending from adjacent P-base regions. This phenomenon has been termed the *JFET action*. Finally, the main body of the drift region contributes a large series resistance R_D, especially for high-voltage devices. The analysis of each of these components of the on-resistance is provided in this section.

6.3.2.1. MOS Physics.

In this section, the charge created at the surface of the semiconductor as a result of the gate bias applied to the MOS structure is analysed. For this analysis, a one-dimensional structure is assumed with no electric field applied parallel to the surface. To perform the analysis, a P-type semiconductor region is assumed. The treatment is applicable to the analysis of current transport in n-channel power MOSFETs. A similar analysis can be performed for an N-type semiconductor region by appropriate changes in the polarity of the voltage, electric field, and free-carrier charge. In the analysis presented here, the oxide layer is assumed to be a perfect insulator that does not allow the transport of any charge between the gate and the semiconductor.

The energy-band diagrams for an ideal MOS structure with a P-type semiconductor are shown in Fig. 6.12 for four examples of different bias potentials on the metal electrode. Here an ideal MOS structure is defined as one that satisfies the following conditions: (1) the insulator has infinite resistivity, (2) charge can exist only in the semiconductor and on the metal electrode, and (3) there is no energy difference between the work function of the metal and the semiconductor. Under these conditions, there is no band bending in the absence of a gate bias and the band structure assumes the form shown in Fig. 6.12a. This is known as the *flatband* condition. In this figure, ϕ_m is the metal work function, χ and χ_o are the semiconductor and oxide electron affinities, ϕ_B is the barrier height between the metal and the oxide, and ψ_B is the potential difference between the intrinsic and Fermi levels in the semiconductor. Under flatband conditions

$$q\phi_m = q\chi + \frac{E_g}{2} + q\psi_B = q\phi_B + q\chi_o \qquad (6.2)$$

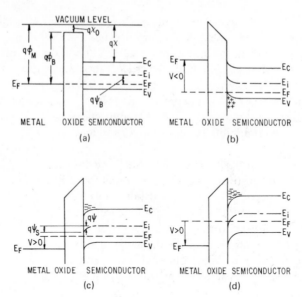

Fig. 6.12. Energy band diagrams for a MOS structure with *P*-type semiconductor region under (*a*) flatband (zero gate bias), (*b*) accumulation (negative gate bias), (*c*) depletion (positive gate bias), and (*d*) inversion (positive gate bias).

When a negative bias is applied to the metal electrode, holes are attracted to the surface, resulting in the bending of the valence band closer to the Fermi level as shown in Fig. 6.12*b*. This condition of excess majority carriers at the surface is referred to as *accumulation*. In contrast, when a positive bias is applied to the metal electrode, the holes are repelled from the surface. At small negative baises, this results in the bending of the valence band away from the Fermi level, resulting in the formation of a surface depletion layer as shown in Fig. 6.12*c*. At larger negative biases on the metal, the band bending increases and the intrinsic level crosses the Fermi level as shown in Fig. 6.12*d*. At this point, the density of electrons at the surface exceeds the density of holes (majority carriers) in the bulk. This condition is defined as *inversion*.

The charge in the inversion layer plays a key role in the determination of current transport in MOSFET devices. To determine this charge, the potential distribution $\psi(x)$ in the semiconductor must first be determined. The electron and hole concentrations can then be obtained from

$$n_p = n_{p0}\, e^{q\psi/kT} \tag{6.3}$$

and

$$p_p = p_{p0}\, e^{-(q\psi/kT)} \tag{6.4}$$

where n_{p0} and p_{p0} are, respectively, the equilibrium electron and hole concentrations in the bulk. The potential distribution $\psi(x)$ can be derived by solving

Poisson's equation:

$$\frac{d^2\psi}{dx^2} = -\frac{\rho(x)}{\epsilon_s} \tag{6.5}$$

where the charge density is given by

$$\rho(x) = q(N_D^+ - N_A^- + p_p - n_p) \tag{6.6}$$

The solution of the Poisson's equation must satisfy the boundary condition that charge neutrality exists in the bulk of the semiconductor far from the surface. In the bulk, charge neutrality requires

$$(N_D^+ - N_A^-) = (n_{po} - p_{po}) \tag{6.7}$$

Using Eqs. (6.3), (6.4), and (6.6) in Poisson's Eq. (6.5), it can be shown that:

$$\frac{d^2\psi}{dx^2} = -\frac{q}{\epsilon_s}[p_{po}(e^{-(q\psi/kT)} - 1) - n_{po}(e^{(q\psi/kT)} - 1)] \tag{6.8}$$

Integration of this equation from the bulk toward the surface gives the electric field distribution [7]:

$$\mathscr{E}(x) = -\frac{d\psi}{dx} = \frac{\sqrt{2}\,kT}{qL_D}F\left(\frac{q\psi}{kT}, \frac{n_{po}}{p_{po}}\right) \tag{6.9}$$

where

$$L_D = \sqrt{\frac{kT}{q^2}\frac{\epsilon_s}{p_{po}}} \tag{6.10}$$

is called the *extrinsic Debye length for holes*, and

$$F\left(\frac{q\psi}{kT}, \frac{n_{po}}{p_{po}}\right) = \left\{\left[e^{-(q\psi/kT)} + \left(\frac{q\psi}{kT}\right) - 1\right] + \frac{n_{po}}{p_{po}}\left[e^{(q\psi/kT)} - \left(\frac{q\psi}{kT}\right) - 1\right]\right\}^{1/2} \tag{6.11}$$

The space charge required per unit area to create this electric field can be obtained from Gauss's law:

$$Q_S = -\epsilon_s\mathscr{E}_s \tag{6.12}$$

where $\mathscr{E}_s$ is the surface electric field. If the surface potential is called ψ_s, the surface charge can be obtained from Eqs. (6.9) and (6.12):

$$Q_S = \frac{\sqrt{2}\,\epsilon_s kT}{qL_D}F\left(\frac{q\psi_s}{kT}, \frac{n_{po}}{p_{po}}\right) \tag{6.13}$$

The variation of the surface charge with shift in the surface potential is illustrated in Fig. 6.13.

For negative values of the surface potential, the bands are bent upward near the surface, creating a positively charged surface accumulation layer. In

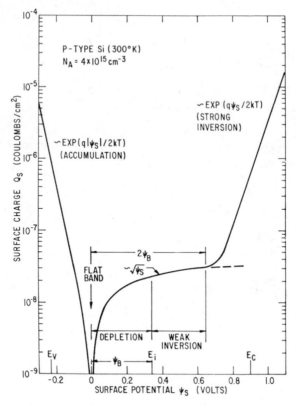

Fig. 6.13. Dependence of the surface space-charge density on the surface potential for P-type silicon.

this regime of operation, the charge in the accumulation layer varies exponentially with the surface potential because the first term on the right-hand side of Eq. (6.11) becomes dominant:

$$Q_S \simeq \frac{\sqrt{2}\,\epsilon_s kT}{qL_D}\, e^{(q|\psi_s|/kT)} \tag{6.14}$$

where $|\psi_S|$ is the magnitude of the surface potential. As the magnitude of ψ_S decreases, the accumulation layer charge decreases exponentially until it becomes equal to zero at the flatband condition shown in Fig. 6.12a. Note that under accumulation conditions, the electric field is essentially confined to the oxide and the entire gate voltage is supported by the oxide.

When the surface potential increases from zero in the positive direction, the bands are bent downward, as illustrated in Fig. 6.12c. As long as the surface potential ψ_S remains below the bulk value ψ_B, the intrinsic level E_i does not cross the Fermi level. The surface is then still p-type and contains a depletion layer. The negative charge in the depletion layer can also be calculated from Eq. (6.13). In this case the second term of the function F [Eq. (6.11)] becomes dominant and the surface charge is given by

$$Q_S \simeq \frac{\epsilon_s}{L_D} \sqrt{2 \frac{kT}{q}} \psi_S \tag{6.15}$$

Using Eq. (6.10) for L_D, we obtain

$$Q_S \simeq \sqrt{2\epsilon_s q p_{p0} \psi_S} \tag{6.16}$$

The surface charge increases gradually as the square root of the surface potential as indicated in Fig. 6.13.

When the surface potential ψ_S exceeds the bulk value ψ_B, the intrinsic level E_i crosses the Fermi level. At this point, a mobile negative charge begins to form at the surface. The concentration of this charge is small as long as ψ_S is less than $2\,\psi_B$. This regime is called *weak inversion*. In this region of operation, the surface charge continues to rise as the square root of the surface potential. When the surface potential exceeds $2\psi_B$, the surface charge begins to rise rapidly as shown on the right-hand side of Fig. 6.13. This occurs because the fourth term in Eq. (6.11) for the function F now becomes dominant. This regime of operation is known as *strong inversion*. It occurs when

$$\psi_S > 2\psi_B = 2\frac{kT}{q} \ln\left(\frac{N_A}{n_i}\right) \tag{6.17}$$

The surface charge under strong inversion can be derived from Eq. (6.13) by assuming that the fourth term in Eq. (6.11) is dominant:

$$Q_S \simeq \sqrt{\frac{2n_{p0}}{p_{p0}}} \frac{\epsilon_s kT}{qL_D} e^{q\psi_S/2kT} \tag{6.18}$$

Substituting for L_D from Eq. (6.10), this expression can be rewritten as:

$$Q_S \simeq \sqrt{2\epsilon_s n_{p0}}\, e^{(q\psi_S/2kT)} \tag{6.19}$$

In the strong inversion region, the surface charge again increases exponentially with surface potential. This charge is an important quantity because it determines the conductivity of the channel in power MOSFETs.

In the depletion region of operation, the applied gate voltage is shared across the gate oxide and the semiconductor depletion layer. With the onset of strong inversion, any further increase in the gate voltage is dropped across the oxide and the voltage across the semiconductor depletion layer remains essentially constant. As a consequence, the depletion layer width in the semiconductor expands until strong inversion occurs, and then remains constant with further increase in gate voltage. The maximum depletion layer width is approximately given by

$$W_m \simeq \sqrt{\frac{2\epsilon_s}{qN_A}(2\psi_B)} \tag{6.20}$$

Use of Eq. (6.17) for ψ_B gives

$$W_m \simeq \sqrt{\frac{4\epsilon_s kT}{q^2 N_A} \ln\left(\frac{N_A}{n_i}\right)} \tag{6.21}$$

A plot of the variation in the maximum depletion depth with background doping concentration is provided in Fig. 6.14. It is worth pointing out that the maximum depletion layer width for the MOS structure is considerably smaller than the maximum depletion layer width in the semiconductor determined by avalanche breakdown.

6.3.2.2. Threshold Voltage.

The voltage on the gate electrode at which strong inversion begins to occur in the MOS structure is an important design parameter for power MOSFETs because it determines the minimum gate bias required to induce an *n*-type conductance in the channel. This voltage is called the *threshold voltage*. For proper device operation, its value can be neither too large nor too small. If the threshold voltage is large, a high gate bias voltage will be needed to turn on the power MOSFET. This imposes problems in the design of the gate drive circuitry. It is also important that the threshold voltage not be too low. Because of the existence of charge in the gate oxide, it is possible for the threshold voltage to be negative for *n*-channel power MOSFETs.

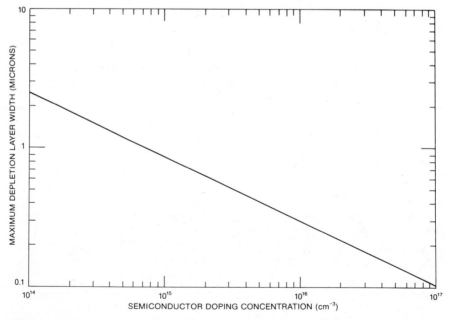

Fig. 6.14. Maximum depletion layer width of an MOS structure as a function of the doping level of the semiconductor.

This is an unacceptable condition because a conductive channel will now exist at zero gate bias voltage; that is, the devices will exhibit normally-on characteristics. Even if the threshold voltage is above zero for an n-channel power MOSFET, its value should not be too low because the device can then be inadvertently triggered into conduction either by noise signals at the gate terminal or by the gate voltage being pulled up during high-speed switching. Typical power MOSFET threshold voltages are designed to range between 2 and 3 V.

In the absence of any difference in the work function of the metal and the semiconductor, the applied gate voltage is given by

$$V_G = V_{ox} + \psi_S \tag{6.22}$$

where V_{ox} is the voltage across the oxide. This voltage is related to the surface charge:

$$V_{ox} = \mathscr{E}_{ox} t_{ox} \tag{6.23}$$

$$= \frac{Q_S}{\epsilon_{ox}} t_{ox} \tag{6.24}$$

$$= \frac{Q_S}{C_{ox}} \tag{6.25}$$

At the point of transition into the strong inversion condition, the gate bias voltage is equal to the threshold voltage:

$$V_G = V_T = \frac{Q_S}{C_{ox}} + 2\psi_B \tag{6.26}$$

Using Eq. (6.16) for Q_S at the point of transition into strong inversion, it can be shown that

$$V_T \simeq \frac{\sqrt{4\epsilon_s q N_A \psi_B}}{C_{ox}} + 2\psi_B \tag{6.27}$$

$$\simeq \frac{\sqrt{4\epsilon_s k T N_A \ln(N_A/n_i)}}{(\epsilon_{ox}/t_{ox})} + \frac{2kT}{q} \ln\left(\frac{N_A}{n_i}\right) \tag{6.28}$$

From this expression, it can be concluded that the threshold voltage will increase linearly with gate oxide thickness and approximately as the square root of the semiconductor doping. The calculated threshold voltage for silicon MOS structures with gate oxide thicknesses ranging from 100 to 10,000 Å are provided in Fig. 6.15 for doping levels ranging from 10^{14} to $10^{18}/cm^3$. These curves apply to a uniformly doped semiconductor and do not take into account the presence of any charge in the oxide.

In actual MOS structures, the threshold voltage is altered due to (1) an unequal work function for the metal and the semiconductor—if the barrier

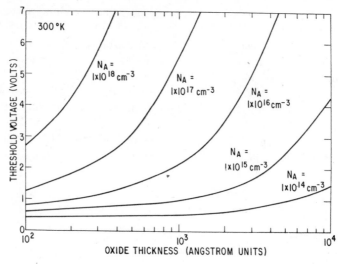

Fig. 6.15. Threshold voltage of ideal silicon MOS structures as a function of oxide thickness.

height between SiO_2 and the metal is ϕ_B, the difference between the metal and the semiconductor work function can be obtained as

$$q\phi_{ms} = q\phi_B + q\chi_0 - \left(q\chi + \frac{E_g}{2} + q\psi_B\right) \tag{6.29}$$

(2) the presence of fixed surface charge Q_{FC} at the oxide–silicon interface, (3) the presence of mobile ions in the oxide with charge Q_I, and (4) the presence of charged surface states at the oxide–silicon interface with charge Q_{SS}. All these charges cause a shift in the threshold voltage:

$$V_T = \phi_{ms} + 2\psi_B - \left(\frac{Q_S + Q_{SS} + Q_I + Q_{FC}}{C_{ox}}\right) \tag{6.30}$$

The fixed surface charge Q_{FC} is located within 20 Å of the oxide–silicon interface and cannot be charged or discharged over a wide range of surface potentials such as those normally encountered in device operation. The density of this charge is dependent on the oxide growth conditions and its subsequent heat treatment. The presence of excess silicon or a deficiency of oxygen has been postulated as the origin of the fixed charge, but the influence of impurities cannot be excluded.

The presence of mobile ions (charge Q_I) has been related to the shift in the threshold voltage under bias stressing at elevated temperatures or under long periods. Sodium ions are the most prominent mobile species in the oxide. The concentration of these ions is dependent on the cleanliness of the wafers during processing and the purity of the gate metal.

Interface states (charge Q_{SS}) are energy levels in the silicon band gap located close to the surface that can charge and discharge with changes in the surface potential. The origin of interface states on a clean free surface in vacuum has been shown to arise from unsaturated bonds at the surface. The interface states at the Si–SiO$_2$ boundary are usually several orders of magnitude lower (10^{10}–10^{12}/cm^2) than on the free surface. The origin of these surface states is not completely understood, but the effect of processing on their density has been extensively studied. This is discussed in the section on device fabrication.

The dependence of the threshold voltage of an aluminum gate MOS capacitor fabricated on a 1000- and 15,000-Å oxide layer grown on P-type silicon is provided in Fig. 6.16 as a function of the doping level. The 1000-Å oxide thickness is typically used to fabricate the MOS gate of the FETs, whereas the 15,000-Å oxide thickness represents the thick field oxide in the inactive portions of the device where the influence of the gate bias must be minimized. In these figures, curves are shown for various values of the charge in the oxide that may arise from either fixed positive interface charge or positive ions at the oxide–silicon interface or from positively charged surface states. For a typical peak channel doping on the order of 5×10^{16} atoms/cm^3 and a charge density of 10^{11}/cm^2, a threshold voltage at 2–3 V can be obtained with an oxide thickness of 1000 Å. This threshold voltage is convenient for controlling the device with integrated drive circuits. For this doping level and interface charge, the threshold voltage of the field oxide is over 30 V, which provides sufficient margin for overdriving the gate without inverting the regions outside the device active area.

As indicated by Eq. (6.30), the threshold voltage for inversion in an MOS capacitor is affected by the barrier height between the metal and the oxide.

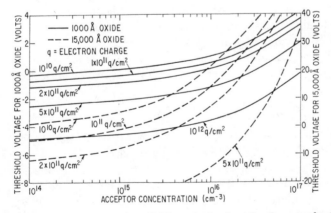

Fig. 6.16. Threshold voltage as a function of P-base doping level for the case of a 1000-Å gate oxide and 15,000-Å field oxide. Curves are shown for charge in the oxide ranging from 10^{10} to 10^{12}/cm^2.

TABLE 6.1. Barrier Heights of Metals on Silicon Dioxide

Metal	Ag	Al	Au	Cu	Mg	Ni	Pd	Pt	W
Barrier height ϕ_m (eV)	4.2	3.2	4.1	3.8	2.3	3.7	4.2	4.4	3.6

Table 6.1 provides a list of metals and the measured barrier height between them and silicon dioxide [8]. As can be seen from this table, the threshold voltage of a device can be tailored to a limited extent by choosing the proper metal.

A greater control over the threshold voltage can be achieved by substituting polycrystalline silicon for the metal. By doping the polycrystalline silicon either N or P type, the work function difference in the poly Si–SiO$_2$–Si system can be varied. Measurements have shown that the Fermi level in heavily doped polycrystalline silicon corresponds exactly to that in monocrystalline silicon [9]. On this basis, the work function difference between N-type polysilicon and P-type monocrystalline silicon is given by the expression

$$q\phi_{\text{PS—Si}} = \frac{kT}{q} \ln\left(\frac{N_{\text{DPS}}N_{\text{A}}}{n_{\text{i}}^2}\right) - E_{\text{g}} \tag{6.31}$$

where N_{A} is the substrate doping concentration and N_{DPS} is the donor concentration in the polycrystalline silicon.

Most power MOSFETs are fabricated by using heavily doped polycrystalline silicon as the gate electrode. This provides a refractory gate material that is compatible with silicon device processing and allows the fabrication of large-area devices with a two-level "metal" process as illustrated in Fig. 6.11. The conductivity of polycrystalline silicon is an order of magnitude lower than that for metals. In very-high-frequency power MOSFETs, the RC gate charging time constant becomes too large for polycrystalline silicon gate electrodes. For these devices, aluminum has been used as the gate electrode, with source and gate metal being interdigitated on the wafer surface. With the development of refractory gate silicide or polycide technology for integrated circuits, it is expected that very-high-frequency power MOSFETs with two-level electrode structures will be forthcoming.

6.3.2.3. Channel Conductance.
Consider a MOSFET having a P-type base region with N^+ source and drain located on either side as shown in Fig. 6.17. The current flow between source and drain is controlled by the electron charge available for transport in the surface inversion layer of the P base as well as the surface mobility of these electrons. According to the MOS analysis, when the surface potential ψ_S exceeds twice the bulk potential ψ_B, a strong inversion layer begins to form. Since the band bending is small beyond this point, the

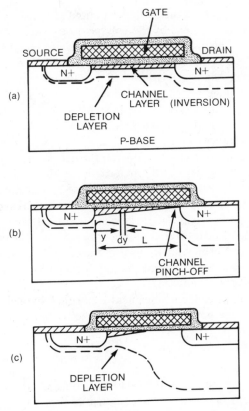

Fig. 6.17. Basic MOS transistor with inversion layer charge transport.

inversion layer charge available for current conduction is

$$Q_n \simeq C_{ox}(V_G - V_T) \tag{6.32}$$

Thus the channel resistance at low drain voltages, where the voltage drop along the channel is negligible, is given by

$$R_{ch} = \frac{L}{Z\mu_{ns}C_{ox}(V_G - V_T)} \tag{6.33}$$

where Z and L are the width and length of the channel, respectively, and μ_{ns} is the surface mobility of electrons. The surface mobility and its dependence on doping, temperature, and surface orientation are discussed extensively in Chapter 2 because of their importance in determining the conductance of the channel in the MOSFET.

As the drain current increases, the voltage drop along the channel between drain and source becomes significant. The positive drain potential opposes the gate bias voltage and reduces the surface potential in the channel. The channel charge near the drain is reduced by the voltage drop along the channel as illustrated in Fig. 6.17b. When the drain voltage becomes equal to $(V_G - V_T)$, the charge in the channel at the drain end becomes equal to zero. This condition is called *channel pinch-off*. At this point, the drain current saturates. Further increase in drain voltage results in no further increase in drain current. The drain voltage is now supported across an extension of the drain under the gate as shown by Fig. 6.17c, resulting in a reduction in the effective channel length.

The channel resistance can be derived as a function of the gate and drain voltages under the following assumptions: (1) the gate structure is an ideal MOS structure as discussed earlier; (2) the free-carrier mobility is a constant, independent of the electric field strength; (3) the base region is uniformly doped; (4) current transport occurs exclusively by drift; (5) the leakage current is negligible; and (6) the longitudinal electric field along the surface is small compared with the transverse electric field resulting from the gate bias. Assumption 6 allows treatment of an elemental segment of the channel dy at location y under conditions similar to the gradual channel approximation used to analyze the impedance of JFETs in Chapter 4.

The resistance of an elemental segment dy of the channel is dependent on the inversion layer charge per unit area and the mobility of the free carriers:

$$dR = \frac{dy}{Z\mu_{ns}Q_n(y)} \tag{6.34}$$

The charge in the inversion layer depends not only on the gate voltage, but also on the drain current because of the potential drop along the channel:

$$Q_n(y) = C_{ox}[V_{GS} - V_T - V(y)] \tag{6.35}$$

where $V(y)$ is the voltage drop along the channel. The voltage drop in segment dy is given by

$$dV = I_D\,dR \tag{6.36}$$

From these expressions

$$\int_0^L I_D\,dy = -Z\mu_{ns}C_{ox}\int_0^{V_D}(V_G - V_T - V)\,dV \tag{6.37}$$

Since the drain current I_D must remain constant throughout the channel:

$$I_D = \frac{\mu_{ns}C_{ox}Z}{2L}[2(V_G - V_T)V_D - V_D^2] \tag{6.38}$$

When the drain voltage is small, the drain current increases linearly with drain voltage:

$$I_D \simeq \mu_{ns}C_{ox}\frac{Z}{L}(V_G - V_T)V_D \tag{6.39}$$

In this linear region, the channel resistance is given by

$$R_{ch} = \frac{L}{Z\mu_{ns}C_{ox}(V_G - V_T)} \tag{6.40}$$

which agrees with Eq. (6.33), representing a homogeneous inversion layer extending between drain and source. In this linear regime, the transconductance can be obtained by differentiating Eq. (6.39) with respect to the gate voltage:

$$g_m = \frac{dI_D}{dV_G} = \mu_{ns}C_{ox}\frac{Z}{L}V_D \tag{6.41}$$

As the drain voltage and current increase, the second term in Eq. (6.38) becomes increasingly important, resulting in an increase in the channel resistance. Physically, this corresponds to the reduction in the channel inversion layer charge near the drain with increasing drain voltage. Ultimately, the channel resistance becomes infinite, corresponding to drain current saturation. The drain voltage at which current saturation occurs is called the *channel pinch-off point*:

$$V_{DS} = (V_G - V_T) \tag{6.42}$$

Substituting for V_{DS} in Eq. (6.38), an expression for the saturated drain current can be derived:

$$I_{DS} = \frac{\mu_{ns}C_{ox}}{2}\frac{Z}{L}(V_G - V_T)^2 \tag{6.43}$$

The saturated drain current is an important parameter because it determines the maximum current that the channel will support. In actual devices, the saturated drain current will be lower than that indicated by Eq. (6.43) because the surface mobility μ_{ns} is a function of the longitudinal electric field. The transconductance of the device in the saturated current region of operation is given by

$$g_{ms} = \mu_{ns}C_{ox}\frac{Z}{L}(V_G - V_T) \tag{6.44}$$

The basic features of the current–voltage characteristics of a MOSFET as described by Eq. (6.38) are shown in Fig. 6.18.

According to the preceding analysis, the drain current will saturate above a certain drain voltage. In actual practice, the drain current does not remain constant in the saturation region because when the drain voltage increases beyond V_{DS}, the length of the channel is reduced as illustrated in Fig. 6.17(c). This decreases the effective channel length and produces a finite drain output conductance. For devices with higher P-base doping levels, the channel shrinkage ΔL can be approximated by treating the gate–drain depletion region like a N^+–P step junction [10]:

$$\Delta L = \sqrt{\frac{2\epsilon_s}{qN_A}(V_D - V_{DS})} \tag{6.45}$$

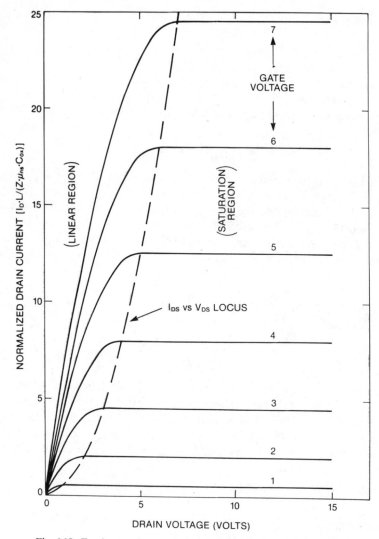

Fig. 6.18. Fundamental current–voltage characteristics of a MOSFET.

The drain output resistance can then be derived:

$$
r_{ds} = 12 \left[L - \sqrt{\frac{2\epsilon_s(V_D - V_{DS})}{qN_A}} \right]^2 \sqrt{V_D - V_{DS}}
$$

$$
\times \left\{ Z\mu_{ns}C_{ox} \sqrt{\frac{2\epsilon_s}{qN_A}} \left[(V_{DS} + 2\phi_B)^2 + 2V_G(V_{DS} + 2\phi_B) \right. \right.
$$

$$
\left. \left. - 12\,\phi_B \left\{ \left(V_G - \phi_B - \frac{4}{3} \sqrt{\frac{\epsilon_s qN_A\phi_B}{C_{ox}^2}} \right) \right\} \right] \right\}^{-1}
\tag{6.46}
$$

A fairly good agreement between the calculated drain output conductance with this expression and the measured output conductance of n-channel devices has been reported. Better results can only be obtained by resorting to a two-dimensional numerical analysis of current transport in these devices [11, 12].

6.3.2.4 On-Resistance.

The on-resistance of a power MOSFET is the total resistance between the source and drain terminals in the on-state. The on-resistance is an important device parameter because it determines the maximum current rating. The power dissipation in the power MOSFET during current conduction is given by

$$P_D = I_D V_D = I_D^2 R_{on} \qquad (6.47)$$

Expressed in terms of the chip area A, this is

$$\frac{P_D}{A} = J_D^2 R_{on,sp} \qquad (6.48)$$

where (P_D/A) is the power dissipation per unit area; J_D is the on-state current density; and $R_{on,sp}$ is the specific on-resistance, defined as the on-resistance per unit area. These expressions are based on the assumption that the power MOSFET is operated in its linear region (low V_D) during current conduction. The maximum power dissipation per unit area is determined by the maximum allowable junction temperature and the thermal impedance. For typical power packages, the maximum power dissipation per unit area is in the range of 100 W/cm^2. The maximum operating current density will thus vary inversely as the square root of the specific on-resistance.

The specific on-resistance of the power MOSFET is determined by several components as illustrated in Fig. 6.11 for the DMOS structure:

$$R_{on} = R_{N+} + R_{ch} + R_A + R_J + R_D + R_S \qquad (6.49)$$

where R_{N+} is the contribution from the N^+-source diffusion, R_{ch} is the channel resistance, R_A is the accumulation layer resistance, R_J is the contribution from the drift region between the P-base regions, R_D is the drift region resistance, and R_S is the substrate resistance. Additional resistances can arise from poor contact between the source–drain metal and the N^+ semiconductor regions as well as from the leads used to connect the device to the package.

The contributions from the N^+ source and drain regions are generally negligible for high-voltage power MOSFETs. For low-voltage devices with breakdown voltages below 50 V, however, these regions can contribute significantly to the on-resistance. This is especially true for the substrate because it must be thick to impart adequate strength to the wafers during device fabrication. For a typical substrate with thickness of 20 mils and resistivity of 0.01 $\Omega\cdot$cm, the substrate resistance per unit area is then given by

$$R_{S,sp} = 0.01 \ \Omega\cdot cm \times 5 \times 10^{-2} \ cm = 5 \times 10^{-4} \ \Omega\cdot cm^2 \qquad (6.50)$$

This resistance can be reduced by using 0.001 $\Omega \cdot$cm arsenic-doped wafers and by lapping the substrate after device fabrication.

To analyze the contribution of the N^+-source resistance, consider the DMOS cell structure shown in Fig. 6.19. In this cell, $2m$ is the cell diffusion window and L_G is the length of the gate electrode between the adjacent cells. The cell diffusion window $2m$ is determined by the photolithographic design tolerances used for device fabrication. These tolerances also determine the N^+-emitter length L_E. If a linear cell is considered with a 1-cm extension perpendicular to the cross section shown in Fig. 6.19, the resistance per square centimeter due to the N^+ emitter regions is given by

$$R_{N^+,\text{sp}} = \frac{1}{2} \rho_{\Box N^+} L_E(L_G + 2m) \tag{6.51}$$

where $\rho_{\Box N^+}$ is the sheet resistance of the N^+ diffusion. In a typical device, $\rho_{\Box N^+} = 10\ \Omega \cdot$cm, $L_E = 10\ \mu$m, and the cell repeat spacing $(L_G + 2m) = 40\ \mu$m. The resulting specific resistance of the N^+-emitter regions $(2 \times 10^{-5}\ \Omega \cdot \text{cm}^2)$ is negligible compared with all other resistances in the structure.

The channel resistance was discussed in the previous section for the basic MOSFET structure. The contribution from the channel depends on the ratio (L/Z), the gate oxide thickness (through C_{ox}), and the gate drive voltage V_G. The contribution from the channel can be minimized by making the channel length L small and keeping its width Z large. This requires a high cell density and good control over the P-base and N^+-emitter diffusion profiles to keep the channel short without causing punch-through breakdown as discussed earlier in Section 6.3.1. To calculate the contribution from the channel, consider the DMOS cell structure shown in Fig. 6.19. The channel resistance per square centimeter for the linear cell structure is given by

$$R_{\text{ch,sp}} = \frac{1}{2} \frac{L(L_G + 2m)}{\mu_{ns} C_{\text{ox}}(V_G - V_T)} \tag{6.52}$$

For a typical device with channel length L of 2 μm, a gate oxide thickness of 1000 Å, and a gate drive of 10 V, the channel resistance per square centimeter is found to be $2.5 \times 10^{-3}\ \Omega \cdot \text{cm}^2$. Note that the channel resistance decreases when the cell repeat spacing is reduced. The channel resistance can also be reduced by decreasing the gate oxide thickness while maintaining the gate drive voltage.

The resistance of the accumulation layer R_A determines the current spreading from the channel into the drift region. The accumulation layer resistance is dependent on the charge in the accumulation layer and the mobility for free carriers at the accumulated surface. For the linear cell geometry, the accumulation layer resistance per square centimeter is

$$R_A = \frac{K(L_G - 2x_p)(L_G + 2m)}{\mu_{nA} C_{\text{ox}}(V_G - V_T)} \tag{6.53}$$

where K is a factor introduced to account for the two-dimensional nature of the current flow from the accumulation layer into the bulk. Good agreement with experimental results has been observed for $K = 3$ [13]. For the above example, with $L_G = 20$ μm and $x_p = 3$ μm, the accumulation resistance per square centimeter is found to be 6×10^{-3} $\Omega \cdot$cm^2. The accumulation layer resistance can be reduced by decreasing the length L_G of the gate electrode between the cells. However, this has an adverse effect on the resistance R_J of the JFET region.

The resistance of the drift region between the P-base diffusions can be calculated if the voltage drop along the vertical direction is neglected. Under the assumption that the depletion layer width of the P-base–N-drift layer junction is negligibly small and the current is flowing uniformly down from the accumulation layer into the JFET region, the resistance of the JFET region can be analyzed as a resistance with increasing cross section when proceeding downward from the surface. The cross section of the JFET region is

$$L_J(x) = L_G - 2\sqrt{x_p^2 - x^2} \tag{6.54}$$

The resistance of a small segment dx of the JFET region per square centimeter is then

$$dR_J = \frac{\rho_D(L_G + 2m)\,dx}{L_J(x)} \tag{6.55}$$

Using Eq. (6.54) and integrating from $x = 0$ to point x_A [13], the total JFET region resistance is obtained:

$$R_J = 2\rho_D(L_G + 2m)\left[\frac{1}{\sqrt{1 - (2x_p/L_G)^2}}\tan^{-1}(0.414)\sqrt{\frac{L_G + 2x_p}{L_G - 2x_p}} - \frac{\pi}{8}\right] \tag{6.56}$$

where ρ_D is the resistivity of the drift region. In deriving this expression, it has been assumed that the JFET region extends to a depth x_A at which angle $\theta = 45°$.

The drift region is assumed to begin beyond this point. The current spreading into the drift region is shown by the dashed lines in Fig. 6.19. The cross section of the drift region increases when proceeding down from the JFET region. For the current spreading at an angle α shown in Fig. 6.19, the drift region resistance per square centimeter is given by [13]

$$R_D = \rho_D \frac{(L_G + 2m)}{\tan(\alpha)} \ln\left[1 + 2\frac{h}{a}\tan(\alpha)\right] \tag{6.57}$$

where the dimensions h and a are indicated in Fig. 6.19. It has been found that a good approximation for α is provided by [13]

$$\alpha = 28° - \left(\frac{h}{a}\right) \qquad \text{if} \quad h \geq a \tag{6.58}$$

and

$$\alpha = 28° - \left(\frac{a}{h}\right) \qquad \text{if} \quad h < a \tag{6.59}$$

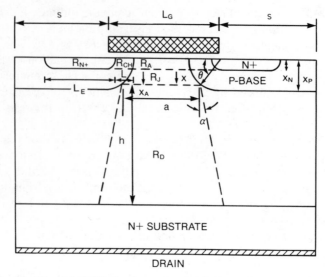

Fig. 6.19. Double-diffused MOSFET cell structure with each component of the on-resistance indicated.

The preceding expression for the drift region resistance is based on the condition of no overlap in the current flow paths. If the drift region thickness is large and the cell window $2m$ is small, the current flow paths will overlap as illustrated in Fig. 6.20. The resistance of the lower region can be calculated by using a uniform current density in this portion. Note that the resistance of the JFET and drift region decrease as the gate length L_G is increased.

Consider the ideal case where the resistances of the N^+-emitter, the N^+ substrate, the n-channel, the accumulation region, and the JFET region are negligible. The specific on-resistance of the power MOSFET will then be determined

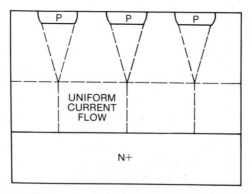

Fig. 6.20. MOSFET cell structure with overlapping current flow in the drift region.

by the drift region. This resistance is identical to that derived for the ideal JFET:

$$\text{(ideal)} \quad R_{on,sp} = 5.93 \times 10^{-9}(BV_{pp})^{2.5} \tag{6.60}$$

for n-channel devices and

$$\text{(ideal)} \quad R_{on,sp} = 1.63 \times 10^{-8}(BV_{pp})^{2.5} \tag{6.61}$$

for p-channel devices.

This treatment is applicable to the DMOS structure. A similar analysis can be performed for the VMOS structure. In this case, no JFET region exists. Instead, the current spreads out from an accumulation layer formed at the tip of the V groove as illustrated in Fig. 6.21. The drift region resistance must now be modified to contain two portions R_{D1} and R_{D2}. The portion at the tip of the V groove has been analyzed [13]:

$$R_{D1,sp} = (0.477)\rho_D(L_G + 2m) \tag{6.62}$$

The resistance due to component R_{D2} is the same as in the case of the DMOS structure for the a and h as illustrated in Fig. 6.21.

Returning to the DMOS device structure, the impact of changing the cell geometry on the on-resistance can best be illustrated by considering some specific cases. Consider the case of a relatively low voltage (100-V) power MOSFET with a drift region thickness W of 4.6 μm and a channel length L of 0.3 μm [14]. The impact of increasing the width of the accumulation region L_G upon the various components of the on-resistance is provided in Fig. 6.22a for two cell window sizes $S = 7$ μm and $S = 30$ μm). It can be seen that the channel and accumulation layer resistances grow as the accumulation layer width L_G increases. Meanwhile, the resistance of the drift region decreases because of improved current spreading. A pronounced minimum in the total on-resistance is observed at an optimum accumulation region width. When the cell

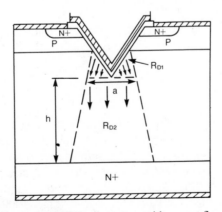

Fig. 6.21. V-Groove MOSFET cell structure with current flow path indicated.

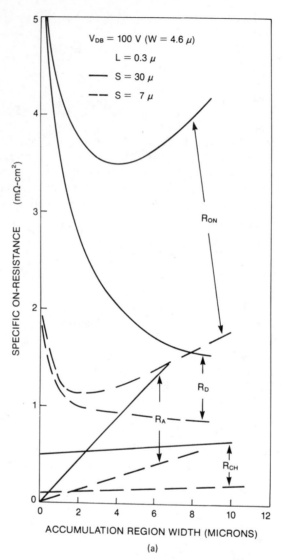

Fig. 6.22. Calculated components of the specific on-resistance for (*a*) 100-V and (*b*) 1000-V power MOSFET structures. (After Ref. 14, reprinted with permission from the IEEE © 1979 IEEE.)

window size is reduced, all components of the resistance undergo a drastic reduction and the point of minimum total resistance shifts to smaller accumulation layer width L_G. This emphasizes the importance of using high-resolution photolithography to fabricate these low-breakdown-voltage power MOSFETs.

Another example of the impact of cell geometry on the on-resistance is provided in Fig. 6.22*b* for the case of a high-voltage (1000 V) device. In this case, the resistance of the drift region becomes predominant and the cell window

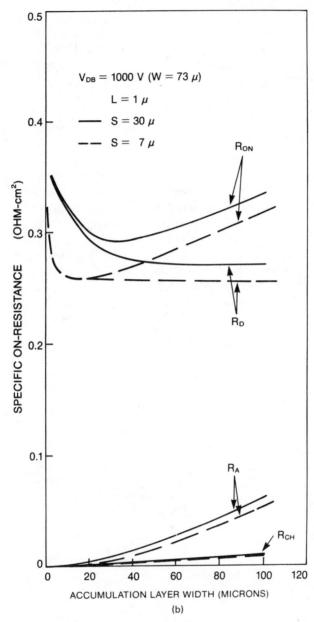

Fig. 6.22. (Continued)

size has only a small impact on the specific on-resistance. It is also significant that the minimum on-resistance is observed to occur at a much larger accumulation layer width L_G than for the low-voltage devices. The high-voltage power MOSFET can, consequently, be fabricated with less stringent photolithographic requirements than in the case of low-voltage devices.

For a given lithographic design rule, power MOSFETs with on-resistances approaching the ideal case can be realized at higher breakdown voltages. The specific on-resistance for state-of-the-art (circa 1981) power MOSFETs are compared with the ideal case in Fig. 6.23 [15]. High-voltage devices with breakdown voltages above 500 V approach within a factor of 2 of the ideal breakdown voltage. In contrast, the low voltage devices with breakdown voltages below 100 V exhibit a specific on-resistance of an order of magnitude higher than the ideal. Recent progress with the application of higher-resolution processing to low-voltage power MOSFETs is improving their on-resistance, making it closer to the ideal case.

The preceding analysis of on-resistance is based on a homogeneously doped drift region. All power MOSFETs are fabricated by using uniformly doped epitaxial layers. It has been proposed that a lower on-resistance can be obtained by using a nonuniform epitaxial doping profile to obtain the same breakdown

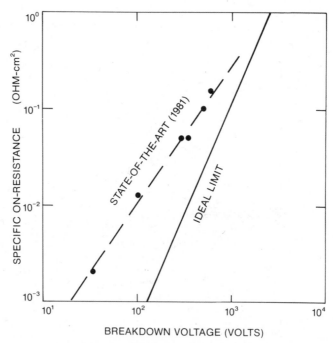

Fig. 6.23. Comparison of actual power MOSFET on-resistances with the ideal case. (After Ref. 15, reprinted with permission from the IEEE © 1981 IEEE.)

voltage [16]. Consider a doping profile $N(x)$ that will minimize the on-resistance of the drift layer

$$R_{\text{on}} = \int \frac{1}{q\mu N(x)} \, dx \tag{6.63}$$

under the condition that the electric field distribution satisfy the breakdown voltage requirement

$$V_{\text{B}} = \int \mathscr{E} \, dx \tag{6.64}$$

Using Poisson's equation to relate the doping and electric field, it can be shown that

$$\frac{d\mathscr{E}}{dx} = \frac{qN}{\epsilon_s} \tag{6.65}$$

These equations can be solved to obtain the optimum doping profile:

$$N(x) = \frac{\epsilon_s \mathscr{E}_{\text{C}}^2}{3qV_{\text{B}} \sqrt{1 - (2\mathscr{E}_{\text{C}} x / 3V_{\text{B}})}} \tag{6.66}$$

A comparison of the optimum profile with the uniformly doped case for the same breakdown voltage is provided in Fig. 6.24. According to this doping profile, the specific on-resistance is

$$R_{\text{on,min}} = \frac{3V_{\text{B}}^2}{\epsilon_s \mu_n \mathscr{E}_{\text{C}}^3} \tag{6.67}$$

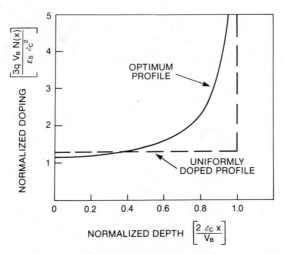

Fig. 6.24. Optimum doping profile for minimizing drift region resistance. (After Ref. 16, reprinted with permission from the IEEE © 1979 IEEE.)

where $\mathscr{E}_C$ is the critical electric field for breakdown. This specific on-resistance is only 12.5% lower than that for the uniformly doped profile. As a result of out-diffusion from the substrate during processing and epitaxial growth with uniformly doped epitaxial layers, the doping profile of low-voltage (100-V) power MOSFETs approaches the ideal profile without the need to take a special effort to tailor the epitaxial layer doping.

6.4. FREQUENCY RESPONSE

The power MOSFET is inherently capable of operation at high frequencies because of the absence of minority carrier transport. Its frequency response can be treated in a manner similar to that used for the power JFET in Chapter 4. Two limits to operation at high frequencies are set by the transit time across the drift region and the rate of charging of the input gate capacitance. The transit time limited frequency response was already treated in Chapter 4. It is a function of breakdown voltage:

$$f_T = \frac{6.11 \times 10^{11}}{(1 + L/d)(BV_{PP})^{7/6}} \tag{6.68}$$

In this formula, L is the channel length and d is the drift region thickness, including the JFET region. To achieve a high frequency response, it is important to keep the channel length small. As in the case of power JFETs, the frequency response diminishes with increasing breakdown voltage. This limit to the frequency response is seldom encountered since it lies in the gigahertz range.

Another limit to high-frequency operation arises from the need to charge and discharge the input gate capacitance. A simple equivalent circuit for the power MOSFET is shown in Fig. 6.25. In addition to the gate–source capacitance, a significant gate–drain capacitance must be included in the analysis because of the overlap of the gate electrode over the drift region. This capacitance is amplified by the Miller effect into an equivalent input gate capacitance of

$$C_M = (1 + g_m R_L) C_{GD} \tag{6.69}$$

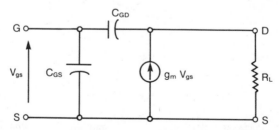

Fig. 6.25. Simple equivalent circuit of the power MOSFET.

The total input capacitance is

$$C_{in} = (C_{GS} + C_M) \qquad (6.70)$$

To analyze the frequency response, it is necessary to evaluate each of the components of the input capacitance arising in the power MOSFET structure. Consider the conventional DMOS structure, shown in Fig. 6.26, with the gate electrode extending between adjacent DMOS cells. The input gate–source capacitance for this structure contains several components: (1) the capacitance C_{N+} arising from the overlap of the gate electrode over the N^+ emitter region, (2) the capacitance C_P arising from the MOS structure created by the gate electrode over the P-base region, and (3) the capacitance C_M arising from running the source metal over the gate electrode. The total gate–source capacitance is given by

$$C_{GS} = C_{N+} + C_P + C_M \qquad (6.71)$$

The capacitance between the source and gate electrodes (C_M) is determined by the dielectric constant and thickness of the intervening insulator

$$C_M = \frac{\epsilon_I A_o}{t_I} \qquad (6.72)$$

where A_o is the area of the overlap between the source and gate electrodes. To reduce this capacitance, a thick insulating layer is generally used during device fabrication. The other components of the capacitance require analysis of the MOS structure.

6.4.1. MOS Capacitance

Consider the basic MOS structure shown in Fig. 6.27. This structure contains two capacitances in series, the oxide capacitance and the semiconductor

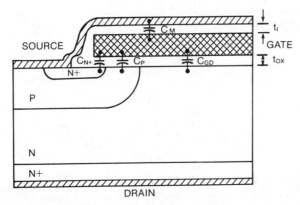

Fig. 6.26. Conventional DMOS structure with capacitances indicated.

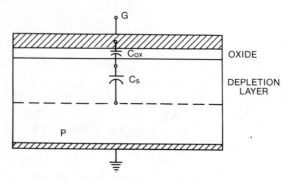

Fig. 6.27. Basic MOS structure with capacitances indicated.

capacitance:

$$\frac{1}{C_G} = \frac{1}{C_{ox}} + \frac{1}{C_S} \tag{6.73}$$

The oxide capacitance per unit area is determined by the thickness of the gate oxide:

$$C_{ox} = \frac{\epsilon_{ox}}{t_{ox}} \tag{6.74}$$

The semiconductor capacitance is determined by the width of the depletion layer. For a P-type substrate, if a negative gate bias is applied, an accumulation layer will form at the surface. Any changes in the gate voltage cause a corresponding change in the accumulation layer charge with a response time corresponding to the dielectric relaxation time. The capacitance is then equal to the oxide capacitance:

$$C_{G,ACC} = C_{ox} \tag{6.75}$$

When a positive bias is applied, a depletion layer forms in the semiconductor. The capacitance of the depletion layer is related to the space charge in the semiconductor:

$$C_S = \frac{dQ_S}{d\psi_S} \tag{6.76}$$

Using Eq. (6.13), it can be shown that

$$C_S = \frac{\epsilon_s}{\sqrt{2}\,L_D} \frac{[1 - e^{-(q\psi_S/kT)} + (n_{p0}/p_{p0})[e^{(q\psi_S/kT)} - 1]]}{F(q\psi_S/kT, n_{p0}/p_{p0})} \tag{6.77}$$

The MOS gate capacitance will decrease with increasing positive gate voltage as a result of widening of the depletion layer. Beyond a certain voltage, a surface inversion layer will form. Once the inversion layer forms, the depletion layer

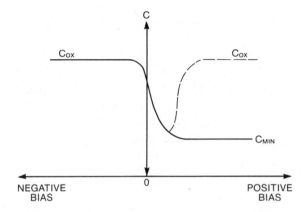

Fig. 6.28. The $C-V$ characteristics of a MOS structure with p-type substrate.

reaches its maximum value and the semiconductor capacitance attains its lowest value. The resulting $C-V$ curve is illustrated in Fig. 6.28 by the solid line. This curve is called the *high-frequency response curve* (typically measured at 1 MHz) because it is assumed that the inversion layer charge cannot follow the AC signal used to measure the capacitance. If either the measurement frequency or the semiconductor lifetime is reduced until the inversion layer charge can respond to the AC signal, the capacitance will become equal to the oxide capacitance as illustrated by the dashed line.

In the case of the power MOSFET, the $C-V$ curve shown by the solid line in Fig. 6.28 is not observed because the inversion layer formation in the P-base region is accompanied by transport of carriers between the N^+ emitter and the inversion layer. Consequently, even at high operating frequencies, the inversion layer charge can respond to the applied high-frequency signal by the rapid transport of electrons from the N^+ emitter into the inversion layer. Consequently, the measured input gate $C-V$ curve of power MOSFETs follows the dashed lines rather than the solid line shown in Fig. 6.28.

6.4.2. Input Capacitance

From the preceding analysis of the MOS capacitance for the gate structure, the various components of the input capacitance can be determined as follows. For the heavily doped N^+-emitter region, the capacitance is determined by the gate oxide thickness

$$C_{N^+} = \frac{\epsilon_{ox} A_{N^+o}}{t_{ox}} \tag{6.78}$$

where A_{N^+o} is the area of overlap of the gate electrode over the N^+ emitter. For the DMOS structure with a channel width Z, the overlap area over the

N^+ emitter is given by

$$A_{N^+o} = X_{N^+}Z \qquad (6.79)$$

where X_{N^+} is the depth of the N^+-emitter diffusion. It is assumed here that the lateral diffusion of the emitter is equal to its depth and that the gate fringing capacitance can be neglected. When taken together, these are reasonable approximations because they compensate for each other.

The capacitance of the gate over the P base (C_P) is dependent on the gate bias. It goes through a minimum with increasing gate bias as shown in Fig. 6.28 by the dashed lines. This capacitance can be reduced by keeping the channel length small. It represents a fundamental contribution to the input capacitance of the power MOSFET.

The gate–drain capacitance also varies with gate and drain voltage. It has a high value during the on-state because the surface of the drift region is under accumulation. As the drain voltage increases and the device supports high voltages, the gate–drain capacitance decreases. As a result of the amplification of the gate–drain capacitance by the Miller effect, this capacitance can severely reduce the frequency response.

The gate–drain overlap capacitance can be drastically reduced by altering the gate structure by eliminating the gate overlap over the drift region as shown in Fig. 6.29. In this structure, the source electrode is also confined to the diffusion window to eliminate the capacitance C_M. A significant improvement in the high-speed switching performance of power MOSFETs has been reported by taking these measures during device design and fabrication [6]. It is noteworthy that elimination of the gate–drain overlap does adversely impact the on-resistance.

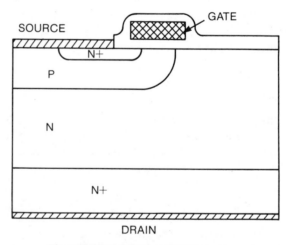

Fig. 6.29. High-frequency DMOS structure.

6.4.3. Gate Series Resistance

In the power MOSFET equivalent circuit shown in Fig. 6.25, no input gate series resistance was assumed to exist. In practice, a series gate resistance is always present, arising from either the resistance of the gate electrode internal to the device or from the gate drive circuit. The frequency response limited by the RC charging time constant of the input gate circuit is given by

$$f_{in} = \frac{1}{2\pi C_{in} R_G} \tag{6.80}$$

In the case of conventional polysilicon gate power MOSFETs, the sheet resistance of the gate electrode is typically over $10\ \Omega/\square$. This has been found to seriously lower the frequency response [17]. By changing to a molybdenum gate structure, an increase in frequency response by an order of magnitude has been reported, demonstrating that this is the limiting factor to device high-frequency performance. Power MOSFETs capable of delivering 100 W at 900 MHz have been developed by using a molybdenum silicide gate process [18].

The maximum frequency of operation at which the input current becomes equal to the load current is given by

$$f_m = \frac{g_m}{2\pi C_{in}} \tag{6.81}$$

This limit to operation is generally avoided because of the high power dissipation in the gate circuit.

6.5. SWITCHING PERFORMANCE

Because of the inherent high-speed turn-on and turn-off capability of power MOSFETs, these devices are often used as power switches. When used in this manner, they are maintained in either the on-state or forward blocking state for most of the time and must rapidly switch between these modes of operation. The power dissipation is determined by the on-resistance during the conducting state and the leakage current in the forward blocking state. These parameters have been discussed earlier. In this section, the parameters that govern the transition between the on and off states are analyzed. In addition, high-speed switching is accompanied by a very high rate of change of the drain voltage, which can cause an undesirable turn-on of the parasitic bipolar transistor in the power MOSFET. This can limit the switching speed and safe operating area of power MOSFETs.

6.5.1. Transient Analysis

A simple analytical procedure for predicting the switching transients experienced with power MOSFETs can be achieved by using the circuit model shown in Fig. 6.30 [19]. As a result of the interaction between the device and the circuit,

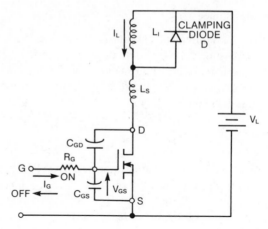

Fig. 6.30. Inductive switching circuit using power MOSFET for controlling load current.

it is necessary to consider a specific type of load for analysis. Here a clamped, inductive load L_1 is considered with a steady-state current I_L flowing through it. The inductance L_S is the stray circuit inductance not clamped by the diode D.

First, consider the case of a step voltage V_{GA} applied at the gate terminal G. The gate voltage V_{GS} that controls the drain current flow is determined by the voltage across the gate–source capacitance C_{GS}. This voltage is governed by the charging of the capacitances C_{GS} and C_{GD} by resistor R_G. As long as the gate voltage V_{GS} remains below the threshold voltage V_T, no drain current will flow. The time taken for the gate voltage V_{GS} to reach the threshold voltage V_T represents a turn-on delay period. During this period, the input capacitance is simply $(C_{GS} + C_{GD})$ because the Miller effect is operative only when the transistor is in its active region. The gate voltage then rises exponentially:

$$V_{GS} = V_{GA}\{1 - e^{-[t/R_G(C_{GS}+C_{GD})]}\} \tag{6.82}$$

The turn-on delay time obtained from this expression is

$$t_1 = R_G(C_{GS} + C_{GD}) \ln\left[\frac{1}{1 - (V_T/V_{GA})}\right] \tag{6.83}$$

Drain current flow begins to occur beyond the turn-on delay time. Consider the case of a MOSFET with linear transfer characteristics as shown in Fig. 6.31. In this case, the drain current will increase in proportion to the gate voltage. Since the device is now in its active region, the Miller effect will determine the gate–drain capacitance that is being charged by the gate circuit. If the stray inductance L_S is small, the Miller effect is small and the gate voltage will continue to rise exponentially as described by Eq. (6.82). The drain current will then take the form

$$I_D = g_m\{V_{GA}[1 - e^{-[t/R_G(C_{GS}+C_{GD})]}] - V_T\} \tag{6.84}$$

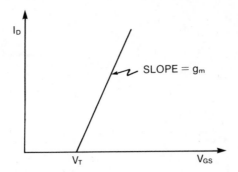

Fig. 6.31. Linear transfer characteristics of a power MOSFET used for transient analysis.

The drain voltage will be nearly constant during this time. This period will end when the drain current reaches the load current I_L, that is, when all the circulating diode current is transferred to the power MOSFET. Beyond this point the drain voltage will fall from V_L to the on-state voltage of the power MOSFET. Since the drain current is now constant, the gate voltage must be

$$V_{GS} = V_T + \frac{I_L}{g_m} \tag{6.85}$$

Since the gate voltage V_{GS} is constant, all the input current flows into the Miller capacitance C_{GD} during this period. The input current is given by

$$I_G = \frac{V_G - V_{GS}}{R_G} \tag{6.86}$$

Thus

$$\frac{dV_{GD}}{dt} = \frac{I_G}{C_{GD}} = \frac{V_G - (V_T + I_L/g_m)}{R_G C_{GD}} \tag{6.87}$$

The rate of change of the drain–source voltage is equal to the rate of change of the gate–drain voltage because the gate–source voltage is constant during this period. Thus

$$\frac{dV_D}{dt} = \frac{V_G - (V_T + I_L/g_m)}{R_G C_{GD}} \tag{6.88}$$

Integration of this equation yields

$$V_D = V_L - \left[\frac{V_G - (V_T + I_L/g_m)}{R_G C_{GD}}\right] t \tag{6.89}$$

The drain voltage will then decrease linearly with time. The composite gate and drain waveforms are shown in Fig. 6.32. Note that the gate voltage may continue to rise beyond time t_3, but this will have no influence on the drain current or voltage because they have reached their steady-state values.

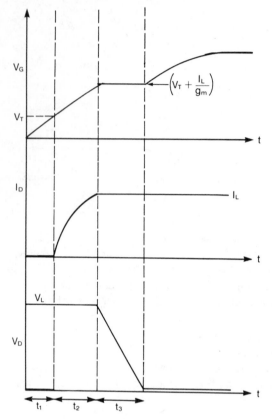

Fig. 6.32. Turn-on transients of a power MOSFET for the case of small stray inductance [low L_S/R_G) ratio].

The preceding analysis is valid only when the ratio (L_S/R_G) is small. Other cases can be treated in a similar manner [19]. The high-power dissipation due to the device turn-on transient occurs during the interval t_2 and t_3, where both high current and voltage are sustained by the device simultaneously. During this period the device current–voltage locus must stay within its safe operating area.

A corresponding analysis can be performed for device turn-off. In this case, an abrupt drop in applied gate voltage from V_G to zero is assumed to occur at the terminal G. The voltage V_{GS} then decreases exponentially with time as a result of discharging of the gate capacitance, until the gate voltage reaches the value required to obtain a saturated drain current equal to the load current I_L. During this period

$$V_{GS} = V_G \, e^{-(t/R_G C_G)} \tag{6.90}$$

The duration of this turn-off delay phase is given by the condition

$$V_{GS}(t_4) = V_T + \frac{I_L}{g_m} = V_G\, e^{-(t_4/R_GC_G)} \tag{6.91}$$

From this equation, the turn-off delay time is obtained:

$$t_4 = R_GC_G \ln\left[\frac{V_G}{V_T + (I_L/g_m)}\right] \tag{6.92}$$

Beyond this time, the drain current will remain at I_L whereas the drain voltage will begin to rise toward V_L because the load current cannot be diverted from the MOSFET into the diode until V_D exceeds V_L. As long as the drain current is constant, the gate voltage V_{GS} will also remain constant and the gate current I_G will drive the gate–drain capacitance C_{GD}. During this period

$$V_{GS} = V_T + \frac{I_L}{g_m} \tag{6.93}$$

and

$$I_G = \frac{V_{GS}}{R_G} = \frac{V_T + (I_L/g_m)}{R_G} \tag{6.94}$$

Since this current charges the gate–drain capacitance C_{GD} and the rate of change of drain–source voltage is equal to the rate of change of the drain–gate voltage, it follows that

$$\frac{dV_{DS}}{dt} = \frac{dV_{DG}}{dt} = \frac{I_G}{C_{GD}} \tag{6.95}$$

Thus

$$V_{DS} = V_{on} + \frac{I_G}{C_{GD}}\, t \tag{6.96}$$

$$= V_{on} + \frac{1}{R_GC_{GD}}\left(\frac{I_L}{g_m} + V_T\right)t \tag{6.97}$$

The drain voltage will rise linearly from the on-state voltage drop V_{on} to the load voltage during this period. The duration of this period can be obtained by equating V_{DS} to V_L:

$$t_5 = \frac{R_GC_{GD}(V_L - V_{on})}{(I_L/g_m) + V_T} \tag{6.98}$$

At this point the freewheeling diode D will turn on. To achieve this, it is essential that the MOSFET drain voltage exceed the load supply voltage V_L. If the stray inductance L_S is small, the overshoot in the drain voltage will be small. However, if L_S is large and the rate of change of the drain current becomes large, the voltage on the MOSFET drain can exceed its breakdown voltage. If the stray inductance is small, the drain voltage can be assumed to remain relatively constant. The gate voltage will then continue to decrease exponentially:

$$V_G = \left(\frac{I_L}{g_m} + V_T\right) e^{-(t/R_G C_G)} \tag{6.99}$$

The drain current will follow this change in the gate voltage:

$$I_D = (I_L + g_m V_T) e^{-(t/R_G C_G)} - g_m V_T \tag{6.100}$$

This period will extend until the gate voltage reaches the threshold voltage V_T and the drain current is reduced to zero. The time interval for this period is

$$t_6 = R_G C_G \ln\left(\frac{I_L}{g_m V_T} + 1\right) \tag{6.101}$$

After this interval, the gate voltage will continue to decay exponentially to zero. Since the drain current and voltage are at their steady-state voltages (off state), this will have no influence on them.

The composite waveforms corresponding to the preceding discussion of turn-off are shown in Fig. 6.33. These waveforms apply to the case of a small ratio (L_S/R_G). Other cases can be treated in a similar manner [19]. The power dissipation during the turn-off transient occurs primarily during the periods t_5 and t_6, where both the current and the voltage are large. Once again the current–voltage locus must remain within the safe operating area of the device. The drain voltage overshoot during period t_6 must especially be noted. Its magnitude depends on the size of the stray inductance. If the stray inductance is large, the MOSFET can be forced into avalanche breakdown, causing destructive failure. The analysis presented here also indicates the need to keep the gate–drain capacitance small. This can be achieved by avoiding the overlap of the gate electrode with the drift region as shown in Fig. 6.29 or by using a thick oxide in the portion where the drift region comes to the surface.

6.5.2 (dV/dt) Capability

It is well known that thyristors are prone to turn-on under the application of a high rate of change of the anode voltage. When a high (dV/dt) occurs at the drain terminal, power MOSFETs can also be forced into current conduction. In certain cases, this will lead to destructive failure of the devices. The various mechanisms that can lead to (dV/dt)-induced turn-on are discussed here. Circuit and device design approaches to raising the (dV/dt) are provided.

To analyze the (dV/dt)-induced turn-on in power MOSFETs, consider the equivalent circuit of the device shown in Fig. 6.34 with a ramp applied between drain and source [20]. In addition to the device capacitances, the equivalent circuit shows the parasitic N–P–N bipolar transistor with a base–emitter

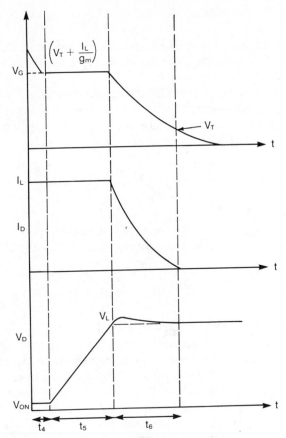

Fig. 6.33. Turn-off transients of a power MOSFET for the case of small stray inductance [low L_S/R_G) ratio].

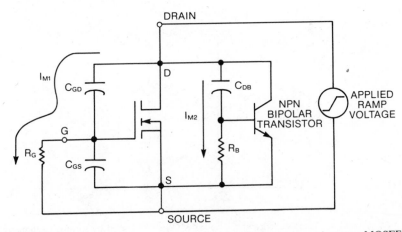

Fig. 6.34. Equivalent circuit used for analysis of (dV/dt)-induced turn-on in power MOSFETs.

shunting resistance R_B. This shunting resistance is present in the device structure as a result of the overlapping of the source metal over both the N^+-emitter
and the P-base regions.

    Mode 1:   For the first mode of (dV/dt)-induced turn-on, consider the current
I_{M1} flowing by means of the gate-to-drain capacitance C_{GD} into the gate circuit
resistance R_G. If the drain voltage is much larger than the gate voltage, the
voltage drop across the gate resistance will be approximately given by

$$V_{GS} = I_{M1}R_G \tag{6.102}$$

$$= R_G C_{GD} \left[\frac{dV}{dt} \right] \tag{6.103}$$

If the gate voltage V_{GS} exceeds the threshold voltage of the power MOSFET, the
device will be forced into current conduction. The (dV/dt) capability set by this
mode is

$$\left[\frac{dV}{dt} \right] = \frac{V_T}{R_G C_{GD}} \tag{6.104}$$

The (dV/dt) capability can be increased by using a very-low-impedance gate
drive circuit and raising the threshold voltage. Since the threshold voltage decreases with increasing temperature, this mode of turn-on can become aggravated with increasing power dissipation within the device. In general, it is nondestructive because the gate voltage does not rise much above the threshold
voltage and the device current is limited by the high device resistance.

    Mode 2:   This mode of (dV/dt)-induced turn-on occurs as a result of the
existence of the parasitic bipolar transistor. At high rates of change of the drain
voltage, a current I_{M2} flows by means of the capacitance C_{DB} into the base short-

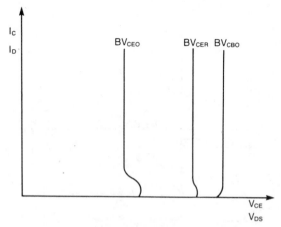

Fig. 6.35. Breakdown characteristics of the power MOSFET induced by turn-on of the parasitic
$N–P–N$ bipolar transistor.

circuiting resistance R_B. If this current is sufficient for the base–emitter junction of the bipolar transistor to become forward-biased, it will turn on the transistor. For low values of R_B, when the emitter junction is inactive, the breakdown voltage of the transistor, and hence the power MOSFET, approaches the collector–base breakdown voltage BV_{CBO} as shown in Fig. 6.35. But under a high (dV/dt) and large values of R_B, the emitter–base junction becomes strongly forward-biased. The breakdown voltage then collapses to the level of the open-base transistor breakdown voltage BV_{CEO}, which is about 60% of the collector–base breakdown voltage BV_{CBO}. If the applied drain voltage is greater than BV_{CEO}, the device will enter avalanche breakdown or may be destroyed by second breakdown if the drain current is not externally limited.

The (dV/dt)-induced turn-on in this mode is dependent on the internal device structure. Consider the DMOS structure shown in Fig. 6.36 with one end of the N^+ emitter short-circuited to the P-base region. The applied (dV/dt) creates a displacement current flow in capacitance C_{DB}. This current must flow laterally in the P-base region to the source metal. It produces a voltage drop along the resistance R_B that forward-biases the edge A of the N^+ emitter. When the edge A of the N^+ emitter becomes forward-biased by the lateral current flow, the bipolar transistor will turn on, thus precipitating further current flow. An estimate of the (dV/dt) at which this will occur can be obtained by assuming that the bipolar transistor will turn on at an emitter–base forward-bias voltage of V_{BE}:

$$\frac{dV}{dt} = \frac{V_{BE}}{R_B C_{DB}} \qquad (6.105)$$

where R_B is a distributed base resistance. To obtain a high (dV/dt) capability, it is important to keep the resistance R_B small. This can be achieved by increasing the doping level of the P base and keeping the length L_{N^+} of the N^+ emitter

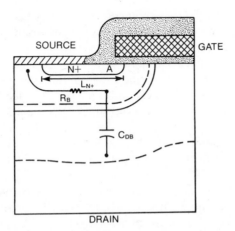

Fig. 6.36. Cross section of power MOSFET with capacitance C_{DB} and base resistance R_B shown.

as small as possible within the constraints of the lithography used for device fabrication.

It should be noted that the P-base sheet resistance increases with increasing drain voltage as result of the extension of the depletion layer, resulting in lowering of the dV/dt capability. In addition, increasing the temperature will lower the voltage V_{BE} at which the emitter will begin to inject. The P-base resistance will also increase with temperature because of a reduction in the mobility. These factors will cause a reduction in the (dV/dt) capability as temperature increases. Experimental confirmation of this mode of (dV/dt)-induced breakdown has been obtained on special structures, where the P base was not short-circuited to the N^+ emitter, allowing the use of large external base resistance R_B to make the effect pronounced [21]. In the case of practical power MOSFETs designed for operation at high switching speeds, the (dV/dt) capability is over 10,000 V/μsec.

6.6. SAFE OPERATING AREA

The safe operating area defines the limits of operation of the device. It is well known that the maximum current at low drain voltages is limited by power dissipation if the leads are sufficiently rated to prevent fusing. The maximum voltage at low drain currents is determined by the avalanche breakdown phenomena as discussed in Section 6.3.1. Under the simultaneously application of high current and voltage, the device may be susceptible to destructive failure even if the duration of the transient is too small to prevent excessive power dissipation. This failure mode has been referred to as *second breakdown*.

6.6.1. Bipolar Second Breakdown

The term *second breakdown* refers to a sudden reduction in the blocking voltage capability when the drain current increases. This phenomenon has been observed in power MOSFETs. It originates from the presence of the parasitic bipolar transistor in the device structure. When the drain voltage is increased to near the avalanche breakdown voltage, current flows into the P-base region in addition to the normal current flow within the channel inversion layer. The avalanche current collected within the P-base flows laterally along the P-base to its contact. The voltage drop along the P base forward-biases the edge of the N^+ emitter furthest from the base contact. When the forward bias on the emitter exceeds 0.6–0.7 V, it begins to inject carriers. The parasitic bipolar transistor is no longer capable of supporting the P-base–N-drift layer breakdown voltage BV_{CBO}. Its breakdown voltage is instead drastically reduced to BV_{CEO}, which is typically 60% of BV_{CBO}.

To analyze second breakdown in power MOSFETs, consider the equivalent circuit of the device with the parasitic bipolar transistor as shown in Fig. 6.37a. A base–emitter resistor R_B is shown in this figure corresponding to the lateral resistance in the P-base region illustrated in Fig. 6.37b. The current flow in the

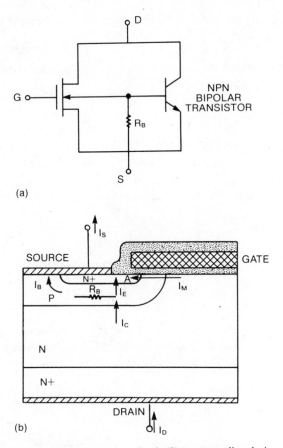

Fig. 6.37. (*a*) Power MOSFET equivalent circuit; (*b*) corresponding device cross section.

device takes two paths—one through the MOS channel and the other through the active bipolar transistor. These currents must satisfy the following conditions:

$$I_D = I_C + I_M \tag{6.106}$$

$$I_S = I_E + I_M \tag{6.107}$$

and

$$I_B = I_C - I_E \tag{6.108}$$

The emitter and collector currents are related by

$$I_C = \gamma_E \alpha_T M I_E \tag{6.109}$$

where γ_E is the injection efficiency, α_T is the base transport factor, and M is the avalanche multiplication factor. The first two parameters can be assumed to be equal to unity for a typical power MOSFET structure. Furthermore, during second breakdown, the emitter current is caused by the forward bias at point A:

$$I_E = I_0\, e^{(qV_B/kT)} \tag{6.110}$$

where V_B is the forward bias caused by the lateral base current flow:

$$V_B = R_B I_B \tag{6.111}$$

Combination of these equations gives

$$I_E = I_0 \exp\left[\frac{qR_B}{kT}(M-1)I_E\right] \tag{6.112}$$

If the first-order expansion of the exponential is used to evaluate the initiation of the second breakdown effect, it can be shown that

$$I_E = \frac{I_0}{[1 - (qR_B/kT)(M-1)I_0]} \tag{6.113}$$

The multiplication factor is related to the drain voltage by

$$M = \frac{1}{[1 - (V_D/BV)^n]} \tag{6.114}$$

with $n \simeq 4$ for electrons. As the drain voltage increases, the emitter current, and hence the source current, can rise catastrophically in accordance with Eq. (6.113). The voltage at which this will occur can be obtained from the preceding equations:

$$V_{D,SB} = \frac{BV}{[1 + qR_B I_0/kT]^{1/n}} \tag{6.115}$$

A reduction of the second breakdown voltage is predicted with increasing base resistance R_B by this analysis.

6.6.2. MOS Second Breakdown

Another phenomenon that causes second breakdown occurs because of the effect of the lateral voltage drop in the P base on the channel current. This is called the *MOSFET body bias effect* [22]. In this case, it is necessary to define a body bias coefficient:

$$\gamma = \frac{\Delta I_D}{\Delta V_B} \tag{6.116}$$

where

$$V_B = R_B I_B \tag{6.117}$$

The source current is then given by

$$I_S = I_M + \gamma V_B = I_M + \gamma R_B I_B \tag{6.118}$$

At high drain voltages, the large electric field causes avalanche multiplication of the channel current. The base current is now given by

$$I_B = I_D - I_S = (M - 1)I_S \tag{6.119}$$

Using Eq. (6.118), it can be shown that

$$I_B = (M - 1)(I_M + \gamma R_B I_B) \tag{6.120}$$

and from this equation

$$I_B = \frac{(M - 1)I_M}{1 - \gamma R_B(M - 1)} \tag{6.121}$$

and

$$I_S = \frac{I_M}{1 - \gamma R_B(M - 1)} \tag{6.122}$$

which is of the same form as Eq. (6.113) derived earlier. As the drain voltage increases, the multiplication factor increases and causes a catastrophic increase in the source current. The drain voltage at which this second breakdown occurs is

$$V_{D,SB} = \frac{BV}{(1 + \gamma R_B)^{1/n}} \tag{6.123}$$

Once again a reduction of the second breakdown voltage occurs with increasing base resistance. An exaggerated example of a power MOSFET I–V characteristic is shown in Fig. 6.38 with the safe operating area indicated.

As temperature increases, the breakdown voltage BV and the coefficient n will increase. This will tend to raise the second breakdown voltage. However,

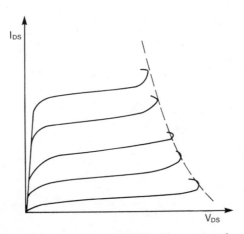

Fig. 6.38. Output characteristics of a power MOSFET with exaggerated second breakdown limit shown by the dashed lines.

the P-base resistance R_B also increases with temperature as a result of a reduction in the mobility. This has a compensating effect. It has been found that the second breakdown voltage and current are only weak functions of temperature [22]. In addition, p-channel power MOSFETs exhibit a smaller collapse in voltage in comparison with n-channel devices because of the lower N-base resistance and the lower hole impact ionization coefficient.

In conclusion, unless the emitter–base junction of the parasitic bipolar transistor in the power MOSFET structure is adequately short-circuited, the devices can exhibit second breakdown. Although this problem was observed in early devices, advances in design and process technology have eliminated this problem in modern devices.

6.6.3. Modern Device Characteristics

Because of the rugged design with very efficient short-circuiting of the emitter–base junction of the parasitic bipolar transistor, modern commercial power MOSFETs are rated with excellent safe operating area. As an example, the safe operating area of a typical 500-V power MOSFET [23] is shown in Fig. 6.39. The curves shown in this figure are simply based on a power dissipation that will keep the junction temperature below 150°C. The ability of a power

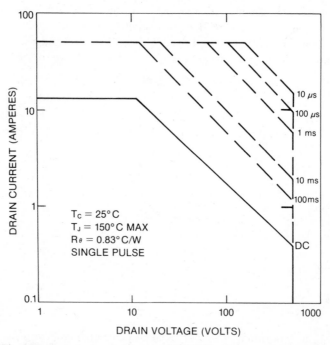

Fig. 6.39. Typical commercial power MOSFET safe-operating-area limits.

MOSFET to operate along these constant power contours substantiates the absence of second breakdown, by adequate short-circuiting of the internal parasitic bipolar transistor, in modern commercially available devices.

6.7. INTEGRAL DIODE

Some power switching circuits require reverse current flow through the active power switching devices. Even though the power switching devices do not have to exhibit any reverse blocking capability, in these applications they must be capable of reverse conduction. Some examples of these types of circuit are DC to AC inverters for adjustable-speed motor drives, switching power supplies, and DC to DC choppers for motor speed control with regenerative breaking. When using bipolar transistors it is necessary to use a diode connected across the transistor (often called an *antiparallel diode*) to conduct the reverse current. This diode must exhibit good reverse recovery characteristics with a small reverse recovery charge to keep the power dissipation low and a reverse recovery waveform that does not contain an abrupt change in current (i.e., snap-off). Any abrupt current changes create very high transient voltages that can damage the power devices.

When the power MOSFET is used in place of the bipolar transistor as the power switching device, the possibility of utilizing the reverse conducting diode inherent in the structure becomes attractive because it eliminates the complexity and cost of adding an external diode. The existence of the integral reverse conducting diode is illustrated in Fig. 6.40. This diode is formed across the

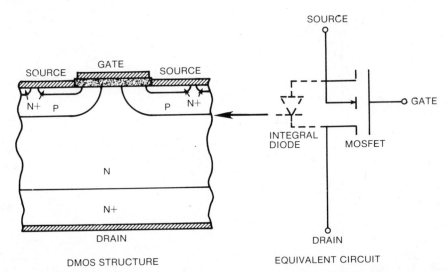

Fig. 6.40. Power MOSFET structure with integral diode formed by *P*-base and *N*-drift layer junction.

P-base–N-drift layer junction with its anode current flowing through the source contact to the P-base region. It should be noted that the applications under consideration here require high-voltage operation. The power MOSFET forward drop in these cases exceeds the junction knee voltage discussed earlier with reference to Fig. 6.3, and it is not practical to gate the power MOSFET and utilize it as a synchronous rectifier.

In the as-fabricated form, the lifetime in the N-drift region of power MOSFETs is generally high. The integral diode can be treated as a $P–i–N$ rectifier between the source and drain terminals. It will conduct current very efficiently with a forward voltage drop in the range of 1 V as the result of the injection of a high concentration of minority carriers into the drift region. The theory of current conduction in a $P–i–N$ rectifier has been discussed in Chapter 5. In the case of the power MOSFET integral diodes, the maximum current carried by the diode must match the power MOSFET rated current because in these applications, the current in one MOSFET invariably flows through the integral diode of another device in the circuit. Because of the high on-resistance of high-voltage power MOSFETs, their forward conduction current density is limited to below 100 A/cm². At these relatively low current densities for the integral diode, the forward voltage drop will be low (about 1 V). It can, therefore, easily sustain the current requirements of the intended applications.

The primary difficulty with using the integral diodes of power MOSFETs is with their reverse recovery characteristics. Because of the high lifetime in the drift region of as-fabricated devices, these diodes exhibit very slow reverse recovery with large recovery charge, so that a very large reverse recovery current is observed. The peak reverse recovery current I_p will increase with increasing (di/dt) (i.e., with higher switching speed) as illustrated in Fig. 6.41. Unfortunately,

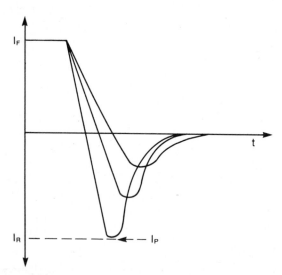

Fig. 6.41. Increase of peak reverse recovery current of $P–i–N$ rectifier with increasing (di/dt).

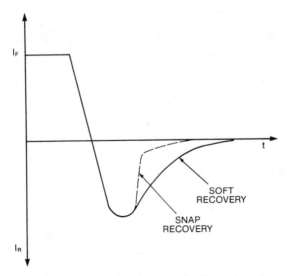

Fig. 6.42. Comparison of the snap-reverse recovery of a diode with soft reverse recovery.

this current flows through the transistors in the circuit imposing power dissipation and thermal stress on them. In addition, some diodes can go through the reverse recovery with an abrupt change in current. This behavior is called *snappy recovery* and is compared with the preferred soft recovery characteristics in Fig. 6.42. The high rate of change of current during the snap–reverse recovery generates a very high voltage ($L_S \, di/dt$) across the stray inductance L_S, which was discussed in Section 6.5.1. This voltage appears across the power MOSFET terminals and can take the device outside its safe operating area.

6.7.1. Control of Switching Speed

Improvement of the reverse recovery characteristics of the integral diode in power MOSFETs was first accomplished by using electron irradiation [24]. Electron irradiation can be used to introduce recombination centers into the drift region as discussed in Chapter 2. During reverse recovery, the resulting lifetime reduction can greatly reduce both the reverse recovery charge and the recovery time. The improvement that can be achieved by using the electron irradiation process is illustrated for a typical device in Fig. 6.43. An important feature of the process is the excellent control over the reverse recovery characteristics obtained by the ability to choose the radiation dose precisely. The reverse recovery time can be tailored from 350 nsec for the as-fabricated devices to less than 100 nsec at a dose of 16 Mrads. Since even good discrete $P-i-N$ rectifiers exhibit a reverse recovery time of over 200 nsec, the radiation process can improve the power MOSFET integral diode to the level that it can replace the external flyback diode. Another important observation with regard to the

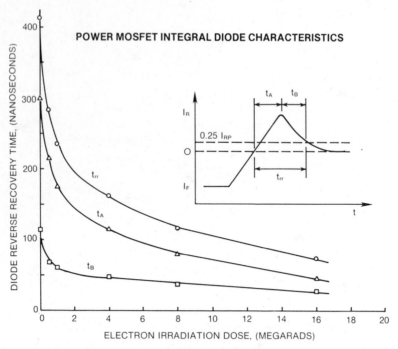

Fig. 6.43. Measured improvement in reverse recovery time with electron irradiation. The inset shows the reverse recovery waveform with the times t_A and t_B indicated. (After Ref. 24, reprinted with permission from *Solid State Electronics.*)

electron irradiation process is that the ratio of the time intervals for the two segments (t_A/t_B) shown in the inset of Fig. 6.43 remains approximately constant at a value of less than 2 for all radiation doses. This is an indication that no snappy recovery phenomenon occurs as the diode speed is increased by the electron irradiation.

It should be noted that the electron irradiation process introduces a positive charge in the gate oxide [24]. This causes a reduction in the threshold voltage of *n*-channel devices and an increase in the threshold voltage of *p*-channel devices. It has been observed that the threshold voltage can become negative for *n*-channel devices after electron irradiation, causing a drastic reduction in the forward blocking capability. Fortunately, the oxide charge due to electron irradiation can be annealed out at relatively low temperatures (150–200°C). At these temperatures, the recombination centers in the bulk that improve the reverse recovery of the integral diode are stable [25]. Other power MOSFET parameters remain essentially unaffected by the electron irradiation process. Consequently, it is an excellent method for obtaining power MOSFETs with high-quality integral diodes because electron irradiation, and subsequent annealing, can be performed after complete device fabrication in a highly controlled manner.

The speed of the integral diode of power MOSFETs can also be improved by gold or platinum doping [26]. A problem observed when using this process is an increase in the on-resistance at high gold and platinum diffusion temperatures. This is caused by compensation because of deep levels arising from the gold or platinum doping. The compensation becomes apparent when the concentration of the deep level becomes comparable to the background doping level. Since the capture cross section for the platinum deep level responsible for minority carrier lifetime reduction is larger than for gold, it produces less compensation for a given reverse recovery speed. Experimental comparison of gold and platinum doping [26] indicates that platinum doping will provide a 50% lower reverse recovery time when compared with gold for the same on-resistance. It is worth pointing out that the gold or platinum doping processes must be performed during device fabrication prior to contact metallization. This provides much less control over the minority carrier lifetime compared with electron irradiation leading to a wider spread in device characteristics. Furthermore, the platinum and gold atoms tend to segregate to regions of the device with strain such as the N^+ substrate and under the gate oxide. This phenomenon can severely degrade the MOS characteristics. The absence of the segregation effect in the case of electron irradiation makes the compensation effect negligible.

6.7.2. Bipolar Transistor Conduction

A potential problem with the use of the integral diode in the power MOSFET arises from the existence of the bipolar transistor in the structure. The equivalent circuit for the integral diode contains this bipolar transistor as shown in Fig. 6.44. The primary diode current flow path is shown by I_D in this figure. This current flow produces a voltage drop across the base resistance R_B that forward-biases the emitter–base junction of the bipolar transistor. If the base resistance

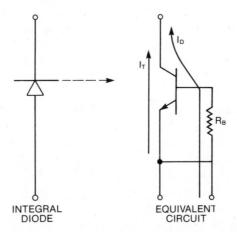

INTEGRAL
DIODE

EQUIVALENT
CIRCUIT

Fig. 6.44. Equivalent circuit of the integral diode showing the bipolar transistor.

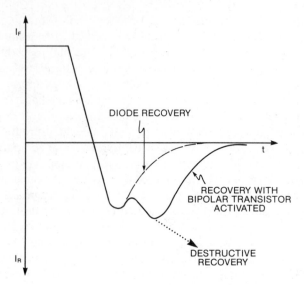

Fig. 6.45. Reverse recovery waveforms observed when the parasitic bipolar transistor is activated.

is not sufficiently low, the bipolar transistor will turn on and a significant current can flow along path I_T because even in the inverse mode the bipolar transistor has some gain. When the diode in the structure goes through its reverse recovery, the high voltage across the transistor can result in its going into second breakdown, leading to catastrophic failure.

The activation of the bipolar transistor during reverse recovery of the integral diode can be observed in the reverse recovery waveform. A typical example of the reverse recovery current waveform with an activated bipolar transistor is provided in Fig. 6.45. If second breakdown occurs, the recovery will proceed according to the dotted line shown in Fig. 6.45, leading to destructive failure. This mode of failure is aggravated by the voltage surge caused by snap recovery of the integral diode because this not only produces a high voltage at the drain but imposes a high (dV/dt) across the device terminals. This high (dV/dt) can precipitate further bipolar action as discussed in Section 6.5.2, leading to destructive failure.

6.8. NEUTRON RADIATION TOLERANCE

Bipolar transistors exhibit severe degradation in electrical performance after being subjected to neutron radiation. In these bipolar devices, the severe reduction in the minority carrier lifetime as a result of the neutron radiation damage, destroys the current gain and drift region modulation. In comparison, power MOSFETs are majority carrier devices for which the minority carrier lifetime is of consequence only in terms of the leakage current. Consequently, these devices can be expected to exhibit much superior neutron radiation tolerance.

6.8.1. On-Resistance

The influence of high energy or fast neutron irradiation on the on-resistance of two typical power MOSFETs is shown in Fig. 6.46 [23]. The on-resistance remains relatively constant until the dose exceeds 10^{13} neutrons/cm^2. In contrast, even a highly neutron radiation resistant bipolar device, such as the field-controlled diode, exhibits a sharp increase in forward voltage drop at neutron influences about 10 times lower than those observed for the power MOSFETs. The reason for this difference in performance is because the on-resistance of power MOSFETs increases only when the deep levels caused by the neutron damage begins to compensate the donors in the drift region and raise its resistivity. The deep-level concentration required to achieve compensation is comparable to the drift region doping level. As a result, power MOSFETs with lower breakdown voltage will exhibit higher neutron radiation tolerance as shown in Fig. 6.46 because their drift region is more heavily doped. In contrast, for a bipolar device such as the FCD, the forward drop will increase sharply when the lifetime (diffusion length) begins to decrease below a critical value. This occurs at neutron fluences well below the point at which compensation is significant.

The observed increase in the power MOSFET on-resistance due to neutron radiation produces an increase in conduction losses for power switching application but has little consequence in linear applications. The sizing of the device

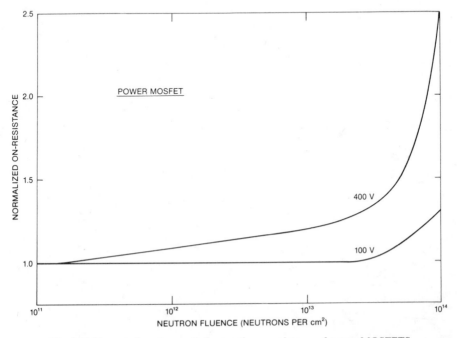

Fig. 6.46. Impact of neutron radiation on the on-resistance of power MOSFETS.

and its heak sink must be initially designed to ensure operation within the maximum safe operating area after radiation. In linear applications, the transconductance is of greater importance. This parameter is affected only when the mobility begins to degrade. The neutron fluences at which the mobility is reduced are about an order of magnitude higher than those at which the on-resistance increases.

6.8.2. Threshold Voltage Shift

The major problem with irradiation of power MOSFETs arises from the introduction of positive charge in the gate oxide by the fast neutron radiation damage. The positive charge in the gate oxide causes decrease in the threshold voltage of n-channel power MOSFETs and increase in the threshold voltage of p-channel power MOSFETs. A typical example of the variation of the threshold voltage with radiation dose is provided in Fig. 6.47. Following the radiation, n-channel power MOSFETs can exhibit negative threshold voltage, that is, normally-on characteristics. It is possible to operate these devices by designing the gate power supply to provide a negative gate drive voltage during the blocking state. Similarly, for p-channel devices, the impact of the radiation on the threshold voltage can be accounted for by providing a high gate drive voltage.

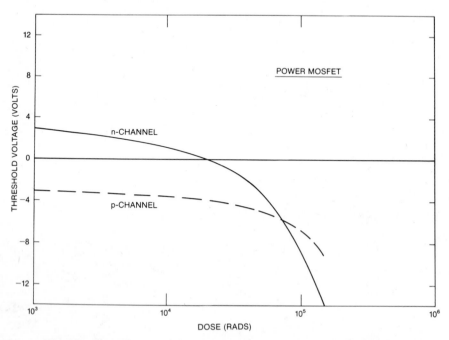

Fig. 6.47. Change in the threshold voltage of power MOSFETs due to radiation.

6.9. HIGH-TEMPERATURE PERFORMANCE

Power MOSFETs exhibit very good high-temperature operating characteristics. Devices are commercially available with peak junction temperatures ratings as high as 200°C. The ability of a power MOSFET to operate at elevated temperatures is related to the absence of minority carrier injection. The effect of increasing temperature on various device parameters is discussed in this section.

6.9.1. On-Resistance

An important merit of the power MOSFET is an increase in its on-resistance with increasing temperature. Although this may appear at first sight to be an undesirable feature because of an increase in power dissipation, it imparts an important benefit in terms of device stability and allowing the operation of devices in parallel. Local variations in drift region resistivity are unavoidable during the fabrication of power devices. Some locations inside the device will contain a lower on-resistance compared with the rest of the device. This promotes localization of current along the paths of lowest resistance. Localization of current produces additional heating. Fortunately, in the case of the power MOSFET, the mobility for holes and electrons decreases with temperature, causing the local resistivity to increase. This will tend to homogenize the current distribution and prevent thermal runaway. Furthermore, when power MOSFETs are connected in parallel, the increase in their on-resistance with temperature promotes good current sharing without the need for external ballasting, which is required for bipolar transistors.

A typical example of the increase in the on-resistance of a power MOSFET with temperature is provided in Fig. 6.48. This curve closely follows the mobility change with temperature discussed in Chapter 2. A general expression that can be used to predict the variation of the on-resistance of p- and n-channel power MOSFETs is

$$R_{on}(T) \simeq R_{on}(25°C)\left(\frac{T}{300}\right)^{2.3} \tag{6.124}$$

where T is the absolute temperature.

6.9.2. Transconductance

The reduction in the mobility with increasing temperature adversely affects the transconductance of power MOSFETs. A typical example of the variation of the transconductance is provided in Fig. 6.49. The transconductance will follow the mobility variation because the other terms that determine its value are relatively unaffected by temperature:

$$g_m(T) \simeq g_m(25°C)\left(\frac{T}{300}\right)^{-2.3} \tag{6.125}$$

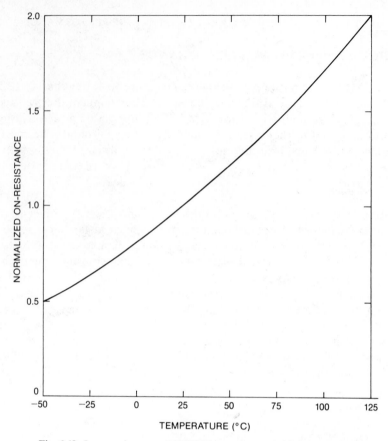

Fig. 6.48. Increase in power MOSFET on-resistance with temperature.

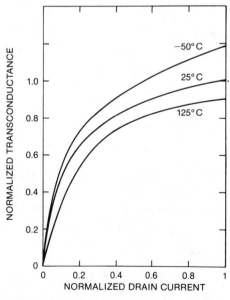

Fig. 6.49. Decrease of power MOSFET transconductance with increasing temperature.

Care must be taken during device design to provide sufficient channel periphery to allow for the degradation in the transconductance if high-temperature operation is anticipated.

6.9.3. Threshold Voltage

In Section 6.3.2 it was shown that the drain current of a power MOSFET will vary as the square of the gate drive voltage ($V_G - V_T$) in the saturated current regime. A plot of the square root of the drain current as a function of the gate voltage can, therefore, be used to obtain a measurement of the threshold voltage. A typical example of this transfer characteristic is shown in Fig. 6.50. The measured drain current indeed varies as the square of the gate voltage, resulting in the straight lines. The intercept of these lines at $I_D = 0$ provides the threshold

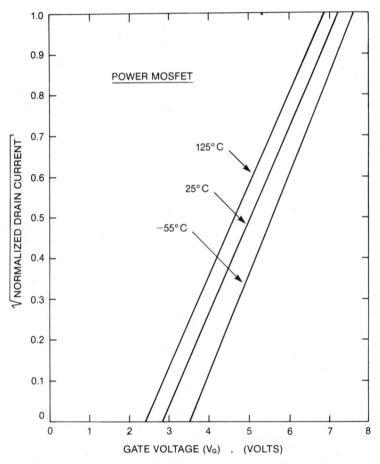

Fig. 6.50. Transfer characteristics of power MOSFET at different temperatures.

voltage. It can be seen that the threshold voltage decreases with increasing temperature.

In the absence of a work-function difference and the oxide charge, it was shown in Section 6.3.2.2 that the threshold voltage is given by

$$V_T = \frac{\sqrt{4\epsilon_s q N_A \psi_B}}{C_{ox}} + 2\psi_B \qquad (6.126)$$

Differentiating this expression with respect to temperature, it can be shown that

$$\frac{dV_T}{dt} = \frac{d\psi_B}{dt}\left[\frac{1}{C_{ox}}\sqrt{\frac{\epsilon_s q N_A}{\psi_B}} + 2\right] \qquad (6.127)$$

This expression is valid even in the presence of a work-function difference and oxide charge because these parameters are not strongly affected by the temperature. The bulk potential ψ_B varies with temperature because the energy gap changes [27]:

$$\frac{d\psi_B}{dt} \simeq \frac{1}{T}\left\{\frac{E_g(T=0)}{2q} - |\psi_B(T)|\right\} \qquad (6.128)$$

From these expressions, the rate of variation of the threshold voltage with temperature can be calculated as a function of the background doping level and oxide thickness. A plot of (dV_T/dT) as a function of background doping is provided in Fig. 6.51 for the Si–SiO$_2$ system using gate oxide thickness as a parameter [27]. Note that the threshold voltage will vary at a higher rate for thicker gate oxides and higher substrate doping levels. During the design and

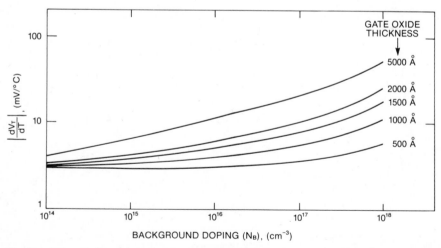

Fig. 6.51. Effect of substrate doping N_B and gate oxide thickness on the rate of change of threshold voltage with temperature. (After Ref. 27, reprinted with permission from the IEEE © 1971 IEEE.)

fabrication of power MOSFETs, it is important to provide an adequate room temperature threshold voltage which will allow operation at high temperatures with sufficient margin to ensure noise immunity and protection from inadvertent turn-on at high (dV/dt) transients.

6.10 DEVICE STRUCTURES AND TECHNOLOGY

Two fundamentally different processes have evolved for the fabrication of discrete vertical-channel power MOSFETs. One of these processes utilizes chemical etching of the silicon surface to form the V-groove structure shown in Fig. 6.1*b*, whereas the other processes relies on the planar diffusion of the *P* base and N^+ emitter from a common window defined by the refractory gate electrode to form the DMOS structure shown in Fig. 6.1*a*. Because of the anisotropic etch used to form the VMOS device structure, the device topology is restricted to the $\langle 110 \rangle$ directions along the surface of the (100)-oriented wafers. This restriction does not apply to the DMOS structure, allowing optimization of surface layout. In this section, the impact of surface topology is discussed after providing a description of the device processing technology.

6.10.1. DMOS (Planar) Structure

For the fabrication of the planar power MOSFET structure shown in Fig. 6.1*a*, the wafer orientation can be chosen to provide the highest surface mobility to achieve a low channel contribution to the on-resistance. For this reason, these devices are fabricated with the use of (100)-oriented wafers. Because of the extremely rapid increase in on-resistance with increasing breakdown voltage, power MOSFETs are confined to blocking voltages of less than 1000 V. The ideal depletion width for these devices is less than 100 μm. They must be fabricated by growing lightly doped epitaxial layers (typically 5–50 μm thick) on heavily doped substrates. The discussion here is confined to the case of *n*-channel devices. The fabrication of complementary (*p*-channel) devices is similar. In the case of *n*-channel devices, the N^+ substrates are antimony doped with a resistivity of 0.01 $\Omega \cdot$cm. Antimony is chosen as the dopant because its autodoping effects are small during epitaxial growth. For low on-resistance devices, arsenic-doped substrates with a resistivity of 0.001 $\Omega \cdot$cm are preferred.

After epitaxial growth, a thick field oxide is grown. This oxide is patterned to create the desired planar diffused edge termination. Sometimes the processing of the diffusion for the edge termination is combined with the DMOS cell fabrication. Following the termination diffusion, the device active area (central region where the DMOS cells are to be fabricated) is opened up. The gate oxide is then grown to a typical thickness of 1000 Å. Recently, device manufacturers have been scaling down the gate oxide thickness to obtain an increase in the transconductance so as to make it possible to drive the power MOSFETs at voltages compatible with low-voltage logic circuits.

A layer of heavily doped polysilicon (or other refractory gate material) is deposited next. After oxidizing the polysilicon, photolithography is used to pattern the polysilicon. A cross section of a small segment of the active region is shown at this point in the process in Fig. 6.52a. At this point in the process, the cell window has been defined. Boron is now implanted through the gate oxide to a dose (typically $10^{14}/cm^2$) that is necessary to obtain the desired threshold voltage in conjunction with the N^+-emitter diffusion. The boron is driven in with a wet oxidation step to obtain the structure shown in Fig. 6.52b. An oxide island is now patterned at the center of each cell window to mask the N^+ emitter diffusion. This oxide island leaves a portion of the P-base ac-

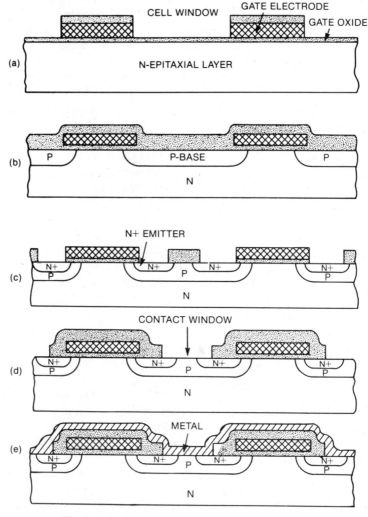

Fig. 6.52. Double-diffused MOSFET process sequence.

cessible at the surface in each cell, allowing good contacts to be subsequently made to it.

The patterning of the oxide island is one of the critical alignment steps during the DMOS process. If the oxide island is misaligned with respect to the edges of the polysilicon in the cell window, the length of the N^+ emitter will increase, causing degradation of the (dV/dt) and the safe operating area. After the oxide island is formed, the N^+ emitter is diffused by using either phosphorus predeposition or ion implantation. The latter process provides greater control over the emitter profile and threshold voltage. A typical dose is $1 \times 10^{15}/cm^2$. The device cross section at this point is shown in Fig. 6.52c. After the N^+ emitter, is driven in a thick layer of oxide is deposited by low pressure chemical vapor deposition (LPCVD). This provides a conformal insulating film over the top and sides of the polysilicon gate electrode.

The oxide layer is now patterned to form the contact windows as illustrated in Fig. 6.52d. This is another critical photolithographic step in the DMOS process because misalignment can cause either poor contact to the P-base region or lead to an overlap of the contact with the polysilicon at the cell edges. This overlap will produce a short circuit between source and gate. The final step in the device process consists of metallization to create the structure shown in Fig. 6.52e.

The DMOS process described here provides the fundamental basis for the fabrication of modern power MOSFETs. Many variations of the process have been explored in an attempt to either eliminate some of the critical alignment steps or to provide self-alignment of the N^+-emitter and contact windows.

6.10.2 VMOS (Nonplanar) Structure

Historically, the VMOS process was applied toward the fabrication of discrete, vertical-channel, power MOSFETs before the development of the DMOS process. This process relies on the anisotropic etching of silicon. It has been found that in certain chemical solutions, the (111) planes in silicon etch at a much slower pace when compared with other surfaces. If (100)-oriented silicon is etched through a window at the surface, these etches will create a V-shaped (or truncated V-shaped) groove as a result of the difference in etch rate between the (100) and (111) planes.

The VMOS (n-channel) process begins with the growth of a lightly doped N-type epitaxial layer on a (100)-oriented N^+ substrate. A boron diffusion is now performed across the entire wafer surface if the device termination is formed by using the V-groove etch step [28]. Alternately, the boron diffusion can be performed over the entire active area but patterned at the edges of the chip to form the desired device termination. A cross section of the active region at this point in the process is shown in Fig. 6.53a. The oxide layer grown during the P-base diffusion step is next patterned to define the N^+ emitter regions as illustrated in Fig. 6.53b. Note that the N^+ emitters extend between the cells where the V groove is be etched.

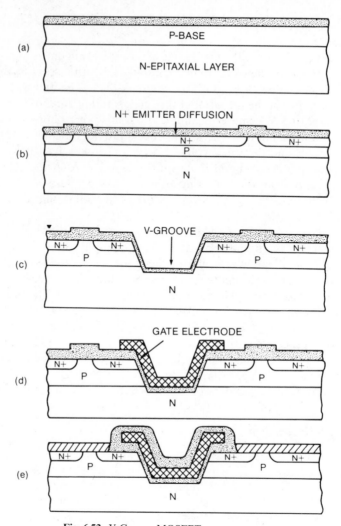

Fig. 6.53. V-Groove MOSFET process sequence.

The oxide layer grown over the wafer surface is now patterned to expose the silicon surface in areas to be etched. This patterning is an important alignment step because it determines the length of the N^+-emitter region and the relative position of the V groove with respect to the P-base contacts. The V groove is formed by using anisotropic etching to create the structure shown in Fig. 6.53c and must extend beyond the bottom of the P-base–N-drift layer junction. The etching is usually performed by using a mixture of potassium hydroxide and isopropanol [29]. Note that the window used to etch the grooves must be aligned along the ⟨110⟩ directions on the wafer surface because the (100) and (111) planes intersect along these lines. The V grooves extend down from the surface at an angle of 57.7° to the surface.

After forming the V grooves, the gate oxide must be grown in them. Since the presence of potassium in the gate oxide can cause severe instability of the threshold voltage, it is important to boil the wafers in hydrochloric acid (HCl) prior to gate oxide growth until all the potassium residue remaining from the V-groove etch step is removed. The polysilicon gate electrode is now deposited and patterned to form the structure shown in Fig. 6.53d. Another important alignment step is involved here because misregistration of the polysilicon with respect to the groove can lead to loss of channel periphery and an increase in the source–gate capacitance. Following the patterning of the polysilicon, an oxide layer is deposited by LPCVD. This layer is patterned to create the contact windows. These windows must also be carefully aligned to the cell structure as described for the DMOS structure. The source metallization is now evaporated and patterned.

Because of the nonplanar surface created by the V groove, the patterning of the polysilicon layer, contact windows and source metallization are more difficult than in the DMOS process. Poor resist coverage at the groove edges can lead to short circuits between source and gate if the source metal overlaps the gate (not shown in Fig. 6.53). It is also mandatory to eliminate even small traces of potassium from the wafer surface after etching the grooves to avoid instabilities in threshold voltage and drift in the breakdown voltages. These technological problems, together with the high electric field at the groove edges, have placed the VMOS structure at a serious disadvantage when compared with the DMOS structure. Although the first commercial power MOSFETs were manufactured with the VMOS process, most modern power MOSFETs are no longer fabricated with this process.

As in the case of the DMOS process, the VMOS process described here is intended to only provide a fundamental sequence for device fabrication. Many innovative techniques for simplifying the process and reduce the alignment problems have been attempted. One interesting variation of the process [30] utilizes isotropic etching of silicon to create a large oxide overhang as illustrated in Fig. 6.54. This overhang is utilized to obtain a selective deposition of the gate contact metal over the channel with a minimum overlapping of the source and drift regions. This is accomplished by evaporating a three-metal system (Cr–Ni–Au) at an angle to the wafer as indicated by the dashed lines in Fig. 6.54. This metal system was chosen because the evaporation of aluminum at

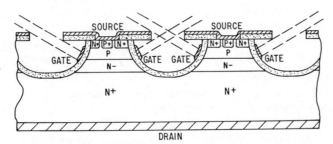

Fig. 6.54. Non-planar power MOSFET structure fabricated by using isotropic etching.

low angles of incidence has been observed to produce a "dull fibrous" film with an electrical resistivity substantially higher than that of normal evaporated aluminum films. Because of the low capacitance in this device structure, very high frequency performance has been achieved.

6.10.3. Gate Oxide Fabrication

Since the electrical characteristics of the MOSFET are modulated by the application of an electric field across the gate oxide, its properties play an important role in determining the device characteristics. A large amount of effort has, consequently, been expended in the semiconductor industry to develop an understanding of oxidation and annealing conditions on the oxide properties. The important process criteria for the growth of the gate oxide are provided in this section.

The silicon dioxide that is used as the insulating barrier between the gate metal and the channel can be grown by subjecting the silicon surface to an oxygen ambient at elevated temperatures. The ambients that have been most commonly used are either dry oxygen or oxygen containing water vapor. Lower interface state densities have been reported when the oxidation ambient contains water vapor [31]. However, the slower oxidation rate in dry oxygen allows greater control over the gate oxide thickness. The higher interface density in this case can be reduced by increasing the oxidation temperature and subsequently annealing the oxide, as described below.

Before considering the effects of annealing on the oxide–silicon interface, it is worth pointing out that the lowest fixed interface charge Q_{FC} and surface state density have been observed for the (100) plane [31, 32]. For the devices made by the anisotropic etching technology, the gate oxide must be formed on a (111) surface since this surface is exposed by the preferential etch. In contrast, the devices made by planar diffusion techniques (DMOS) can be fabricated from (100)-oriented wafers to obtain a lower fixed interface charge and surface state density.

Low-temperature annealing of the oxide films in either a mixture of hydrogen and nitrogen or a nitrogen ambient containing water vapor has proved to be very effective in decreasing the interface state density. A 1000°C anneal in dry nitrogen must be performed prior to the preceding low-temperature annealing to achieve a minimum interface state density [33]. It is believed that during the subsequent low-temperature anneal, the hydrogen in the oxide diffuses to the oxide–silicon interface and ties up the dangling bonds, reducing the surface state density. The optimum annealing temperature is about 400°C, with a low surface state density observed after 60 min of annealing time. A similar annealing of the interface states can be accomplished after the aluminum gate metallization has been applied. A faster annealing rate has been observed in this case. This is believed to arise from the generation of hydrogen at the aluminum by the reduction of water vapor in the oxide. This process has the drawback of the possible introduction of mobile ions into the silicon dioxide from the aluminum, which can lead to an instability in the threshold voltage.

A drift in the threshold voltage due to the migration of mobile ionic species in the oxide has been one of the most serious problems in MOS gated FETs. This is particularly true with devices made by the preferential etching of the V groove with a potassium hydroxide–isopropanol solution. In this case, the complete removal of potassium ions from the silicon surface prior to the growth of the gate oxide is imperative. The high mobility of alkali ions in silicon dioxide at relatively low temperatures (in the range of 200–300°C) has been demonstrated conclusively [34, 35]. The migration of these alkali ions can be suppressed by either doping the oxide with phosphorus or capping it with a layer of silicon nitride. However, polarization effects [36] in the phosphosilicate glass films and the trapping of charge at the oxide–nitride interface have prevented the application of these techniques to the gate area of power MOSFET devices.

A more promising process innovation has been the addition of halogens to the gas stream prior to and during oxidation. The exposure of the oxidation tube to a mixture of HCl and dry O_2 has been found to be effective in reducing the mobile ion contamination of oxides grown subsequently in the tube [37]. The mobile ion contamination in the oxide has also been found to decrease with increasing HCl concentration in the gas stream during oxidation. However, it is important to keep in mind that HCl will etch silicon at the oxidation temperature if its concentration exceeds a few percent. Typically, an HCl concentration of about 1% by volume in the gas stream is effective in suppressing ionic contamination. The addition of chlorine during oxidation has also been reported to lower the interface state density and the fixed oxide charge density and to improve the dielectric breakdown strength of the oxide [37–39]. It is interesting to note that these improvements are observed only during dry oxidation and that the addition of HCl during oxidation with water vapor has been found to have no effect on the oxide properties.

In addition to the effect of the oxide thickness on the gate threshold voltage, the growth of the oxide over the channel can also affect the threshold voltage by the redistribution of the boron in the channel during oxidation. Experiments conducted on the thermal oxidation of silicon doped with gallium, boron, and indium have found that the surface concentration of these impurities decreases after oxidation [40]. This occurs because these impurities segregate into the oxide, thus depleting the silicon surface. Impurity segregation can seriously alter the diffusion profile of boron in the P base of both the DMOS and VMOS devices and, thus, influence the threshold voltage as well as the penetration of the depletion layer into the P base when the device is blocking current flow. Several models have been developed to allow the calculation of the diffusion profile of impurities with inclusion of the impurity redistribution during oxidation [41]. These models have also been applied to the calculation of the effect of oxidation on the threshold voltage of MOSFETs transistors [42]. In addition, the strain at the surface under the polysilicon gate of the DMOS structure

alters the lateral diffusion of impurities. It has been found that the boron diffusion is retarded whereas the phosphorus diffusion is enhanced. This effect can
severely alter the channel length of both n- and p-channel DMOS power
MOSFETs.

6.10.4. Cell Topology

In previous discussions of power MOSFETs in this chapter, cross sections of
devices were shown without consideration of the cell layout on the surface. For
the VMOS device structure, the anisotropic etching of the grooves can be accomplished only when the grooves are oriented along the ⟨110⟩ directions on
the (100) wafer surface. This constraint allows only two design options for the
surface topology of VMOS devices: long stripes or rectangular cells with V
grooves running perpendicular to each other. Even the latter option is impractical because of the poor contour at the corners where the V grooves intersect
at right angles [43]. Therefore, V-groove power MOSFET cells are confined to
long fingers with interdigitation of source and gate [15].

In contrast, the DMOS structure allows any conceivable cell topology as
long as it meets all technological constraints such as alignment tolerances. The
cell windows that are commonly used for the design of power MOSFETs have
square, circular, and hexagonal shapes. These windows can be located in either
a square or hexagonal cell pattern as illustrated in Fig. 6.55. The impact of the
layout of these cells, as well as the linear cell shown in Fig. 6.56(a), on the resistance has been analyzed [44]. It has been shown that the on-resistance of all

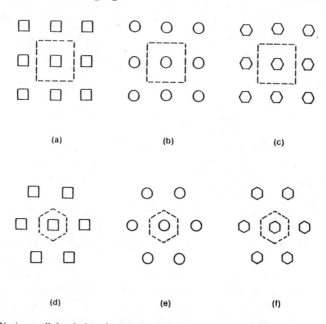

Fig. 6.55. Various cellular designs for power MOSFETs: (a) square window, square cell; (b) circular
window, square cell; (c) hexagonal window, square cell; (d) square window, hexagonal cell; (e)
circular window, hexagonal cell; (f) hexagonal window, hexagonal cell.

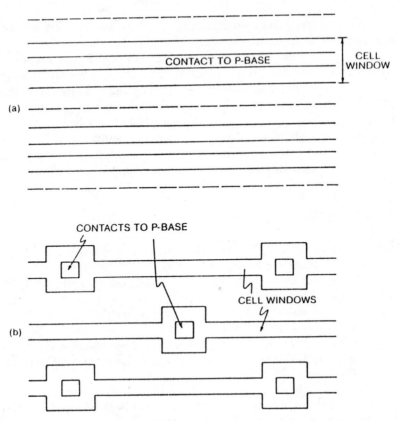

Fig. 6.56. Comparison of linear cell design with (*a*) continuous contact and (*b*) broken contact to P-base region.

cellular designs are equal if the size of the cell window and the ratio of the area of the cell window to the total cell area are kept the same. The on-resistance of the cellular designs shown in Fig. 6.55 is smaller by a factor of 1.6 than a linear cell with the same cell window size. This difference can be made up by breaking up the linear cell into regions where contact to the P base is made as shown in Fig. 6.56b. In all cases, maintenance of the smallest possible cell window size produces a lower on-resistance as long as the cell size is also scaled down to keep the cell window area:cell area ratio constant. Although early power MOSFETs were primarily fabricated by using linear cell geometries, most modern power MOSFETs utilize a cellular topology.

6.11. TRENDS

Power MOSFETs are the most commercially advanced devices discussed in this book. Their attributes are a voltage-controlled characteristic with low input gate power, a stable negative temperature coefficient for the on-resistance, wide safe operating area, and high switching speed. Because of these features, when first introduced, the devices were expected to replace power bipolar transistors in many applications. This displacement has not occurred at a rapid pace as a result of the high on-resistance of power MOSFETs and their relatively high manufacturing cost. With rapid progress along the learning curve, the manufacturing cost of power MOSFETs is now beginning to rival that of bipolar transistors. In addition, for low voltage (< 100 V) and in high-frequency circuits, the power MOSFET clearly offers superior performance. Although power MOSFETs with breakdown voltage of up to 1000 V have been reported, most of the applications for these devices are expected to lie in the low-voltage area such as for automotive circuits.

REFERENCES

1. W. Shockley and G. L. Pearson, "Modulation of conductance of thin films of semiconductors by surface charges," *Phys. Rev.*, **24**, 232–233 (1948).

2. D. Kahng and M. M. Atalla, *IRE–AIEE Solid State Device Research Conference*, 1960.

3. M. S. Adler and V. A. K. Temple, *Semiconductor Avalanche Breakdown Design Manual*, General Electric Technology Marketing Operation, New York, 1979.

4. M. N. Darwish and K. Board, "Optimization of breakdown voltage and on-resistance of VDMOS transistors," *IEEE Trans. Electron Devices*, **ED-31**, 1769–1773 (1984).

5. V. A. K. Temple and P. V. Gray, "Theoretical comparison of DMOS and VMOS structures for voltage and on-resistance," *IEEE International Electron Device Meeting Digest*, Abstract 4.5, pp. 88–92 (1979).

6. Y. Shimada, K. Kato, and T. Sakai, "High efficiency MOS–FET rectifier device," *IEEE Power Electronics Specialists Conference Proceedings*, pp. 129–136 (1983).

7. C. G. B. Garrett and W. H. Brattain, "Physical theory of semiconductor surfaces," *Phys. Rev.*, **99**, 376–387 (1955).

8. B. E. Deal, E. H. Snow, and C. A. Mead, "Barrier energies in metal–silicon dioxide–silicon structures," *J. Phys. Chem. Solids*, **27**, 1873–1879 (1966).

9. W. M. Werner, "The work function difference of the MOS system with aluminum field plates and polycrystalline silicon field plates," *Solid State Electron.*, **17**, 769–775 (1974).

10. V. G. K. Reddi and C. T. Sah, "Source to drain resistance beyond pinch-off in metal-oxide–semiconductor transistors (MOST)," *IEEE Trans. Electron Devices*, **ED-12**, 139–141 (1965).

11. C. T. Sah and H. C. Pao, "The effects of fixed bulk charge on the characteristics of metal-oxide–semiconductor transistors," *IEEE Trans. Electron Devices*, **ED-13**, 393–409 (1966).

12. T. L. Chiu and C. T. Sah, "Correlation of experiments with a two-section model theory of the saturation drain conductance of MOS transistors," *Solid State Electron.*, **11**, 1149–1163 (1968).

13. S. C. Sun and J. D. Plummer, "Modelling of the on-resistance of LDMOS, VDMOS, and VMOS power transistors," *IEEE Trans. Electron Devices*, **ED-27**, 356–367 (1980).

14. C. Hu, "A parametric study of power MOSFETs," *IEEE Power Electronics Specialists Conference Record*, pp. 385–395 (1979).

15. B. J. Baliga, "Switching lots of watts at high speed," *IEEE spectrum*, **18**, 42–48 (1981).

16. C. Hu, "Optimum doping profile for minimum ohmic resistance and high-breakdown voltage," *IEEE Trans. Electron Devices*, **ED-26**, 243–244 (1979).

17. H. Ikeda, K. Ashikawa, and K. Urita, "Power MOSFETs for medium-wave and short-wave transmitters," *IEEE Trans. Electron Devices*, **ED-27**, 330–334 (1980).

18. H. Esaki and O. Ishikawa, "A 900 MHz, 100 watt, VD-MOSFET with silcide gate self-aligned channel," *IEEE International Electron Devices Meeting Digest*, Abstract 16.6, pp. 447–449 (1984).

19. S. Clemente and B. R. Pelly, "Understanding the power MOSFET switching performance," *Proceedings of IEEE Industrial Application Society Meeting*, Abstract 32.B, pp. 763–776 (1981).

20. R. Severns, "dV/dt effects in MOSFETs and bipolar junction transistor switches," *IEEE Power Electronics Specialists Conference Record*, pp. 258–264 (1981).

21. D. S. Kuo, C. Hu, and M. H. Chi, "dV/dt breakdown in power MOSFETs," *IEEE Electron Device Lett.*, **EDL-4**, 1–2 (1983).

22. C. Hu and M. H. Chi, "Second breakdown of vertical power MOSFETs," *IEEE Trans. Electron Devices*, **ED-29**, 1287–1293 (1982).

23. B. R. Pelly, "Power MOSFETs—a status review," *1983 International Power Electronics Conference Record*, pp. 19–32 (1983).

24. B. J. Baliga and J. P. Walden, "Improving the reverse recovery of power MOSFET integral diodes by electron irradiation," *Solid State Electron.*, **26**, 1133–1141 (1983).

25. A. O. Evwaraye and B. J. Baliga, "The dominant recombination center in electron irradiated semiconductor devices," *J. Electrochem. Soc.*, **124**, 913–916 (1977).

26. Y. Ohata, "New MOSFETs for high power switching applications," *Proc. Powercon 11*, Abstract C5, pp. 1–11 (1984).

27. R. Wang. J. Dunkley, T. A. DeMassa, and L. F. Jelsma, "Threshold voltage variations with temperature in MOS transistors," *IEEE Trans. Electron Devices*, **ED-18**, 386–388 (1971).

28. K. P. Lisiak and J. Berger, "Optimization of non-planar power MOS transistors," *IEEE Trans. Electron Devices*, **ED-15**, 1229–1234 (1978).

29. J. B. Price, "Anisotropic etching of silicon with $KOH-H_2O$–isopropyl alcohol," *Electrochemical Society Meeting Digest*, pp. 339–353 (1973).

30. J. G. Oakes, R. A. Wickstrom, D. A. Tremere, and T. M. S. Heng, "A power silicon microwave MOS transistor," *IEEE Trans. Microwave Theory Techniques*, **MTT-24**, 305–311 (1976).

31. P. V. Gray D. M. Brown, "Density of SiO_2–Si interface states," *Appl. Phys. Lett.*, **8**, 31–33 (1966).

32. E. Arnold, J. Ladell, and G. Abowitz, "Crystallographic symmetry of surface state density in thermally oxidized silicon," *Appl. Phys. Lett.*, **13**, 413–416 (1968).

33. Y. T. Yeow, D. R. Lamb, and S. D. Brotherton, "An investigation of the influence of low-temperature annealing treatments on the interface state density at the Si–SiO$_2$ interface," *J. Phys. D: Appl. Phys.*, **8**, 1495–1506 (1975).

34. E. Yon, W. H. Ko, and A. B. Kuper, "Sodium distribution in thermal oxide on silicon by radiochemical and MOS analysis," *IEEE Trans. Electron Devices*, **ED-13**, 276–280 (1966).

35. S. R. Hofstein, "Stabilization of MOS devices," *Solid State Electron.*, **10**, 657–670 (1967).

36. E. H. Snow and B. E. Deal, "Polarization phenomena and other properties of phosphosilicate glass films on silicon," *J. Electrochem. Soc.*, **113**, 263–269 (1966).

37. R. J. Kriegler, Y. C. Cheng, and D. R. Colton, "The effect of HCl and Cl$_2$ on the thermal oxidation of silicon," *J. Electrochem. Soc.*, **119**, 388–392 (1972).

38. M. C. Chen and J. W. Hile, "Oxide charge reduction by chemical gettering with trichloroethylene during thermal oxidation of silicon," *J. Electrochem. Soc.*, **119**, 223–225 (1972).

39. C. M. Osburn, "Dielectric breakdown properties of SiO$_2$ films grown in halogen and Hydrogen containing environments," *J. Electrochem. Soc.*, **121**, 809–815 (1974).

40. A. S. Grove, O. Leistiko, and C. T. Sah, "Redistribution of acceptor and donor impurities during thermal oxidation of silicon," *J. Appl. Phys.*, **35**, 2695–2701 (1964).

41. W. G. Allen and C. Atkinson, "Comparison of models for redistribution of dopants in silicon during thermal oxidation," *Solid State Electron.*, **16**, 1283–1287 (1973).

42. H. G. Lee, J. D. Sansbury, R. W. Dutton, and J. L. Moll, "Modelling and measurement of surface impurity profiles of laterally diffused regions," *IEEE J. Solid State Circuits*, **SC-13**, 445–461 (1978).

43. K. E. Bean, "Anisotropic etching of silicon," *IEEE Trans. Electron Devices*, **ED-25**, 1185–1193 (1978).

44. C. Hu, M. H. Chi, and V. M. Patel, "Optimum design of power MOSFET's," *IEEE Trans. Electron Devices*, **ED-31**, 1693–1700 (1984).

PROBLEMS

6.1. The P-base region of a power MOSFET is fabricated by using a diffusion with a surface concentration of $10^{18}/cm^3$ and a depth of 3 μm. Determine the minimum achievable channel length for a drift layer doping concentration of $10^{15}/cm^3$.

6.2. Consider an N-type source diffusion into the P-base region described in Problem 6.1 so that the peak base doping concentration is $10^{17}/cm^3$. Determine the threshold voltage of the resulting power MOSFET, assuming that a gate oxide of 1000 A is used. Ignore any work function differences and fixed charge in the oxide.

6.3. What is the effect of a positive fixed charge density of 10^{11} and $10^{12}/cm^2$ on the threshold voltage for the device described in Problem 6.2?

6.4. Consider the power MOSFET structure described in Problem 6.2 with a channel length of 2 μm. Determine the channel width required to achieve a channel resistance of 1 mΩ, assuming that a gate drive voltage of 15 V is used.

6.5. Calculate the drift layer doping concentration and thickness for a power MOSFET designed with a breakdown voltage of 100 V. Assume that ideal parallel-plane breakdown is achievable.

6.6. What is the ideal specific on-resistance for the device described in Problem 6.5?

6.7. Recompute the drift region parameters when the breakdown voltage is limited to 80% of the ideal value by the edge termination of the device.

6.8. A power MOSFET is fabricated by using the drift region obtained in Problem 6.7 with the DMOS process with a linear cell geometry. The polysilicon gate width is 10 μm and the polysilicon window 16 μm. The source diffusion depth is 1 μm and the length is 5 μm. Its sheet resistance is 10 $\Omega/\square$. The base diffusion depth is 3 μm, resulting in the channel parameters obtained in Problem 6.4. Calculate the specific on-resistance of the device neglecting accumulation layer and substrate contributions.

6.9. What are the percentage contributions to the on-resistance from the channel and drift regions for the device described in Problem 6.8?

6.10. Compare the specific on-resistance of the device structure in Problem 6.8 with the ideal specific on-resistance computed in Problem 6.6.

6.11. What is the impact of using an antimony doped substrate with resistivity of 0.01 $\Omega\cdot$cm and thickness of 20 mils on the specific on-resistance?

6.12. What is the impact of using an arsenic doped substrate with resistivity of 0.002 $\Omega\cdot$cm and thickness of 10 mils on the specific on-resistance?

6.13. Estimate the frequency response of a power MOSFET with the DMOS structure as described in Problem 6.8 with an active area of 0.1 cm^2. Assume that the gate capacitance is determined by the oxide alone and that the gate drive circuit has a resistance of 50 Ω.

6.14. The device described in Problem 6.13 is operated in a switching circuit with a gate drive voltage of 15 V. Determine the turn-on delay times for a drive circuit impedance of 50 Ω.

6.15. Calculate the transconductance of the device described in Problem 6.8 in the current saturation region.

7 POWER MOS—BIPOLAR DEVICES

Previous chapters have discussed bipolar devices such as the FCD and unipolar MOS-controlled devices such as the power MOSFET. It was shown that in bipolar devices, the injection of minority carriers into their drift region reduces the resistance to forward current flow. These devices are capable of operation at relatively high current densities (typically 200–300 A/cm^2), even when designed to support high voltages. This reduces device size and cost. Their principal drawback has been the need to supply large gate currents during forced gate turn-off.

The power MOSFET was developed with the objective of addressing this gate drive problem. The MOS gate structure has a very high steady-state impedance. This allows control of the device by a voltage source, since only relatively small gate drive currents are required to charge and discharge the input gate capacitance. Their gate drive circuitry can be integrated on a single chip with the use of recently developed high-voltage integrated-circuit technology. This provides a large reduction in cost, a significant increase in reliability and an improvement in overall system performance. Unfortunately, the ease of gating the power MOSFET is offset by its high on-resistance, which arises from the absence of minority carrier injection. The drift region of the power MOSFET represents a very high series resistance during current conduction, which limits its operating forward current density to relatively low values (typically in the range of 10 A/cm^2 for a 600-V device). On the basis of these features of power bipolar and MOSFET devices, it is apparent that the combination of the bipolar current conduction with MOS-controlled current flow would provide the ideal features of high operating forward current density and low gate drive power.

One approach to combining these features is to use discrete bipolar transistors and power MOSFETs connected either in the Darlington configuration as shown in Fig. 7.1a or in the emitter switched Darlington configuration shown in Fig. 7.1b. In these implementations, most of the forward current flows occurs in the bipolar transistor whereas the high voltage power MOSFET (transistor Q1) provides its base drive currrent. These circuits present an MOS input to the gate drive circuitry. Since a combination of discrete high-voltage bipolar and MOSFET elements are being used here, the overall (average) current density during forward conduction is between that observed in a bipolar transistor (typically 50 A/cm^2 for a 600-V device) and that of a power MOSFET (typically 10 A/cm^2).

The devices shown in Fig. 7.1 can be simultaneously fabricated on a single chip to provide the user with a three-terminal "device" with characteristics superior to those of either the bipolar transistor or the power MOSFET [1, 2]. This is fundamentally a circuit solution to the problem of combining bipolar and MOS characteristics. Since no new physics is involved, this approach is not discussed any further.

A more powerful and innovative approach is based on a combination of the physics of bipolar current conduction with MOS gate control. The first successful combination of the physics of bipolar and MOS operation was reported in 1979 [3]. This device, called the *MOS gated thyristor*, is discussed in Section 7.1. Although providing MOS gate control to turn on the device, these devices did not provide forced gate turn-off capability. An additional conceptual breakthrough was required to realize that the regenerative action of the four-layer thyristor structure in these devices could be defeated to achieve fully MOS gate-controlled output characteristics with gate turn-off capability [4]. These devices, called *insulated-gate transistors* (IGTs), are discussed in Section 7.2. They are an important new class of power devices that are expected to replace the bipolar transistor in many of its applications. The concept of devices based on combining the physics of bipolar and MOS operation has led to several other possible structures such as the MOS-controlled thyristor (MCT). These device concepts are also discussed at the end of this chapter.

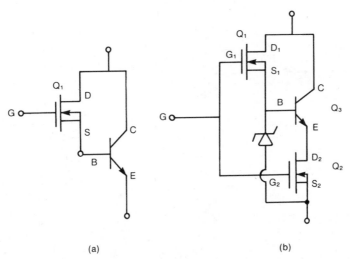

(a)   (b)

Fig. 7.1. Combinations of discrete power bipolar and MOSFET devices to create MOS gated composite circuits.

7.1. MOS GATED THYRISTOR

The MOS gated thyristor represents the first successful combination of the physics of bipolar and MOS devices [3]. In this device, and MOS gate structure is combined with the four-layer thyristor structure to create a device that can be turned on by using a control voltage applied to the gate terminal. To understand the benefits resulting from MOS gating, it is necessary to first discuss the conventional thyristor structure and its shortcomings.

7.1.1. Conventional Thyristor Structure

The operation of a conventional thyristor has been treated in detail elsewhere [5] and is not discussed in detail here. Only a brief analysis is performed to provide a comparison with the MOS gated structure.

The four-layer structure of a conventional thyristor and its equivalent circuit are shown in Fig. 7.2. This P–N–P–N structure can be regarded as two transistors—an upper N–P–N transistor and a lower P–N–P transistor—that are internally connected in such a fashion as to obtain regenerative feedback between each other. During forward blocking, with positive anode voltage, junction J2 becomes reverse-biased. This junction serves as a common collector between the upper N–P–N and lower P–N–P transistors and provides the coupling between them. If the cathode short circuit at point B is ignored for the moment, an anode current flow I_A produces a hole current $\alpha_{PNP}I_A$ across junction J2. The current gain α_{PNP} of the lower P–N–P transistor is governed mainly by the base transport factor of this very-wide-base transistor. Similarly, a cathode current I_K produces an electron current $\alpha_{NPN}I_K$ across junction J2. The current

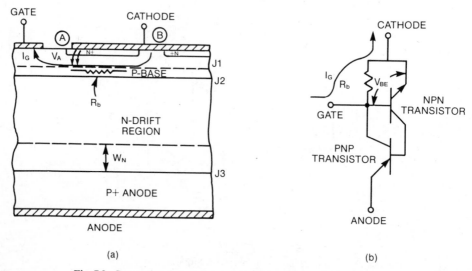

(a)   (b)

Fig. 7.2. Conventional thyristor structure (*a*) and its equivalent circuit (*b*).

gain α_{NPN} of the upper N–P–N transistor is governed mainly by the emitter injection efficiency of this narrow-base transistor. In addition to these currents, a leakage current I_L flows across junction J2 as the result of space-charge generation in the depletion layer and thermal generation in the neutral regions. The total anode current is given by

$$I_A = \alpha_{PNP}I_A + \alpha_{PNP}I_K + I_L \qquad (7.1)$$

On the basis of current conservation

$$I_K = I_G + I_A \qquad (7.2)$$

Combining these equations, it can be shown that

$$I_A = \frac{\alpha_{NPN}I_G + I_L}{1 - \alpha_{PNP} - \alpha_{NPN}} \qquad (7.3)$$

From this expression it can be inferred that the device will switch from its forward blocking state to the forward conduction state when the sum of the current gains $(\alpha_{PNP} + \alpha_{NPN})$ approaches unity.

The current gains of the coupled transistors are a strong function of the current flow. This is especially true of the short-circuited cathode structure shown in Fig. 7.2a. The presence of the cathode short circuit at point B provides a path for current flow directly from the P base into the cathode terminal; this short-circuits the cathode junction J1. The effectiveness of the cathode short circuit is represented in the equivalent circuit shown in Fig. 7.2b by a base resistance R_b. When the emitter–base junction of the upper N–P–N transistor is forward-biased to a voltage V_{BE}, the injected current across the N^+-cathode–P-base junction is given by

$$I_{N^+} = I_0[e^{qV_{BE}/kT} - 1] \qquad (7.4)$$

Meanwhile, the current flow through the shorting resistor R_b is given by

$$I_R = \frac{V_{BE}}{R_b} \qquad (7.5)$$

The input gate current is the sum of these two currents:

$$I_G = I_{N^+} + I_R \qquad (7.6)$$

If the current gain of the N–P–N transistor in the absence of the short-circuiting resistor $(R_b = \infty)$ is α_{NPN}, the current gain with the short circuit is given by

$$\alpha_{NPN,\text{short circuit}} = \frac{\alpha_{NPN}}{1 + (I_R/I_{N^+})} \qquad (7.7)$$

This short-circuited transistor current gain is substantially lower than that for a device without the short circuit, until the N^+ cathode–P-base junction becomes forward biased, that is, V_{BE} approaches 0.7 V.

The thyristor breakover point occurs when the sum of the transistor current gains equals unity. This is equivalent to

$$\alpha_{NPN} = (1 - \alpha_{PNP}) \tag{7.8}$$

where α_{PNP} is related to the anode voltage by the following equation

$$\alpha_{PNP} = \frac{1}{\cosh(W_n/L)} \tag{7.9}$$

where W_n is the undepleted N-drift region width and L is the minority carrier diffusion length. The undepleted drift region width is related to the anode voltage by

$$W_n = W_D - \sqrt{\frac{2\epsilon_s}{qN_D} V_A} \tag{7.10}$$

where V_A is the applied anode voltage, W_D is the N-drift region thickness, and N_D is its doping level. The current gain of the upper N–P–N transistor can be increased by the application of a gate current I_G. This current flow between gate and cathode terminals results in a voltage drop across the P-base resistance R_b, which, in turn, causes an increase in the gain of the N–P–N transistor. When the gate current is sufficient to raise α_{NPN} to the level defined by Eq. (7.8), the thyristor will switch to the on-state. From this description, it can be concluded that a high base resistance R_b is desirable in order to reduce the gate drive current. As a rough approximation, it can be assumed that the breakover point is reached when the forward bias at point A of the N^+ cathode–P base junction becomes 0.7 V. The gate current required to achieve this is

$$I_G = \frac{0.7}{R_b} \tag{7.11}$$

if the injected current I_{N^+} is neglected. Thus a low gate drive current can be achieved by designing the device with short circuits that are spaced far apart so as to increase R_b.

The thyristor structure is susceptible to turn-on under a high applied (dV/dt). The high (dV/dt) across the anode–cathode terminals produces a displacement current across junction J2 to charge the junction capacitance. This displacement current can also flow into the short circuits by way of the P base. The displacement current flow forward-biases the N^+ cathode–P-base junction and has the same effect as the gate current. As a first approximation, the turn-on of the thyristor under a high applied (dV/dt) can be assumed to occur when the cathode junction becomes forward-biased at point A by 0.7 V. The (dV/dt) at which this will occur is given by

$$\frac{dV}{dt} = \frac{0.7}{C_J R_b} \tag{7.12}$$

where C_J is the junction capacitance. This expression indicates the need to use a high short-circuiting density to achieve a low P-base resistance R_b in order to maximize the (dV/dt) capability. The design of a thyristor with high (dV/dt) capability will necessitate providing a large base drive current. Thus the design of a large (dV/dt) capability and a high gate sensitivity for the conventional thyristor imposes conflicting requirements on device design.

7.1.2. MOS Gated Structure

The operation of the MOS gated thyristor structure is based on controlling the gain of the upper $N-P-N$ transistor by means of a voltage applied to an MOS gate structure. The MOS gate structure can have either the DMOS or the VMOS configuration. Since the first MOS gated thyristors were fabricated by the VMOS process, this structure is illustrated in Fig. 7.3. The MOS gate structure must extend into the depletion region of the P-base–N-drift region junction J2. In the absence of a gate voltage, the depletion layer edge in the P base will remain essentially flat and parallel to junction J2 as shown in Fig. 7.3 by the dashed lines. The current gain of the upper transistor will then be low even at high anode voltages as a result of the short circuit at point B. When a positive gate bias is applied, the surface of the P base is depleted as a result of the gate electric field. This promotes the extension of the depletion layer in the P base along the surface of the V groove as shown by the dotted lines in Fig. 7.3. The undepleted base width W_u of the $N-P-N$ transistor is reduced by the application of the gate voltage. If the gate voltage is sufficient to allow the depletion layer in the P base to extend to the N^+ cathode, strong injection of electrons can occur because the junction J1 will become forward-biased at point A. This method of forward-biasing the emitter junctions can be used to increase the

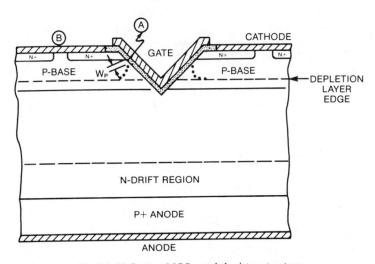

Fig. 7.3. V-Groove MOS gated thyristor structure.

gain of the upper N–P–N transistor by the application of the gate voltage, causing regenerative turn-on of the thyristor structure.

The current gain of the upper N–P–N transistor in the presence of the cathode short circuit is given in Eq. (7.7). This current gain is low in the absence of gate voltage because of the low base resistance R_b. However, the gain of the upper transistor is also related to the undepleted base width W_u of the P-base region:

$$\alpha_{NPN} = \frac{1}{\cosh(W_u/L_n)} \tag{7.13}$$

The undepleted base width W_u can be reduced to zero by the applied gate bias, allowing the current gain at point A to increase dramatically as the result of forward biasing of the N^+-cathode–P-base junction after depletion layer punchthrough.

The gate voltage required to achieve turn-on of the MOS gated structure is dependent on the doping level of the P-base region. A higher turn-on gate voltage will be required to deplete the P base as its doping level increases. The gate voltage required to produce the turn-on is approximately equal to the threshold voltage of the MOSFET as discussed in Chapter 6. It is not dependent on the N^+-cathode short-circuiting density and the short-circuiting resistance R_b. Thus the (dV/dt) capability can be increased in the MOS gated thyristor by providing a high short-circuiting density while retaining a low gate drive power.

The MOS gated thyristor structure allows the decoupling of the (dV/dt) capability and the gate sensitivity for the first time. It also enables the use of a higher cathode short-circuiting density for improvement of the high-temperature operating capability of thyristors. In addition, the power MOSFET process when applied to the MOS gated thyristor allows the fabrication of devices with very long gate widths. Since this eliminates the relatively slow current spreading that is typically observed in conventional thyristor structures, very high speed turn-on can be obtained by using the MOS gated thyristor structure. Devices that exploit these features have been commercially introduced for use in applications that need fast turn-on such as laser modulators and particle accelerators. These devices can block up to 600 V with (dV/dt) capability of over 1,000 V/μsec and turn-on time of less than 200 nsec. It should be noted that these devices do not have gate turn-off capability and must be turned off by reversal of the anode voltage. In addition, lateral devices have also been explored for integrated-circuit applications [6].

7.2. INSULATED-GATE TRANSISTOR

Among the recently created family of MOS–bipolar devices, the IGT [4a] is the most commercially advanced device. It has also been called the *conductivity-modulated field-effect transistor* (COMFET) [4b]. Because of its significantly superior characteristics for low- and medium-frequency applications, when compared with both bipolar transistors and power MOSFETs, a substantial amount of work has been undertaken to obtain a good understanding of the operating characteristics. A vigorous effort is also underway to improve its power ratings by increasing both current and voltage handling capability. Devices with a forward blocking capability of 500-V and 25-A average current rating are already available [7]. These devices can be used in circuits to turn off currents of up to 150 A. These ratings are substantially better than those for any available power MOSFET and represent an important extension of power MOS devices that threaten to replace power bipolar transistors. Further, it has been established both theoretically and experimentally that p-channel IGTs have performance comparable to that of n-channel devices, thus allowing the use of complementary devices in power systems [8].

In this section of the chapter, the basic physics of operation of the IGT is discussed, including the important issue of regenerative latch-up of the devices during both steady-state conduction and dynamic switching. The reduction in switching speed arising from minority carrier injection is analyzed, and techniques developed to improve the switching speed are described.

7.2.1. Basic Structure and Operation

At first sight, the structure of the IGT appears to be identical to that of the MOS gated thyristor shown in Fig. 7.3. Its operation is, however, fundamentally different in that the IGT structure is designed to prevent the regenerative turn-on inherent to the four-layer thyristor structure. The conceptual breakthrough required to create the IGT was the realization that the MOS gate can be used to create a (inversion layer) channel linking the N^+-emitter region to the N-drift layer without regenerative latch-up. Since the lower junction is forward-biased in the on-state, current flow in the IGT can now occur via this channel. This allows fully gate-controlled output characteristics with forced gate turn-off capability.

Consider the DMOS cross-sectional structure of the IGT shown in Fig. 7.4. In this structure, current flow cannot occur when a negative voltage is applied to the collector with respect to the emitter because the lower junction (J3) will become reverse-biased. This provides the device with its reverse blocking capability as illustrated in Fig. 7.5. When a positive voltage is applied to the collector terminal with the gate short-circuited to the emitter terminal, the upper junction (J2) becomes reverse-biased and the device operates in its forward blocking mode. With positive collector terminal voltages, if a positive gate bias is applied of sufficient magnitude to invert the surface of the P-base region under the gate, the device operates in its forward conducting state because electrons can now flow from the emitter N^+ region to the N-base region. In this forward conducting state, junction J3 is forward-biased and the substrate P^+ region injects holes into the N-base region. When the forward bias is increased, the injected hole concentration increases until it exceeds the background doping level of the N base. In this regime of operation, the device characteristics are similar to

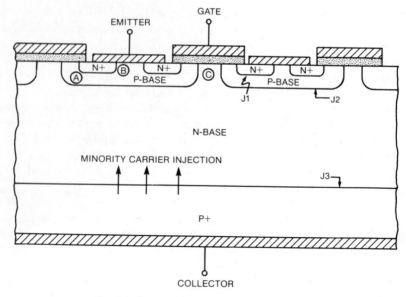

Fig. 7.4. Cross section of DMOS–IGT structure.

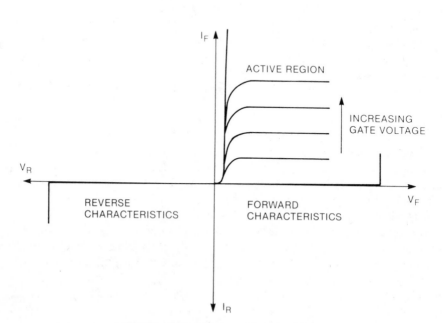

Fig. 7.5. Output characteristics of an IGT.

those of a forward-biased $P-i-N$ diode. These devices can be operated at high current densities even when designed to support high blocking voltages. As long as the gate bias is sufficiently large to produce enough inversion layer charge for providing electrons to the N-base region, the IGT forward conduction characteristics will resemble those of a $P-i-N$ diode. However, if the inversion layer conductivity is low, a significant voltage drop will occur across this region, as is observed in conventional MOSFETs. At this point, the forward current will saturate and the device will operate in its active region. The composite electrical characteristics are shown in Fig. 7.5.

To switch the IGT from its on-state to the off-state, it is necessary to discharge the gate by short-circuiting it to the emitter terminal. In the absence of a gate voltage, the inversion region at the surface of the P base under the gate cannot be sustained. Removal of the gate bias cuts off the supply of electrons to the N-base and initiates the turn-off process. In the presence of a high concentration of minority carriers injected into the N-base region during forward conduction, the turn-off does not occur abruptly. The collector current instead decays gradually with a characteristic time constant determined by the minority carrier lifetime.

The features offered by the IGT are its high forward conduction current density, low drive power due to its MOS gate structure, fully gate controlled output characteristics with gate turn-off capability, and a unique reverse blocking capability. These characteristics approach those of an ideal power switch suitable for many DC and AC power control circuits.

As already pointed out, the IGT contains a parasitic $P-N-P-N$ thyristor structure between the collector and the emitter terminals. If this thyristor latches up, the current can no longer be controlled by the MOS gate. It is important to design the device in such a manner that this thyristor action will be suppressed. This can be achieved by preventing the injection of electrons from the N^+ emitter region into the P base during device operation. The N^+ emitter region will begin to inject electrons into the P-base if the N^+-P junction becomes forward-biased by more than 0.7 V as the result of any lateral current flow in the P-base. This injection can be suppressed by designing narrow N^+ emitter regions and keeping the P-base sheet resistance low.

7.2.2. Device Analysis

As a result of the combination of an MOS gate structure with bipolar current conduction, the analysis of IGTs differs from that of other power devices. To understand its operation, a combination of the physics of the power MOSFET, the bipolar transistor, and the $P-i-N$ rectifier is necessary. In this section, the static and dynamic switching characteristics are analyzed, on the basis of the assumption that the regenerative turn-on of the parasitic thyristor structure does not take place during device operation. However, during device design it is necessary to ensure that this latch-up phenomenon is suppressed. This necessitates an analysis of the parasitic thyristor structure under both static and dynamic conditions.

7.2.2.1. Reverse Blocking Capability.

The reverse blocking capability of the IGT is provided by junction J3 (see Fig. 7.4) when a negative voltage is applied to the collector terminal. For negative collector terminal voltages, junction J3 becomes reverse-biased and its depletion layer extends primarily into the lightly doped N-base region. It is important to note that the breakdown voltage during reverse blocking is determined by an open-base transistor formed between the P^+ substrate, the N-base region, and the P-base region. This structure is prone to punch-through breakdown if the N base is too lightly doped.

The analysis of the breakdown voltage of an open-base transistor was provided in Chapter 3. To obtain the desired reverse blocking capability, it is essential to design the N-base resistivity and thickness optimally. The design is also affected by the minority carrier diffusion length. As a general guideline, the N-base width is chosen so that its thickness is equal to the depletion width at the maximum operating voltage plus one diffusion length. Since the forward voltage drop increases with increasing N-base width, it is important to perform an optimization of the breakdown voltage with the objective of maintaining a narrow N-base width.

When the blocking voltage requirement increases, the N-base width must be correspondingly increased:

$$2d \simeq \sqrt{\frac{2\epsilon_s V_m}{qN_D}} + L_p \tag{7.14}$$

where $2d$ is the N-base width, V_m is the maximum blocking voltage, and L_p is the minority carrier (holes in the n-channel device) diffusion length. At large blocking voltages, the depletion layer width becomes much larger than the diffusion length. The N-base width then increases approximately as the square root of the blocking voltage.

The reverse blocking capability is also impacted by the method of terminating the reverse blocking junction J3. In a typical IGT, the N-base width is in the range of 100 μm. Such devices require the use of a thick P^+ substrate on which the N-base region is epitaxially grown. The junction (J3) extends across the entire plane of the wafer. When preparing individual devices, it is necessary to cut through the reverse blocking junction and passivate this surface to achieve reverse blocking capability. As discussed in Chapter 3, a positive-bevel angle at junction J3 would be highly desirable to reduce surface electric field. This can be achieved by using a V-shaped blade during sawing of the wafers, followed by chemical etching and surface passivation to suppress surface leakage.

7.2.2.2. Forward Blocking Capability.

During operation of the IGT with positive collector voltage, the forward blocking capability of the IGT is provided by the P-base–N-base junction (J2). To operate in the forward blocking mode, the gate must be short-circuited to the emitter to prevent the formation of the surface inversion layer under the gate. Under a positive voltage applied to the collector, the junction (J2) becomes reverse-biased. A depletion layer extend from this junction on both sides. The breakdown voltage of this junction is limited by the considerations discussed in Chapter 6 for the power MOSFET with the additional impact of the presence of the lower junction (J3) in the IGT.

During design of the forward blocking capability, it is first necessary to examine the P-base profile. The P-base profile must be tailored with the objective of controlling the MOS channel threshold voltage while making sure that the depletion layer of junction J2 in the P base does not punch through to the N^+ emitter region. This limits the minimum channel length that can be achieved as discussed earlier in Section 6.3.1.1 for power MOSFETs. The design considerations for the P base and N^+ emitter depth for the IGT are identical to those already discussed in that section for the power MOSFET.

Beyond the P-base profile, the spacing between the DMOS cells also affects the forward blocking capability. The effect of cell spacing on the breakdown voltage of the DMOS structure was treated in Section 6.3.1.2 for power MOSFETs. As the DMOS cell spacing is increased, the breakdown voltage will decrease as a result of depletion layer curvature, leading to field crowding at the edges of the P base. In the case of the IGT, the N-base doping level is lower than that for a power MOSFET with the same breakdown voltage. A larger DMOS cell spacing can, consequently, be tolerated in the IGT.

In addition to these design considerations, the presence of the lower junction (J3) must be taken into account. The IGT breakdown voltage can be severely degraded by punch-through of the depletion layer of junction J2 to the lower junction (J3). For symmetrical devices, that is, devices with equal forward and reverse blocking capability, which are used in AC circuits, the N-base width must be chosen in accordance with Eq. (7.14), used to design the reverse blocking capability. In many DC circuit applications, where power MOSFETs and bipolar transistors are used, the IGT is not required to support reverse voltage. This offers the opportunity to reconfigure the device structure to optimize the forward conduction characteristics for a given forward blocking capability without consideration for the reverse blocking capability. The doping profile and electric field distribution for such asymmetric IGTs is compared with those of the symmetrical device in Fig. 7.6. In the asymmetric IGT structure, the uniformly doped N-base of the symmetrical IGT is replaced by a two-layer N-base region. This alters the electric field distribution as illustrated in the right-hand side of the figure. If the critical electric field for breakdown is assumed to be independent of the N-base doping level and the N-base doping in layer 1 is very low, the electric field distribution changes from the triangular case in the symmetrical IGT to a rectangular case in the asymmetric structure. Under these circumstances, if the N-buffer layer thickness d_2 is assumed to be approximately equal to one diffusion length L, the forward blocking capability of the asymmetric device will be twice as high as for the symmetrical device. In actual practice, the maximum electric field decreases with reduced N-base

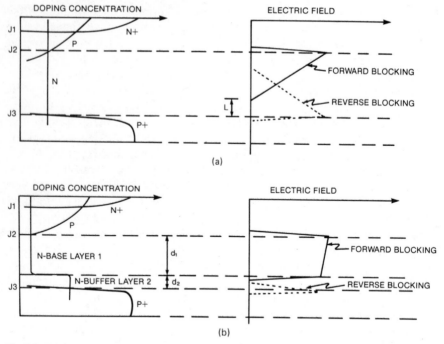

Fig. 7.6. Doping profile and electric field distribution for (a) symmetrical IGT and (b) asymmetric IGT.

doping as a result of redistribution of the field over a larger distance. This factor, in combination with the finite N-base doping level required to optimize the DMOS cell structure, results in an increase of forward blocking voltage by a factor of 1.5–2.0 for the same total N-base width.

During the design of the N-base profile for the asymmetrical IGT structure, it is important to keep the thickness of the N-buffer layer as small as possible. Since the depletion layer of the forward blocking junction (J2) must not punch through to the lower junction (J3), the charge Q_B in the buffer layer must be sufficient to allow the electric field to reduce to zero within it:

$$Q_B = (d_2 N_B) > \frac{\epsilon_s \mathscr{E}_c}{q} = 1.3 \times 10^{12} \tag{7.15}$$

where $\mathscr{E}_c$ is the critical electric field at breakdown in volts per centimeter and N_B is the buffer layer doping per cubic centimeter. This equation indicates that very narrow buffer layers can be used by arbitrarily raising the buffer layer doping concentration. An upper limit to the N-buffer layer doping concentration is imposed by a reduction of the injection efficiency of the junction (J3), which degrades the forward conduction characteristic. The optimum doping concentration and thickness of the N-buffer layer are in the range of 10^{16}–10^{17}/cm^3 and 10 μm, respectively.

7.2.2.3. Forward Conduction.

To operate the IGT in its forward conduction mode, it is necessary to create an inversion layer under the MOS gate that connects the N^+ emitter to the N-base region. As in the case of a power MOSFET, the gate voltage must be sufficiently above the threshold voltage to make the channel resistance small during current flow. Because of the large reduction in the drift layer resistance in the IGT due to the conductivity modulation arising from the injected carriers, the channel resistance must be designed to be much lower than that for a power MOSFET with the same breakdown voltage. Once the channel is formed, forward current flow will occur by the injection of minority carriers across the forward-biased junction (J3). The injected carrier density is typically 100–1000 times greater than the N-base doping level, resulting in a drastic reduction of its series resistance. This feature allows operation of the IGT at high current densities during forward conduction.

The analysis of the forward conduction characteristics can be performed by using two approaches. To understand these approaches, consider the equivalent circuit of the IGT shown in Fig. 7.7. This equivalent circuit consists of a coupled $P-N-P$ and $N-P-N$ transistor pair representing the four-layer parasitic thyristor structure with a MOSFET shunting the upper $N-P-N$ transistor. Also note the short-circuiting resistance R_s between the base and the emitter of the $N-P-N$ transistor. The magnitude of the short-circuiting resistance R_s is determined by the sheet resistance of the P-base region and the distance between the edge of the N^+ emitter at point A and its contact at point B (see Fig. 7.4).

If the short-circuiting resistance R_s is so small that the N^+ emitter does not become forward-biased above 0.6 V during forward conduction, the upper $N-P-N$ transistor can be assumed to be inactive for purposes of device analysis.

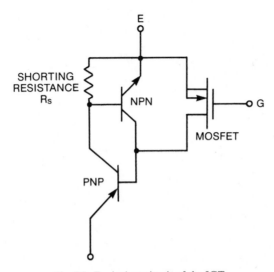

Fig. 7.7. Equivalent circuit of the IGT.

The forward conduction characteristics of the IGT can then be analyzed by using either of the two simplified equivalent circuits shown in Fig. 7.8. In the case of the circuit shown in Fig. 7.8a, the IGT is regarded as a $P-i-N$ rectifier in series with a MOSFET, whereas for the circuit shown in Fig. 7.8b, the IGT is regarded as a wide-base $P-N-P$ transistor driven by a MOSFET in a Darlington configuration. It is worth pointing out that these circuits are valid only for aiding device analysis and cannot be used to emulate IGT characteristics by using discrete devices. The model shown in Fig. 7.8b, based on a bipolar transistor driven by a power MOSFET, offers a more complete description of the IGT, but the $P-i-N$ rectifier/MOSFET model shown in Fig. 7.8a can be used to understand device behavior in many cases.

P–i–N Rectifier/MOSFET Model.   In the analysis of the forward conduction characteristics by using this model, the device is treated as composed of two sections, shown by the dashed lines in Fig. 7.9. A single current flow path is assumed to exist through the $P-i-N$ rectifier and MOSFET connected in series. The current-voltage relationship for the IGT can be developed by coupling the equations derived earlier in Chapter 5 for the current conduction in the $P-i-N$ rectifier and Chapter 6 for the MOSFET, using their common potential at region C in the structure.

According to the analysis of the forward conduction characteristics of a $P-i-N$ rectifier discussed in Section 5.2.1.3 on the FCD, the voltage drop across the $P-i-N$ rectifier ($V_{F,PiN}$) is related to its forward conduction current density $J_{F,PiN}$ by

$$J_{F,PiN} = \frac{2qD_a n_i}{d} F\left(\frac{d}{L_a}\right) e^{qV_{F,PiN}/kT} \tag{7.16}$$

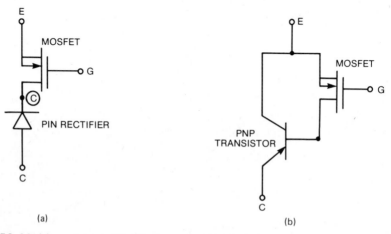

(a)   (b)

Fig. 7.8. Models used to analyze the forward conduction characteristics: (a) $P-i-N$ rectifier/ MOSFET model; (b) $P-N-P$ transistor/MOSFET model.

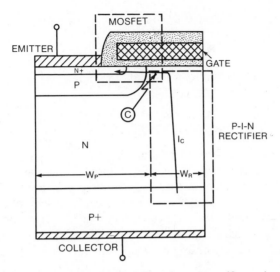

Fig. 7.9. Cross section of IGT with MOSFET and $P-i-N$ rectifier sections delineated.

The current density in the $P-i-N$ rectifier is related to the current by

$$J_{F,PiN} = \frac{I_C}{W_R Z} \tag{7.17}$$

where z is the length perpendicular to the cross section of the linear cell shown in Fig. 7.9. From these equations

$$V_{F,PiN} = \frac{kT}{q} \ln\left[\frac{I_C d}{2q W_R Z D_a n_i F(d/L_a)} \right] \tag{7.18}$$

The same current I_C flows through the MOSFET channel. The voltage drop across the MOSFET is related to the current flowing through it and the gate bias voltage by the relationship derived in Section 6.3.2.3 on power MOSFETs:

$$I_C = \frac{\mu_{ns} C_{ox} Z}{2 L_C} [2(V_G - V_T) V_{F,MOS} - V_{F,MOS}^2] \tag{7.19}$$

where the term $V_{F,MOS}$ is used in place of V_D to indicate that this expression applies to the MOSFET portion of the IGT. Here, L_C is the channel length. In the forward conduction mode, sufficient gate voltage is applied such that the forward voltage drop across the device is low. Under these conditions, $V_{F,MOS} \ll (V_G - V_T)$ and the MOSFET section of the devices is operating in its linear region. Thus

$$I_C = \frac{\mu_{ns} C_{ox} Z}{L_C} (V_G - V_T) V_{F,MOS} \tag{7.20}$$

The voltage drop across the MOSFET section is then given by

$$V_{F,MOS} = \frac{I_C L_C}{\mu_{ns} C_{ox} Z(V_G - V_T)} \tag{7.21}$$

The forward drop across the IGT is simply the sum of the voltage drop across the MOSFET and the P–i–N rectifier:

$$V_F = \frac{kT}{q} \ln\left[\frac{I_C d}{2q W_R Z D_a n_i F(d/L_a)}\right] + \frac{I_C L_C}{\mu_{ns} C_{ox} Z(V_G - V_T)} \tag{7.22}$$

From this equation, the forward conduction characteristics can be computed as a function of the gate bias voltage. Typical forward conduction characteristics will take the form shown in Fig. 7.10. Note the presence of the diode knee below which very little current flow occurs because of lack of injection from the junction (J3). Although the curves shown in Fig. 7.10 have a uniform spacing with increasing gate bias, in actual devices they tend to bunch together at high gate voltages because of a reduction in channel mobility with increasing gate field.

Several important conclusions about IGT performance can be derived by using the P–i–N rectifier/MOSFET model. First, it becomes apparent that there will be a knee in the output conduction characteristic below which very little current flow will occur. Consequently, the IGT is not suitable for applications requiring devices with forward voltage drops of less than 0.7 V. From this model it is also apparent that the IGT forward conduction current density will rise exponentially as in a P–i–N rectifier. This behavior has been verified experimentally in devices as shown in Fig. 7.11 [9]. In this figure the forward conduction characteristics of a bipolar transistor operating at a current gain of 10 and a power MOSFET operating at a gate bias of 20 V are also provided for com-

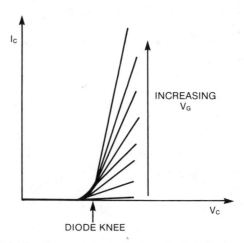

Fig. 7.10. Forward conduction characteristics emphasizing the diode knee and linear gate-controlled region.

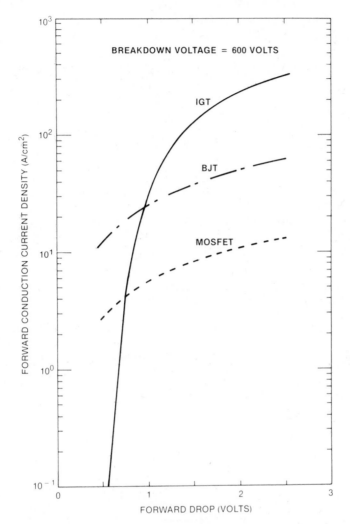

Fig. 7.11. Exponential increase in forward current density of an IGT compared with the linear characteristics of the bipolar transistor and power MOSFET.

parison. Once the forward drop exceeds 1 V, the current density of the IGT surpasses that of the power MOSFET and bipolar transistor. At typical operating forward voltage drops of 2–3 V for devices with breakdown voltages of 600 V, the IGT operates at a current density of approximately 20 times that obtained in the power MOSFET and 5 times that obtained in the bipolar transistor.

Using the $P-i-N$ rectifier/MOSFET model, it is also possible to derive the IGT characteristics under current saturation. When the gate bias is low, such that the voltage drop across the channel becomes significant, the current becomes limited by the MOSFET section. This portion of the device characteristic can be obtained by using Eq. (7.19) to describe the MOSFET portion. As in the case of a MOSFET, the collector current will saturate at

$$I_C = \frac{\mu_{ns}C_{ox}}{2}\frac{Z}{L}(V_G - V_T)^2 \qquad (7.23)$$

The P–i–N diode/MOSFET model can be used to understand the behavior of the forward conduction characteristics as a function of lifetime by taking into account the effect of changes in diffusion length on the P–i–N rectifier. It also allows analysis of the impact of increasing breakdown voltage by accounting for the effect of a wider N-base region on the P–i–N rectifier portion. It also allows analysis of the change in forward conduction characteristics with temperature. These effects are discussed further in other sections. The major shortcoming of this model is that it ignores a component of current in the device that flows into the P-base region. This component is included in the model based on a MOSFET driving a bipolar transistor.

Bipolar Transistor–MOSFET MODEL. The bipolar transistor–MOSFET model is based on the circuit shown in Fig. 7.8b. Here, the MOSFET provides the base drive current for the wide-base P–N–P bipolar transistor. These components of the device are illustrated in its cross section shown in Fig. 7.12. In this figure, the electron current I_e flowing through the MOSFET channel and the hole current I_h flowing across the P–N–P bipolar transistor section are also indicated. These currents are related through the current gain of the

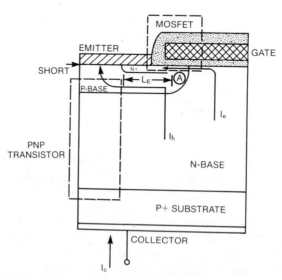

Fig. 7.12. Cross section of an IGT with MOSFET and bipolar transistor sections delineated.

wide-base *P–N–P* transistor:

$$I_h = \left(\frac{\alpha_{PNP}}{1 - \alpha_{PNP}}\right) I_e \tag{7.24}$$

The emitter current is the sum of these components:

$$I_E = I_h + I_e \tag{7.25}$$

$$I_E = \frac{1}{(1 - \alpha_{PNP})} I_e \tag{7.26}$$

Since no gate current component exists because of the very high impedance of the MOS structure, this equation also gives the collector terminal current. From these equations, it can be concluded that the MOSFET channel current I_e is a fraction of the terminal current I_C.

It should be noted that the *P–N–P* transistor must have a large base width to support the forward and reverse blocking voltages. Its current gain α_{PNP} is determined primarily by the base transport factor α_T, given by

$$\alpha_T = \frac{1}{\cosh(W_L/L_a)} \tag{7.27}$$

where W_L is the undepleted base width of the lower *P–N–P* transistor and L_a is the ambipolar diffusion length. The undepleted base width is essentially equal to the thickness of the *N*-base region because the depletion width is small during forward conduction. Because of the high injection level conditions prevalent in the *N*-base region, the ambipolar diffusion length L_a should be used during device analysis. This complicates the analysis because L_a is a function of injection level. The current gain α_{PNP} is typically about 0.5.

To compute the forward conduction characteristics, it is necessary to relate the forward drop across the device to the internal currents I_e and I_h. This can be performed by analyzing the voltage drop along the MOSFET current path I_e in the same manner as described earlier for the *P–i–N* rectifier/MOSFET model. For the bipolar transistor–MOSFET model, Eq. (7.22) must be modified by replacing I_C with I_e:

$$V_F = \frac{kT}{q} \ln\left[\frac{I_e d}{2qW_R Z D_a n_i F(d/L_a)}\right] + \frac{I_e L_C}{\mu_{ns} C_{ox} Z(V_G - V_T)} \tag{7.28}$$

The collector current is related to the MOSFET current I_e through Eq. (7.26). Substitution for I_e in Eq. (7.28) gives the forward conduction characteristic:

$$V_F = \frac{kT}{q} \ln\left[\frac{(1 - \alpha_{PNP})d I_C}{2qW_R Z D_a n_i F(d/L_a)}\right] + \frac{(1 - \alpha_{PNP})L_C I_C}{\mu_{ns} C_{ox} Z(V_G - V_T)} \tag{7.29}$$

The forward conduction characteristic described by this equation is similar to that derived by using the *P–i–N* rectifier/MOSFET model, except that the amplification of the output current by the *P–N–P* bipolar transistor is now accounted for.

The bipolar transistor current component I_h also alters the saturated current. When the gate bias is reduced so that the voltage drop in the MOSFET channel limits current flow, the electron current I_e is given by

$$I_e = \frac{\mu_{ns} C_{ox}}{2} \frac{Z}{L_C} (V_G - V_T)^2 \tag{7.30}$$

The saturated collector current is then

$$I_{C,sat} = \frac{1}{(1 - \alpha_{PNP})} \frac{\mu_{ns} C_{ox}}{2} \frac{Z}{L_C} (V_G - V_T)^2 \tag{7.31}$$

From this equation, an expression for the transconductance of the IGT in the active region can be obtained by differentiation with respect to V_G:

$$g_{ms} = \frac{1}{(1 - \alpha_{PNP})} \mu_{ns} C_{ox} \frac{Z}{L_C} (V_G - V_T) \tag{7.32}$$

The transconductance of the IGT is larger than for a power MOSFET with equal channel aspect ratio (Z/L_C). It depends on the gain of the wide-base bipolar transistor inherent in the IGT structure. Since the current gain α_{PNP} of the P–N–P transistor is typically about 0.5, the transconductance of the IGT is typically a factor of 2 times larger than that for a MOSFET with the same channel aspect ratio.

In the preceding analysis, it was assumed that the current will saturate and remain constant when the voltage drop across the MOSFET channel exceeds $(V_G - V_T)$. Such an assumption would result in an infinite drain output resistance. This feature is not observed in actual devices for two reasons. First, as in the case of power MOSFETs, the effective channel length decreases when the collector voltage increases. But in the IGT, an additional reduction in the drain output resistance arises from the presence of the bipolar transistor current flow. As the collector voltage increases, the current gain α_{PNP} of the bipolar transistor increases because its undepleted base width is reduced. The base transport factor of this transistor controls its current gain. Thus

$$\alpha_{PNP} = \alpha_T = \frac{1}{\cosh(W/L_a)} \tag{7.33}$$

where the undepleted base width W is given by

$$W = 2d - \sqrt{\frac{2\epsilon_s V_C}{q N_D}} \tag{7.34}$$

The impact of the bipolar transistor current flow on the collector output resistance becomes worse with increasing collector voltage. At low collector voltages, the undepleted base width is large and does not change rapidly with increasing collector voltage. Consequently, the collector output resistance is large at low collector voltages. When the collector voltage is raised to the point at which the N-base width becomes small, the current gain of the P–N–P transistor begins to grow rapidly with increasing collector voltage. The collector output resistance decreases rapidly with increasing collector voltage in this portion of the characteristics.

If the decrease in collector output resistance due to channel length reduction is neglected, it can be shown that the collector output resistance arising from an increase in the gain of the bipolar transistor will be given by

$$\frac{1}{r_C} = \frac{dI_C}{dV_C} = \frac{\sinh(W/L)}{[\cosh(W/L) - 2]^2} \sqrt{\frac{\epsilon_s}{2qN_DL^2V_C}} \tag{7.35}$$

A typical set of output characteristics (solid lines) for an IGT at high collector voltages is illustrated in Fig. 7.13. Note the reduction in the collector output resistance with increasing collector voltage. The collector output resistance of an IGT with uniformly doped N-base region will be lower than that for a power MOSFET. To obtain a higher collector output resistance, it is necessary to eliminate the increase in the current gain of the bipolar transistor with increasing collector voltage.

One approach to increasing collector output resistance is to use the asymmetric IGT structure illustrated in Fig. 7.6. Here, the depletion layer of the forward blocking junction (J2) expands at low collector voltages to the width d_1 of the lightly doped portion of the N-base region. The depletion width does not increase significantly with further increase in collector voltage. A pictorial comparison of the depletion width profile as a function of increasing collector voltage for the asymmetric structure with that for the symmetrical structure is provided in Fig. 7.14. It can be seen that the undepleted N-base width W changes rapidly with increasing collector voltage for the symmetrical IGT structure. In

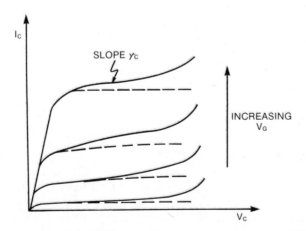

Fig. 7.13. Output characteristics of symmetrical (solid lines) and asymmetric (dashed lines) IGT structures.

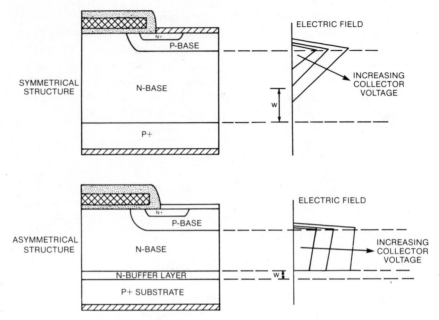

Fig. 7.14. Comparison of depletion layer spreading of the symmetrical and asymmetric IGT structures.

contrast, for the asymmetric IGT structure, the undepleted base width W remains equal to the thickness of the N-buffer layer d_2 for essentially all collector voltages. The current gain of the $P–N–P$ transistor then remains constant at

$$\alpha_{PNP} = \frac{1}{\cosh(d_2/L)} \tag{7.36}$$

for all collector voltages. This results in a higher collector output resistance as indicated by the dashed lines in Fig. 7.13. The collector output resistance of the asymmetric IGT structure will be similar to that of the power MOSFET as governed by the reduction in channel length with increasing collector voltage. It is worth pointing out again that the buffer layer reduces the injection efficiency of the lower junction. This can result in a higher forward drop during current conduction.

Another approach to increasing the collector output resistance is by utilizing its dependence on the minority carrier diffusion length L_a. The current gain α_{PNP} remains small until the depleted N-base width becomes comparable to the diffusion length. For devices with small diffusion lengths, this will occur only at high collector voltages. The collector output resistance will then remain large until the collector voltage reaches close to the breakdown voltage. An increase in the collector output resistance can thus be achieved by reducing the lifetime. This has been observed experimentally following electron irradiation [14].

7.2.2.4. Gate-Controlled Turn-off. An important characteristic of the insulated gate transistor is its gate turn-off capability. Since the current flowing through the MOS channel controls the output characteristics, the collector current flow can be interrupted by removing the gate bias voltage. To perform the transition from the on-state to the forward blocking state, the gate is connected to the source by an external circuit that allows discharging the gate capacitance. When the gate voltage falls below the threshold voltage of the MOS gate structure, the IGT channel inversion layer no longer exists. At this point, the electron current I_e ceases. If the gate turn-off is performed by using a low external resistance so as to abruptly reduce the gate voltage to zero, the collector current will decrease abruptly as the result of the sudden reduction of the channel current I_e to zero. After this point, the collector current will continue to flow because the hole current I_h does not cease abruptly. The high concentration of minority carriers injected into the N-base region supports hole current flow. As the minority carrier density decays as a result of recombination, it leads to a gradual reduction in the collector current.

Typical examples of the IGT voltage and current waveforms during device turn-off are provided in Fig. 7.15. The magnitude of the abrupt drop in the collector current ΔI_C at time t_1 is determined by the current gain of the P–N–P transistor:

$$\Delta I_C = I_e = (1 - \alpha_{PNP})I_C \qquad (7.37)$$

This relationship has been verified experimentally by using devices with two N-base widths processed with a broad range of minority carrier lifetimes [10]. Note that the collector voltage waveform depends on the nature of the load impedance. If the load is resistive, the collector voltage is a reflection of the collector current waveform. In the case of an inductive load, the collector voltage rises abruptly and often overshoots the supply voltage before settling down. The stress on the IGT is substantially higher in this case, but it has been demonstrated that IGTs can be operated under these conditions without the need for snubbers. This is due to the uniform current distribution in the device during turn-off as shown by two-dimensional computer simulation [22].

As the minority carrier lifetime is reduced, not only is the decay time constant of the tail reduced, but the magnitude of the initial abrupt drop ΔI_C is also greater because the current gain of the P–N–P transistor is smaller. Furthermore, for a given lifetime in the N-base region, the magnitude of the abrupt drop ΔI_C increases with increase in breakdown voltage capability because of the need to increase the width of the N-base region. This reduces the current gain of the P–N–P transistor, resulting a larger channel current component I_e.

The rate of change of the initial drop in collector current (ΔI_C) can be altered by controlling the rate of change of the gate voltage during turn-off. As in the case of the turn-off of a power MOSFET described in Section 6.5.1, the gate input capacitance is discharged through an input gate resistance, resulting in both a turn-off delay time and a controlled rate of fall of the collector current during the initial period. A typical set of collector current turn-off waveforms

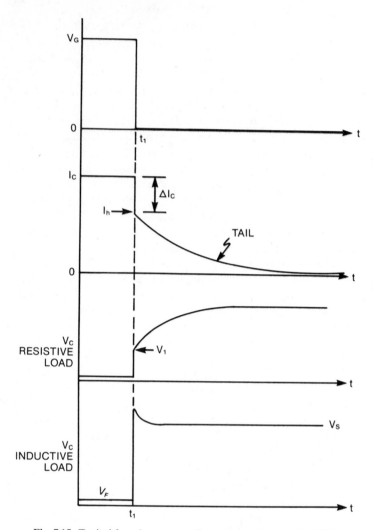

Fig. 7.15. Typical forced gate turn-off waveforms observed for IGTs.

are shown in Fig. 7.16. Note that as the initial drop ΔI_C of collector current slows down because of an increase in the gate resistance, the magnitude of the tail also decreases because some recombination can occur during the initial period. The gate resistance can be used to tailor the (dI/dt) during turn-off. This technique is especially important in the case of high-speed devices that exhibit large initial changes ΔI_C of collector current [11].

7.2.2.5. Latch-up. The presence of the parasitic four-layer thyristor structure in the IGT creates the possibility of the device latching up by regenerative action. This mode of operation is highly undesirable because it leads to loss of

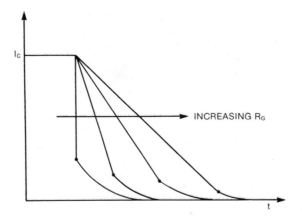

Fig. 7.16. Insulated-gate-transistor collector current turn-off waveforms tailored by changing gate resistance. (After Ref. 11, reprinted with permission from the IEEE © 1983 IEEE.)

control of the collector current by the applied gate voltage. Once the device has latched up, it can be turned off only by either externally turning off the collector voltage or reversing its polarity. In DC circuits, latch-up usually produces catastrophic failure of the devices as a result of excessive heat dissipation.

Several modes of IGT latch-up have been identified. They can be generally classified as the static and dynamic modes. In the static mode, the collector voltage is low and the latch-up occurs when the steady-state current density exceeds a critical value. The dynamic mode of latch-up occurs during switching. This mode involves both high collector current and voltage. The current density at which latch-up occurs in the dynamic mode is lower than that for the static mode. It is dependent on the load inductance. Both cases are treated in the paragraphs that follow.

Static Latch-up. The latch-up of an IGT during steady-state current conduction is called *static latch-up*. It has been found that, when the forward conduction current density exceeds a critical value, the device switches from its gate-controlled output characteristics to a single curve independent of the gate voltage [12]. This is evidence of the regenerative turn-on of the parasitic four-layer thyristor structure. It is easy to detect steady-state latching by the observed reduction in the forward voltage drop. Nondestructive measurements can be performed on devices, indicating that the steady-state latch-up does not occur at a small region of the device but is distributed over most of the active area.

To analyze the latch-up phenomenon, consider the IGT structure shown in Fig. 7.12. In the previous analyses, it was assumed that the upper $N-P-N$ transistor was inactive. During latch-up, this is no longer true. As indicated in Fig. 7.12, the hole current I_h flows into the P-base region and is collected by the emitter metal at the short circuit. To reach the short circuit, the hole current must traverse a path underneath the N^+ emitter region. The lateral hole current flow produces a forward bias across the N^+-P junction (J1) at point A. Although the P base collects holes along its entire surface, a simplified analysis of the latch-up phenomenon can be performed by considering an effective hole current I_h traversing under the entire N^+-emitter length L_E. The forward voltage drop across the N^+-P junction (J1) is then given by

$$V_A = R_P I_h \tag{7.38}$$

where R_P is the resistance of the P base under the N^+ emitter. Using Eqs. (7.24) and (7.26), it can be shown that

$$V_A = \alpha_{PNP} R_P I_C \tag{7.39}$$

where α_{PNP} is the current gain of the $P-N-P$ transistor. To activate the $N-P-N$ transistor and raise its gain to the point at which latch-up can occur, that is

$$\alpha_{NPN} + \alpha_{PNP} = 1 \tag{7.40}$$

it is necessary for the forward-bias on the N^+-P junction (J1) to reach 0.7 V. The steady-state latch-up current is, therefore, given by

$$I_{L,ss} = \frac{0.7}{\alpha_{PNP} R_p} \tag{7.41}$$

From this relationship, it can be concluded that the latching current can be increased by reducing the current gain α_{PNP} and the resistance of the P-base region. Both these effects have been confirmed experimentally.

A verification of the impact of the P-base resistance on the latching current has been performed by fabricating devices with different N^+-emitter lengths L_E. Since these devices were simultaneously fabricated, the P-base sheet resistances for all devices are identical. Thus

$$R_p \propto \rho_{\square P} L_E \tag{7.42}$$

where $\rho_{\square P}$ is the sheet resistance of the P-base. The latching current is then given by

$$I_{L,ss} \propto \frac{1}{\alpha_{PNP} \rho_{\square P} L_E} \tag{7.43}$$

From this expression, it can be concluded that the latching current should decrease inversely with increasing emitter length L_E. In Fig. 7.17, the measured variation (round points) of latching current density is plotted as a function of the length of the N^+ emitter [12]. The latching current density decreases nearly inversely with N^+ emitter length as predicted by Eq. (7.43). A slight deviation from the inverse relationship is observed as a result of the presence of the deep P^+ region diffused into the center of the DMOS cells.

The introduction of a deep, high-concentration P^+ diffusion at the center of each DMOS cell of the IGT is an extremely important processing technique

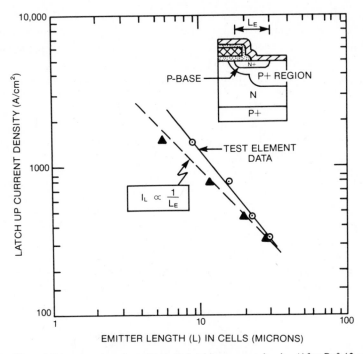

Fig. 7.17. Effect of N^+-emitter length on the static latching current density. (After Ref. 12, reprinted with permission from the IEEE © 1984 IEEE.)

for raising the latching current density. From Eq. (7.41), it is clear that the P-base doping should be increased to raise latching current density. Unfortunately, raising P-base doping increases the threshold voltage to unacceptable levels. To circumvent this problem, a two-layer P-base region has been developed for IGTs, as illustrated in the inset of Fig. 7.17. One of these layers is the original P-base region, which is tailored to obtain the appropriate threshold voltage. The second region is a deeper, higher doped P^+ diffusion performed from the center of each cell, which reduces the resistance of the P base under the N^+ emitter. Ideally, the deep P^+ diffusion should extend to the edge of the DMOS cell, that is, the edge of the N^+ emitter, but must not extend into the channel. For the devices used to obtain the data in Fig. 7.17, the depth of the P^+ region was 5 μm, resulting in a lateral extension of 4 μm. If this portion of the N^+-emitter length is subtracted from the total length, the data shown by the triangular points are obtained. These data are in excellent agreement with the inverse relationship between the latching current density and emitter length derived above.

It was also shown above that the static latching current should be a function of the current gain α_{PNP} of the P–N–P transistor. If the P–N–P transistor current gain is high, the hole current I_h passing through the P base is a large fraction of the collector current. The latching current will then be low. As the current gain α_{PNP} decreases, such as following lifetime reduction in the N-base region, the latching current will increase. An increase in the static latching current, following both electron and neutron irradiation to control the switching speed, has been observed experimentally [7, 13].

A more complete evaluation of the steady-state latching current can be derived by analyzing the current gain of the N–P–N transistor as a function of the hole current I_h flow. Consider the N–P–N transistor of gain α'_{NPN} with a resistor R_P across its base–emitter junction as shown in Fig. 7.18. The resistor R_P shunts a portion of the emitter current. The collector current is related to the emitter–base potential V_{BE} [15]:

$$I_C = I_0(e^{qV_{BE}/kT} - 1) \tag{7.44}$$

where

$$I_0 = \frac{qD_n\bar{n}_p}{L_n} \tag{7.45}$$

The emitter current I_E consists of two components—one (I_S) flowing through the shunting resistance and the other (I'_E) through the transistor:

$$I_E = I_S + I'_E \tag{7.46}$$

Expressing these currents in terms of V_{BE}, it can be shown that

$$I_E = \frac{V_{BE}}{R_p} + \frac{I_0}{\alpha'_{NPN}}(e^{qV_{BE}/kT} - 1) \tag{7.47}$$

The current gain of the transistor in the presence of the shunting resistance is

$$\alpha_{NPN} = \frac{I_C}{I_E} \tag{7.48}$$

$$= \alpha'_{NPN}\left[\frac{e^{qV_{BE}/kT} - 1}{e^{qV_{BE}/kT} - 1 + (V_{BE}\alpha'_{NPN}/I_0R_p)}\right] \tag{7.49}$$

This expression can be used to calculate the variation of the current gain with emitter–base potential. When the forward bias on the emitter is small, the current gain α_{NPN} tends to zero. As V_{BE} increases, the exponential terms in Eq. (7.49)

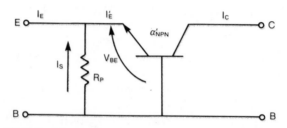

Fig. 7.18. Short-circuited emitter N–P–N transistor structure.

become dominant and the current gain α_{NPN} in the presence of the shunting resistor becomes equal to the current gain α'_{NPN} in the absence of the shunting resistor.

In the IGT, the hole current I_h flowing through the P-base forward-biases the N^+ emitter at its edge A and increases the emitter–base voltage V_{BE}. Consequently, the current gain of the short-circuited N–P–N transistor increases from zero to nearly unity with increase in hole current flow as illustrated in Fig. 7.19. Under steady-state current conduction, the voltage across the junction (J2) remains small and the current gain of the P–N–P transistor can be assumed to remain constant at a value

$$\alpha_{PNP,SS} = \frac{1}{\cosh(2d/L_a)} \tag{7.50}$$

where $2d$ is the width of the N-base region. Static latch-up will then occur when the hole current reaches a value at which

$$\alpha_{NPN,SS} = (1 - \alpha_{PNP,SS}) \tag{7.51}$$

This point is indicated in Fig. 7.19. Since the N-base width is much larger than the diffusion length, the preceding latch-up condition can be rewritten as

$$\alpha_{NPN,SS} = \left[1 - 2e^{-(2d/L_a)}\right] \tag{7.52}$$

When the lifetime in the N base is reduced to increase switching speed, the $\alpha_{PNP,SS}$ also decreases. This raises the current I_h at which α_{NPN} will be sufficiently high to create latch-up. Thus lifetime reduction has two effects on the static latch-up: (1) it reduces the magnitude of the hole current flowing into the P-base region [i.e., the ratio (I_h/I_C) decreases], and (2) it reduces the

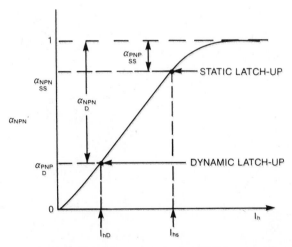

Fig. 7.19. Increase in N–P–N transistor current gain with hole current flow in an IGT.

current gain of the *P–N–P* transistor. These two effects combine to raise the static latching current density. The expected increase in the static latching current after electron and neutron irradiation has been confirmed experimentally [7, 13].

Dynamic Latch-up. Static latch-up occurs when the steady-state forward conduction current density of the IGT exceeds a critical value. Under switching conditions, these devices have been observed to latch up at a lower current density. It has been confirmed experimentally that if the IGT is subjected to rapid turn-on but does not undergo gate-controlled turn-off, its latching current density is identical to the static latching current density. The gate and collector current–voltage waveforms for this case are shown in Fig. 7.20. Note that to maintain this mode of operation, the anode voltage must be independently turned off while the IGT is in its steady-state forward conduction mode prior to turning off the gate voltage. This type of operation is encountered in an AC circuit where the device is turned on during the positive half cycle but does not have to undergo gate-controlled turn-off. In this application, the IGT can be

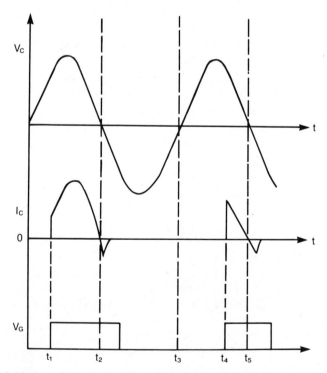

Fig. 7.20. Switching waveforms for an IGT operating in an AC circuit without undergoing forced gate turn-off.

used to control the rate of rise of the collector current (load current) at times t_1 and t_4 to minimize the generation of electrical noise in the power circuits. The gate voltage is maintained on beyond times t_2 and t_5, at which point the collector voltage reverses; this allows the IGT to undergo reverse recovery. It is not subjected to gate-controlled turn-off conditions. The time (t_1 and t_4) at which the IGT is turned on can be adjusted to control the power delivered to the load.

If the IGT is turned off when the collector voltage is positive by switching off the gate drive voltage, the dynamic latching current density is found to be less than the static latching current density. Since this gate-controlled turn-off capability is an important characteristic of the IGT for operation in DC circuits, the analysis of the reduction in latching current due to gate-controlled turn-off is very important.

First, consider the case of a resistive load in a DC circuit. The current and voltage waveforms relevant to this case are shown in Fig. 7.15 for the case where the gate series resistance is small. At time t_1, the collector voltage undergoes an abrupt increase from the forward voltage drop V_F of the IGT to a voltage V_1. This voltage is determined by the magnitude of the abrupt drop in collector current (ΔI_C). Since the IGT forward voltage drop V_F is much smaller than the supply voltage V_S, it follows that

$$V_1 = \Delta I_C R_L \tag{7.53}$$

The voltage V_1 is supported by the P-base–N-base junction (J2) resulting in a sudden reduction in the undepleted base width W of the P–N–P transistor. The current gain of the P–N–P transistor then exhibits a corresponding increase at time t_1. Before turn-off, that is, at $t < t_1$

$$\alpha_{PNP,SS} = \frac{1}{\cosh(2d/L_a)} = 2e^{-(2d/L_a)} \tag{7.54}$$

where $2d$ is the N-base width (the depletion layer width under forward bias being negligible). After turn-off, at $t = t_1$

$$\alpha_{PNP,DR} = \frac{1}{\cosh(W_{DR}/L_a)} \tag{7.55}$$

where DR refers to dynamic turn-off under a resistive load. Here

$$W_{DR} = 2d - \sqrt{\frac{2\epsilon_s V_1}{qN_D}} \tag{7.56}$$

Even though the P–N–P current gain undergoes an abrupt increase in time t_1, the hole current I_h flowing through the P base at this time is identical to its steady-state value prior to turn-off. The condition for latch-up during turn-off with resistive load is

$$\alpha_{NPN,DR} = (1 - \alpha_{PNP,DR}) \tag{7.57}$$

Since $\alpha_{PNP,DR}$ is larger then $\alpha_{PNP,SS}$, the hole current I_h at which latch-up occurs and hence the steady-state collector current I_C prior to turn-off, will be less than for static latch-up as indicated in Fig. 7.19.

If a linear relationship between α_{NPN} and I_h is assumed to exist for the region of interest shown in Fig. 7.19, the reduction in latching current density under dynamic turn-off will be given by

$$\frac{I_{L,DR}}{I_{L,SS}} = \frac{I_{h,DR}}{I_{h,SS}} = \frac{\alpha_{NPN,DR}}{\alpha_{NPN,SS}} \tag{7.58}$$

$$= \frac{(1 - \alpha_{PNP,DR})}{(1 - \alpha_{PNP,SS})} \tag{7.59}$$

Using Eqs. (7.54)–(7.59), it can be shown that

$$\frac{I_{L,DR}}{I_{L,SS}} = \frac{\exp(2d/L_a) - 2\exp(2\epsilon_s V_1/qN_D L^2)^{1/2}}{\exp(2d/L_a) - 2} \tag{7.60}$$

and substituting for V_1 from Eq. (7.53):

$$\frac{I_{L,DR}}{I_{L,SS}} = \frac{\exp(2d/L_a) - 2\exp(2\epsilon_s \Delta I_C R_L/qN_D L^2)^{1/2}}{\exp(2d/L_a) - 2} \tag{7.61}$$

The magnitude of the initial fall in collector current (ΔI_C) depends on the current gain of the P–N–P transistor during steady-state current conduction:

$$\Delta I_C = (1 - \alpha_{PNP,SS})I_C \tag{7.62}$$

Furthermore, if the forward voltage drop of the IGT in the on-state is negligible, then

$$V_S = I_C R_L \tag{7.63}$$

Using these equations, it can be shown that

$$\frac{I_{L,DR}}{I_{L,SS}} = \frac{\exp(2d/L_a) - 2\exp[2\epsilon_s(1 - \alpha_{PNP,SS})V_S/qN_D L_a^2]^{1/2}}{\exp(2d/L_a) - 2} \tag{7.64}$$

From this equation, it can be concluded that the degradation in latching current due to turn-off under a resistive load will become worse as the DC supply voltage increases. The degradation in latching current will also be more pronounced for devices with smaller P–N–P transistor current gain α_{PNP}. Thus, although the use of lifetime reduction techniques to increase the switching speed may at first sight be expected to raise latching current density because of a reduction in the current gains of the coupled transistors, in practice lifetime reduction can even cause a decrease in the dynamic latching current density in accordance with Eq. (7.64). Experimental measurements on n-channel IGTs subjected to increasing doses of electron irradiation have revealed a reduction in the latching current density with increasing dose.

These conclusions apply to the symmetrical IGT structure with a homogeneously doped N-base region. In the case of the asymmetric IGT structure, the N-base region 1 of thickness d_1 will be depleted at essentially all values of V_1, leaving an undepleted base width of d_2 (see Figs. 7.6 and 7.14). The latching current ratio for the asymmetric IGT structure can be derived as done earlier for the symmetrical structure:

$$\frac{I_{L,DR}}{I_{L,SS}} = \frac{1 - [\cosh(d_2/L_a)]^{-1}}{1 - \{\cosh[(d_1 + d_2)/L_a]\}^{-1}} \tag{7.65}$$

Note that it has been assumed that the depletion layer width under steady-state forward current flow is also negligible for the asymmetric IGT structure. As in the case of the symmetrical IGT structure, if the widths d_1 and d_2 are greater than the diffusion length L so as to obtain good forward blocking capability, Eq. (7.65) simplifies to

$$\frac{I_{L,DR}}{I_{L,SS}} = \frac{e^{d_2/L_a} - 2}{e^{d_2/L_a} - 2e^{-(d_1/L_a)}} \tag{7.66}$$

From this equation, it can be concluded that the reduction in latching current density for the asymmetric IGT structure under gate-controlled switching with resistive load will be worse than for the symmetrical IGT structure because d_1 is smaller than the undepleted base width for the symmetrical IGT structure even when $V_1 = V_S$.

In the case of an inductive load, the collector waveform takes a different shape when compared with the resistive load. As shown in Fig. 7.15, the collector voltage now rises abruptly to the power supply voltage at time t_1. The slight overshoot in collector voltage will be neglected to simplify the analysis. Then the current gain of the $P–N–P$ transistor rises to a value corresponding to $V_1 = V_S$ immediately after turn-off:

$$\alpha_{PNP,DI} = \frac{1}{\cosh(W_{DI}/L_a)} \tag{7.67}$$

where

$$W_{DI} = 2d - \sqrt{\frac{2\epsilon_s V_S}{qN_D}} \tag{7.68}$$

Here the subscript DI is used to indicate dynamic turn-off with an inductive load. From the analysis applied for the resistive load, the reduction of the latching current due to forced gate turn-off under an inductive load is given by

$$\frac{I_{L,DI}}{I_{L,SS}} = \frac{1 - \alpha_{PNP,DI}}{1 - \alpha_{PNP,SS}} \tag{7.69}$$

Using Eqs. (7.54) and (7.67), it can be shown that

$$\frac{I_{L,DI}}{I_{L,SS}} = \frac{\exp(2d/L_a) - 2\exp(2\epsilon_s V_S/qN_D L_a^2)^{1/2}}{\exp(2d/L_a) - 2} \tag{7.70}$$

Comparison with Eq. (7.64) shows that, in the case of the inductive load, the decrease in the latching current will be larger than for the resistive load because V_1 is always less than V_s.

The effect of lifetime reduction for the inductive load is quite different when compared with the resistive load. In this case, the ratio of dynamic latching current to the static latching current will be only weakly dependent on the lifetime through the diffusion length L_a in Eq. (7.70). The lifetime reduction will decrease the ratio of the hole current I_h to the collector current I_C and raise the static latching current density. This will result in an increase in the absolute dynamic latching current as shown in Fig. 7.21 [7].

In the preceding analysis it has been assumed that the resistance of the P base does not change with collector voltage. As the reverse voltage across the junction (J2) increases, its depletion layer not only extends into the N-drift layer and increases α_{PNP}, but also extends into the P-base region and raises its sheet resistance. An increase in the shunting resistance R_P occurs with increasing junction voltage. This will degrade the dynamic latching current even more than the amount calculated by using the preceding equations. A quantitative analysis of the effect of the increase in R_P on the dynamic latching current requires numerical analysis.

An additional significant observation is that the dynamic latching current density is a function of the gate series resistance R_G. It has been found experimentally that the dynamic latching current will increase with increasing series gate resistance [7]. When the gate resistance increases, the rate of rise of the collector voltage decreases, as illustrated in Fig. 7.22. This allows the hole cur-

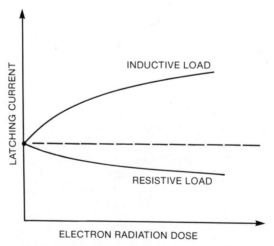

Fig. 7.21. Effect of lifetime reduction due to electron irradiation on the latching current under forced gate turn-off for resistive and inductive loads.

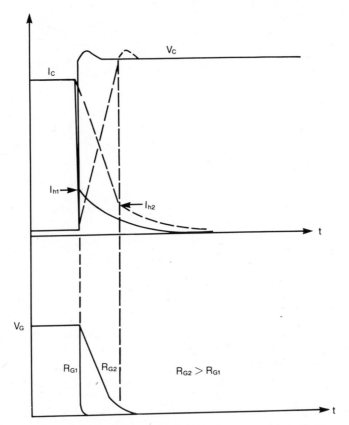

Fig. 7.22. Forced gate turn-off waveforms for two cases of gate series resistance.

rent I_h to decay to a lower value ($I_{h_2} < I_{h_1}$) with increasing gate resistance before the collector voltage (and hence α_{PNP}) increases during the turn-off. The dynamic latching current thus increases as the gate resistance is increased. A typical trend based on experimental data [7] is shown in Fig. 7.23.

Since IGTs are typically used in circuits that take advantage of their gate turn-off capability, it is necessary to design the structure such that the dynamic latching current density exceeds the maximum operating current density. This is a much more stringent design requirement than that needed to meet a static latch-up criterion. An ideal design for the IGT is one whereby the current will saturate at a level three to five times higher than the steady-state operating current level (to allow for surge conditions) for the specified gate drive voltage. At the same time, the dynamic latching current level must exceed the collector saturation current. Such a design would prevent IGT latch-up under all circuit operating conditions and prevent destructive failure of devices. The feasibility of such designs has been verified experimentally [22].

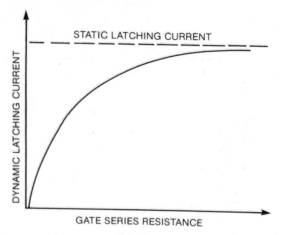

Fig. 7.23. Increase in the latching current of IGTs with increase in gate series resistance.

7.2.3. Frequency Response

When initially developed, the IGT exhibited a relatively slow switching speed. A large forced gate turn-off time was observed to result from the presence of a slowly decaying collector current as shown in Fig. 7.15. The long tail in the collector current waveform arises from the presence of the large minority carrier charge in the N-base region injected during forward conduction. This stored charge must be removed by recombination. It has now been found that the switching speed can be improved at the disadvantage of a reduction in forward conduction capability.

7.2.3.1. Turn-off Time.   The switching speed of the IGT can be increased by the introduction of recombination centers in the N-base region to reduce the minority carrier lifetime. In general, minority carrier lifetime reduction has been achieved by two approaches: either by diffusing transition element impurities into silicon or by means of high-energy particle bombardment to create lattice damage. In both cases, deep levels are introduced in the silicon band gap that are responsible for enhancement of the recombination rate for minority carriers. Among these approaches, electron irradiation offers the advantages of being a clean process that can be carried out at room temperature following complete device fabrication and preirradiation testing. This process offers greater precision over the lifetime control with tighter distribution in device characteristics when compared with the diffusion of deep-level impurities such as gold.

The measured reduction in the forced gate turn-off time with increasing electron irradiation dose is shown in Fig. 7.24. During these measurements, the load resistance was adjusted so as to maintain a constant collector current of 10 A before turn-off for different values of collector voltage. It can be seen that

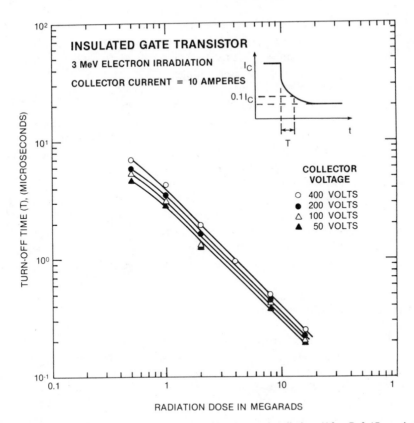

Fig. 7.24. Reduction in turn-off time of IGTs due to electron irradiation. (After Ref. 17, reprinted with permission from the IEEE © 1984 IEEE.)

the turn-off time decreases with increasing radiation dose. Since the turn-off time is dominated by the recombination tail, the turn-off time can be anticipated to decrease in proportion to the minority carrier lifetime. The minority carrier lifetime τ_f after irradiation is dependent on the preirradiation lifetime τ_i and the electron irradiation dose ϕ by the relationship

$$\frac{1}{\tau_f} = \frac{1}{\tau_i} + K\phi \tag{7.71}$$

where K is the radiation damage coefficient. At high doses, the first term in Eq. (7.71) is negligible if the initial lifetime is large, and the lifetime after irradiation decreases in inverse proportion to the radiation dose. In Fig. 7.24, the turn-off time can be observed to decrease inversely with increasing radiation dose at levels above 1 Mrad. Since the gate turn-off time before radiation ranged from 15 to 30 μsec, at doses below 1 Mrads the turn-off time is determined by a combination of both the initial lifetime and the radiation dose, so that the curves exhibit a departure from the inverse relationship observed at the higher doses.

The turn-off time is also a function of the collector voltage and current. In Fig. 7.24, the turn-off time is observed to increase with increasing collector voltage. This phenomenon can be ascribed to the need to establish larger depletion layer widths with increasing collector voltage.

An unusual phenomenon observed for IGTs is the reduction in the turn-off time with increasing collector current. Typical experimental results are shown in Fig. 7.25. Note that this effect occurs both before and after electron irradiation. As the collector current increases, it has been found that the current gain of the $P-N-P$ transistor decreases. This causes an increase in the channel current component ΔI_c in the turn-off waveform (Fig. 7.15) with increasing collector current. For these power devices, the turn-off time is defined as the time taken for the collector current to decay to 10% of its steady-state value. The waveforms in Fig. 7.26 illustrate the impact of increasing collector current on the turn-off time under these circumstances, assuming that the current tail remains invariant. As I_C increases, the turn-off time (as defined above) decreases because the 10% point (A) moves up the decay tail in the collector current wave-

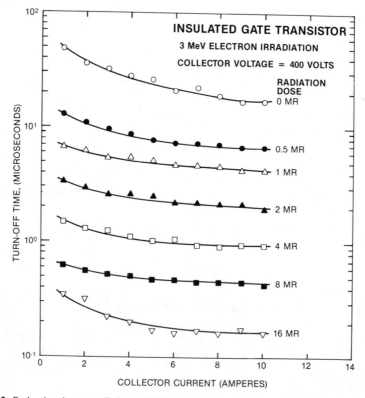

Fig. 7.25. Reduction in turn-off time of IGTs with increasing collector current. (After Ref. 17, reprinted with permission from the IEEE © 1984 IEEE.)

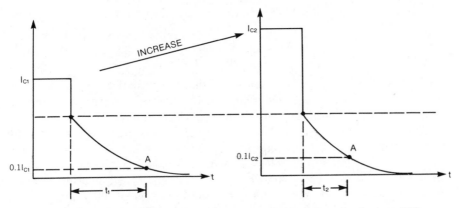

Fig. 7.26. Forced gate turn-off waveforms at two collector currents for the same IGT.

form. Note that even though the turn-off time decreases, the switching losses due to the current tail will be independent of the collector current.

An analysis of the current tail can be performed using the charge control model applied to an open-base bipolar transistor [26]. In this case, it can be shown that

$$i_B(t) = \frac{Q}{\beta \tau_B} + \frac{dQ}{dt} \qquad (7.71a)$$

$$= 0 \qquad (7.71b)$$

where Q is the stored charge in the base, β is the common emitter current gain of the P–N–P transistor, and τ_B is the base transit time given by

$$\tau_B = \frac{[\gamma(b + 1) - 1][1 - \text{sech}(W_B/L_a)]\tau}{2\gamma - [\gamma(b + 1) - 1][1 - \text{sech}(W_B/L_a)]} \qquad (7.71c)$$

Here, γ is the injection efficiency, b is the mobility ratio (μ_n/μ_p), W_B is the width of the N-base region, and τ is the excess carrier lifetime.

The corresponding equation for the output (collector) current of the IGT is

$$i_C(t) = -\frac{Q(1 + \beta)}{\beta \tau_B} - \frac{dQ}{dt} = -\frac{Q}{\tau_B} \qquad (7.71d)$$

$$i_C(t) \simeq i_C(0)\, e^{-(t/\beta \tau_B)} \qquad (7.71e)$$

where $i_C(0)$ is the collector current at the beginning of the current tail.

The current tail is an exponentially varying waveform with a characteristic time constant determined by the free-carrier lifetime τ via Eq. (7.71c). In the case of symmetrical blocking high-voltage devices with fast switching speed, the diffusion length L_a is small compared with the base width W_B and the injection efficiency γ is close to unity. The parameter τ_B is then given by

$$\tau_B = \frac{b\tau}{(2 - b)} \tag{7.71f}$$

From this expression, it can be concluded that the current tail will decay at a rate proportional to the free-carrier lifetime. This is consistent with the observed reduction in the turn-off time with increasing electron radiation dose.

These results demonstrate that electron irradiation is effective in reducing the turn-off time in IGTs. A reduction in the turn-off time from over 20 μsec to less than 200 nsec can be achieved by using a radiation dose of 16 Mrad. It should be noted that when the turn-off tests were conducted by using a power MOSFET at a drain current of 10 A and a drain voltage of 400 V, a gate turn-off time of 150 nsec was measured. This turn-off time was determined by the gate drive impedance (50 Ω) of the circuit used to discharge the considerably higher gate capacitance of the power MOSFET. This indicates that for typical gate drive circuits, the switching speeds in IGTs can approach those of power MOSFETs. Faster switching speeds can be achieved in power MOSFETs at the expense of reducing the gate drive impedance.

7.2.3.2. Forward Conduction.

In bipolar power devices, an improvement in switching speed is generally accompanied by a reduction in forward conduction capability. In the case of the IGT, the device behave like a $P-i-N$ rectifier during forward current conduction, and high-level injection of minority carriers into the drift region drastically lowers its resistance by conductivity modulation. When the diffusion length is reduced by the electron irradiation, the modulation of the drift region is also reduced and the IGT forward current density decreases. The measured reduction in the forward current with increasing electron irradiation dose is shown in Fig. 7.27 [17]. It can be seen that at a fixed forward drop of 2 V, the current decreases from 12 A before radiation to 1.5 A after radiation at a dose of 16 Mrad.

Another more commonly used measure of the impact of lifetime reduction is in terms of an increase in forward voltage drop at a fixed collector current. For example, at a fixed collector current of 10 A, the forward drop can be seen to increase from 1.8 V before radiation to about 5 V after radiation at a dose of 16 Mrad. This can be analyzed by using Eq. (7.29) and taking into account the effect of the lifetime reduction due to electron irradiation on α_{PNP} and L_a. The increase in switching speed of the IGTs is accompanied by a loss in current handling capability. As a point of comparison, a power MOSFET of the same size as these IGTs would have a forward drop of 35 V at a current of 10 A. From this, it can be concluded that even after increasing the speed of the IGT to obtain a turn-off time of 200 nsec, its forward conduction characteristics remain superior to those of the power MOSFET. It should be noted that as the turn-off time is reduced further, the IGT conduction characteristics will approach that of the power MOSFET.

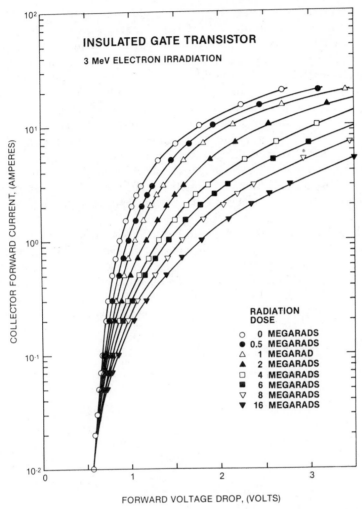

Fig. 7.27. Effect of electron irradiation on the forward conduction characteristics of IGTs. (After Ref. 17, reprinted with permission from the IEEE © 1984 IEEE.)

7.2.3.3. Trade-off Curves.

The results described above demonstrate that electron irradiation can be utilized to control the switching speed of IGTs. The radiation has been shown to be capable of reducing the gate turn-off time from over 20 μsec to 200 nsec. However, this increase in the switching speed is accompanied by an increase in the forward voltage drop. Since a short gate turn-off time is desirable in order to reduce switching losses and a low forward voltage drop is desirable in order to reduce the conduction losses, it becomes necessary to perform a trade-off between these device characteristics. This can be most effectively performed by using a plot of the forward voltage drop versus the gate turn-off time as shown in Fig. 7.28 [17] for a typical device. Depending on the application, the appropriate device characteristics can be selected by choosing the irradiation dose. In the case of circuits operating at low frequencies with large duty cycles where the conduction losses dominate over the switching losses, IGTs with turn-off times in the 5–20-μsec range would be the best. An example of this is line-operated phase-control circuits. For medium-frequency circuits with shorter duty cycles, where the switching losses are comparable to the conduction losses, IGTs with turn-off times in the range of 0.5–20 μsec would be appropriate. An example of this type of application would be for AC motor drives operating at frequencies ranging from 1 to 10 kHz. For high-frequency circuits, the switching losses would become dominant and IGTs with gate turn-off times ranging from 100 to 500 nsec would be required. An example of such circuits would be in switching power supplies operating at between 20 and 100 kHz.

In all applications, it is desirable to have the lowest possible forward voltage drop for the switching speed dictated by the circuit. Through numerical computer analysis, it has been shown that a better trade-off curve can be obtained by changing from a uniform lifetime reduction throughout the N-base region to an optimum carrier lifetime profile [27]. From this analysis, it can be inferred that a higher concentration of deep levels localized at the lower

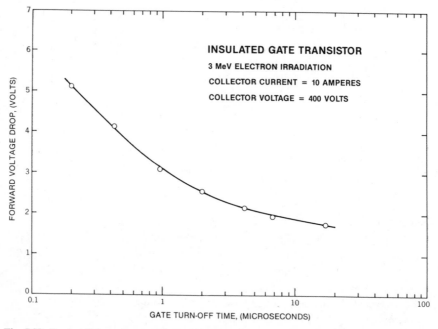

Fig. 7.28. Trade-off curve between forward voltage drop and turn-off time for n-channel IGTs. (After Ref. 17, reprinted with permission from the IEEE © 1984 IEEE.)

junction (J3) of the IGT is preferable. Such a profile for deep levels can be achieved by using high-energy proton implantation. In this case, the lifetime reduction due to the damage occurs primarily near the range of the protons in silicon. By use of 3-MeV protons, the damage can be localized at a depth of 100 μm from the silicon surface. This is close to the lower junction (J3) of a symmetrical IGT designed to support 600 V. Experiments conducted on these devices have confirmed that a better trade-off curve can be achieved when compared with 3-MeV electron irradiation [28]. A reduction in the forward voltage drop by 0.5 V has been reported at a switching speed of 1 μsec.

7.2.4. Complementary Devices

Complementary devices are often needed in power circuits such as adjustable-speed motor drives. For these applications, p-channel power MOSFETs have been developed to complement the n-channel devices. Because of the lower mobility for holes in silicon, p-channel power MOSFETs have a higher specific on-resistance than do n-channel devices. These devices must be made about three times larger in area than n-channel devices to handle the same power rating. It has been demonstrated that this penalty in size is not experienced with IGTs.

In the IGT, the drift region is flooded with minority carriers during forward conduction. Since the concentration of the free carriers greatly exceeds the doping level, the carrier transport is determined by ambipolar diffusion and drift, which is similar for both n-channel and p-channel devices. These devices differ only in terms of the contribution of the channel resistance. For devices with long turn-off times, the lifetime in the drift region is large and the gain of the lower transistor (P–N–P for n-channel and N–P–N for p-channel IGTs) is high. The channel contribution is then negligible. Consequently, complementary IGTs with slow switching speed are found to exhibit nearly identical forward conduction characteristics [8].

When the switching speed is increased by lifetime reduction, the current gain of the wide-base transistor in the IGT structure is reduced. The current contribution from the channel then grows with increasing switching speed. Since the mobility for holes in the channel is smaller than that for electrons, the rate of increase in forward voltage drop with decreasing turn-off time will be worse for p-channel IGTs.

The results of experimental measurements performed on 600-V n- and p-channel IGTs over a range of switching speeds by using electron irradiation are shown in Fig. 7.29 [8]. As expected, the n- and p-channel trade-off curves merge for turn-off times over 10 μsec. At shorter turn-off times, the trade-off curve for the p-channel devices lies above that for the n-channel IGTs. It should be noted that even for turn-off times of 1 μsec, the difference in the forward voltage drop is less than 50% compared with 300% for power MOSFETs. When the turn-off time is reduced to the point at which the channel current becomes

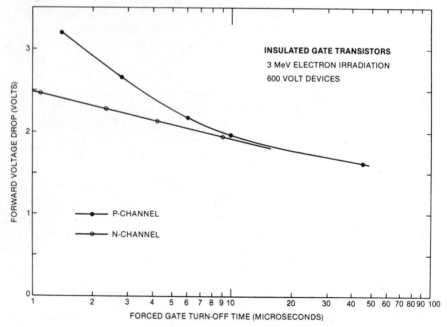

Fig. 7.29. Comparison of the trade-off curves for *n*-channel and *p*-channel IGTs. (After Ref. 8, reprinted with permission from the IEEE © 1984 IEEE.)

a significant proportion of the total collector current, the difference in forward voltage drop will approach the factor of 3 as in the case of power MOSFETs. Nevertheless, *p*-channel IGTs continue to be attractive because their absolute forward drop is still far lower than that for *p*-channel power MOSFETs.

7.2.5. High-Voltage Devices

In the case of the power MOSFET, it was shown that the on-resistance increases sharply with breakdown voltage. This has prevented the development of power MOSFETs with high current carrying capability at high voltages. The increase in on-resistance with breakdown voltage arises from an increase in the resistivity and thickness of the drift region required to support the operating voltage. In the case of the IGT, the drift region resistance is drastically reduced by the high concentration of injected minority carriers. The contribution to the forward drop from the drift region then becomes dependent on its thickness but independent of its original resistivity. The drift region thickness has the additional effect of altering the current gain α_{PNP} of the wide-base transistor. When the blocking voltage capability of the IGT is increased by increasing the drift region width, the current gain α_{PNP} is reduced. More channel current is then observed for these devices, and their forward voltage drop increases.

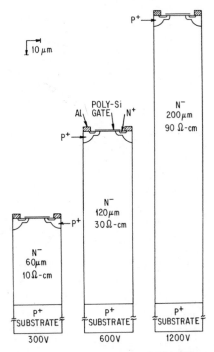

Fig. 7.30. Scaled drawing of the cross section of 300-, 600-, and 1200-V IGTs.

A comparison of the forward conduction characteristic of 300-, 600-, and 1200-V IGTs has been experimentally performed [18]. For visual comparison, a scaled drawing of the device structures is provided in Fig. 7.30. These devices had the symmetrical blocking structure. The drift layer increases from 60 μm for the 300-V devices to 200 μm for the 1200-V devices. The impact of the increase in drift region thickness on the forward conduction characteristics is shown in Fig. 7.31 for the case of IGTs with identical turn-off times. As expected, the forward voltage drop increases with increase in voltage rating. An important point to note is that this effect is much greater for the power MOSFETs, as indicated by the dashed lines. As a consequence, the IGT:power MOSFET current density ratio at a forward drop of 2 V grows from 10 for the 300-V devices to 150 for the 1200-V devices. A similar effect has been reported for the asymmetric IGT structure [19]. These results indicate that the forward conduction current density of the IGT will decrease approximately as the square root of the breakdown voltage. This moderate rate of reduction in current density has allowed the rapid development of devices with both high current and high voltage capability.

The trade-off curve between forward voltage drop and turn-off speed changes when the blocking voltage capability (and hence the drift region width) is altered.

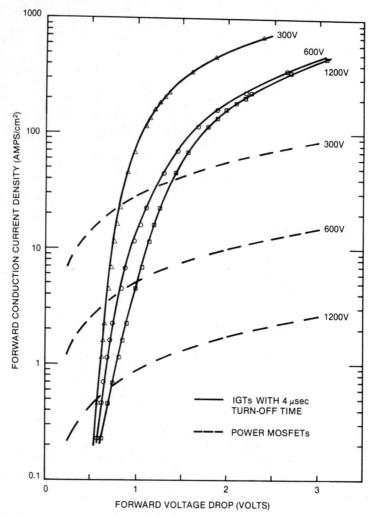

Fig. 7.31. Comparison of the forward conduction characteristics of 300-, 600-, and 1200-V IGTs with a turn-off time of 4 μsec. (After Ref. 18, reprinted with permission from the IEEE © 1985 IEEE.)

A typical set of curves for 300-, 600-, and 1200-V *n*-channel IGTs is provided in Fig. 7.32 for the case when electron irradiation is used as a lifetime control method [18]. The forward voltage drop increases more sharply with decreasing turn-off time for the higher-voltage devices. This behavior is due to the wider drift region in the higher voltage structures. These trade-off curves demonstrate that IGTs with blocking voltage capability of over 1000 V can be developed with turn-off speeds of well below 1 μsec. A 1200-V, 20-A, latch-proof device structure has been reported [19].

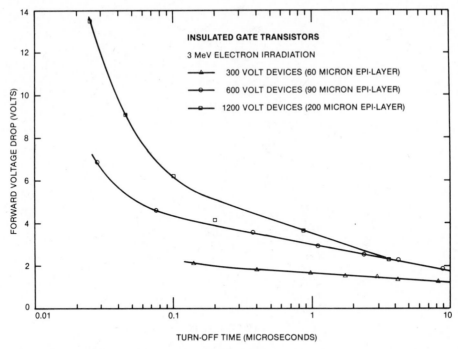

Fig. 7.32. Trade-off curves for 300-, 600-, and 1200-V *n*-channel IGTs. (After Ref. 18, reprinted with permission from the IEEE © 1985 IEEE.)

7.2.6. High-Temperature Performance

One of the important assets of the IGT is its excellent high-temperature forward conduction characteristic. This feature makes the device attractive for applications in which high ambient temperatures may be encountered. Devices have been operated successfully with heat-sink temperatures approaching 200°C. To take advantage of this feature, it is imperative to raise the latching current density even higher than described previously to maintain fully gate controlled operation.

7.2.6.1. Forward Conduction.   When large gate bias voltages are applied to the IGT, the conductivity of the inversion region, in the channel under the MOS gate, becomes very high. Under these conditions, the forward current in the IGT becomes insensitive to the gate bias voltage and the forward *I–V* characteristics are similar to those of a *P–i–N* rectifier. A typical set of forward conduction characteristics measured at temperatures ranging from 25 to 200°C is shown in Fig. 7.33 [20]. It can be seen that the collector current increases exponentially with increasing collector voltage at currents below 0.1 A. This

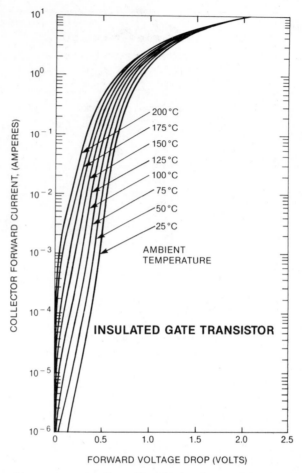

Fig. 7.33. Effect of temperature on the forward conduction characteristics of an IGT under strong gate bias. (After Ref. 20, reprinted with permission from *Solid State Electronics*.).

behavior is similar to that of a $P-i-N$ diode operating at high injection levels where the current is given by

$$I_C = \frac{1}{\alpha_{PNP}} \frac{2qD_aA}{L_a} \, n_i \, e^{qV_C/2kT} \tag{7.72}$$

where D_a and L_a are the ambipolar diffusion coefficient and diffusion length, respectively; A is the diode area, n_i is the intrinsic carrier concentration; q is the electronic charge; k is Boltzmann's constant; and T is the absolute temperature. At anode currents above 0.1 A the anode forward drop increases more rapidly with increasing anode current. This behavior is also similar to that observed in $P-i-N$ diodes and is due to a decrease in the ambipolar diffusion length L_a at high injection levels because of carrier–carrier scattering and Auger recombination. Another factor that can increase the forward drop is a decrease in the anode injection efficiency at higher currents as described earlier in Chapter 4 for the FCD.

The analysis based on current flow at high injection levels in a $P-i-N$ diode can be applied only for forward voltage drops of up to 1 V. Above this value, the voltage drop in the channel can no longer be neglected. This drop is dependent on the electron mobility in the inversion layer, which decreases with increasing temperature. As a result, the series resistance of the channel increases when the temperature is raised. In addition, the voltage drop across the drift region increases because of a reduction in the ambipolar diffusion length.

These effects are most apparent when the IGT forward conduction characteristics are shown on a linear scale to emphasize the high-current region. The forward characteristics shown in Fig. 7.34 can be viewed to consist of two segments: a diode drop portion followed by a resistive portion. The diode voltage drop can be seen to decrease from 0.83 V at room temperature to about 0.5 V at 200°C. At the same time, the resistance of the second segment increases from

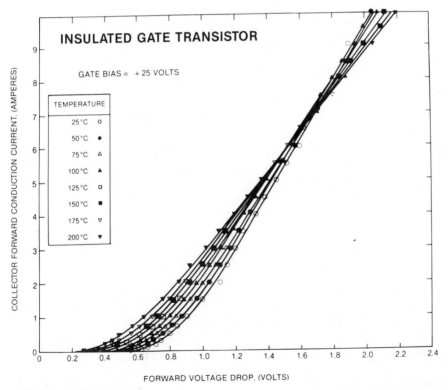

Fig. 7.34. Temperature dependence of the forward conduction characteristics of an IGT in the high-current region. (After Ref. 20, reprinted with permission from *Solid State Electronics*.)

0.124 Ω at room temperature to 0.169 Ω at 200°C. It is important to note that the decrease in the diode forward drop compensates to a large extent for the increase in channel resistance. As a result of this, the IGT forward current carrying capability is relatively unaffected by an increase in the ambient temperature.

The IGT forward voltage drop is shown as a function temperature at various fixed forward currents in Fig. 7.35. It can be seen that a decrease in the forward drop is observed at currents below 5 A and an increase is observed at currents above 9 A, whereas the forward drop at a collector current of 7 A is found to remain independent of temperature. This is a unique feature of the IGT in contrast to other power switching devices. As an example, in the power MOSFET, the decrease in electron mobility with increasing temperature results in an increase in the forward voltage drop by a factor of 3 between room temperature and 200°C. Compared with the IGT, this requires a significant derating of the current handling capability of power MOSFETs with increasing temperature. This feature makes the IGT well suited for applications in which high-temperature ambients are encountered. It is worth pointing out that the small positive temperature coefficient of the forward drop at higher current levels is beneficial in ensuring homogeneous current distribution within chips

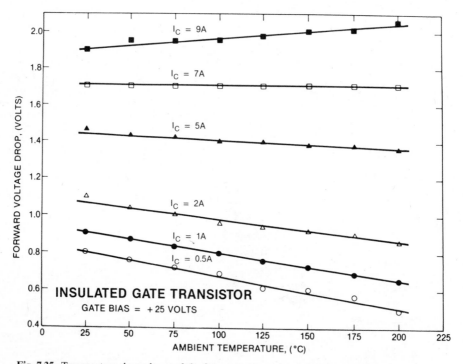

Fig. 7.35. Temperature dependence of the forward voltage drop of IGTs. (After Ref. 20, reprinted with permission from *Solid State Electronics*.)

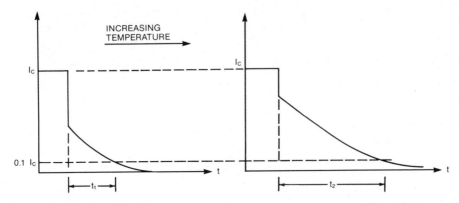

Fig. 7.36. Effect of an increase in temperature on the IGT collector turn-off waveform.

and for achieving good current sharing when paralleling devices. The paralleling of multiple IGTs without matching devices or providing emitter ballasting has been successfully used to achieve high current circuit operation [11].

7.2.6.2. Switching Speed. In the IGT, the gate turn-off time is dominated by the tail in the collector current waveform. As the temperature increases, the minority carrier lifetime in the drift region has been found to increase. This not only slows down the recombination process, but increases the $P-N-P$ transistor gain. The latter effect produces a smaller initial drop ΔI_C in the collector current during turn-off. The resulting change in the collector current waveforms with temperature is illustrated in Fig. 7.36. Both these phenomena cause an increase in the turn-off time with increasing temperature. A typical example of the measured change in turn-off time with temperature is provided in Fig. 7.37 [20]. In this case, the turn-off time shows an approximately linear change with temperature with a 50% increase between room temperature and 200°C.

7.2.6.3. Latching Current. One problem encountered in operation of the IGT at high current levels has been found to be latch-up of the parasitic $P-N-P-N$ thyristor structure inherent in the device structure. Latch-up of this tyristor can occur, causing loss of gate-controlled current conduction. At room temperature, it has been found that, in the early developmental devices, the latching current was about six times greater than the average current level at which the device is expected to operate. Since the current gains of the $N-P-N$ and $P-N-P$ transistors increase with increasing temperature, the latching current has been found to decrease with increasing temperature. This effect is also aggravated by an increase in the resistance R_P of the P-base with temperature.

An example of the measured reduction in the latching current with temperature is shown in Fig. 7.38 [20]. In this figure, the latching current was measured under dynamic conditions with a resistive load. The latching current is observed to decrease by a factor of 2 between room temperature and 150°C. A similar

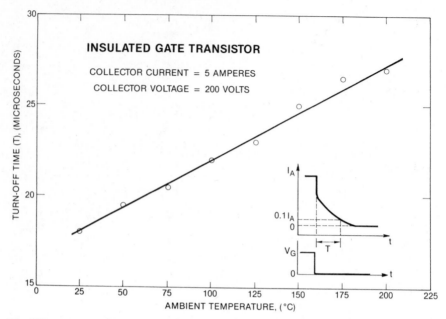

Fig. 7.37. Increase in IGT forced gate turn-off time with increasing temperature. (After Ref. 20, reprinted with permission from *Solid State Electronics.*)

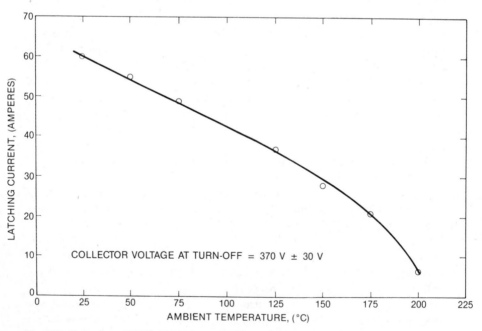

Fig. 7.38. Reduction of IGT latching current with increasing temperature. (After Ref. 20, reprinted with permission from *Solid State Electronics.*)

reduction of the latching current with temperature has been observed under dynamic switching with an inductive load [7]. For an operating current of 10 A, these devices can be operated with a surge current margin of a factor of 2 for junction temperatures of up to 175°C. Recent progress in improving the device structure and processing has allowed raising of this margin to over 10, making the IGTs useful in applications with ambient temperatures of even 200°C. Devices that current limit (instead of latch-up) have been demonstrated to operate at up to 125°C [22].

7.2.7. Device Structures and Fabrication

At first glance, the cross-sectional structure of the IGT appears to be identical to that of the MOS gated thyristor. The difference between these structures, which prevents the latch up phenomenon in the IGT, is the much lower resistance of the P-base region. Another difference between these devices is that in the MOS gated thyristor, turning on the device at one point will result in a spreading of the current throughout the active area of the device, so that its gate structure can be localized. In the IGT, however, it is necessary to design a very large channel width to keep the channel resistance small. In these devices, it is necessary to form thousands of cells that work in parallel as in the case of power MOSFETs. The design of the cell layout also has a significant impact on the latching current density. These factors are discussed in the paragraphs that follow. Since the fabrication procedure is similar to that for the power MOSFETs (discussed in detail in Chapter 6), only salient differences are emphasized here to avoid repetition.

7.2.7.1. P-Base Region.   Among the structural parameters that affect the IGT electrical characteristics, the P-base region (of n-channel devices) has the most prominent impact. As its doping increases, the sheet resistance of the P-base under the N^+ emitter decreases; this raises the latching current density as given by Eq. (7.43). Unfortunately, an increase in the P-base doping results in an increase in the threshold voltage [refer to Eq. (6.28)]. If a single-layer P-base region is used with a threshold voltage of 5 V, the IGT is found to latch up under steady-state conditions at a current density of less than 100 A/cm^2 at room temperature [4]. Such a low latching current density would not allow exploiting the capability of the IGT.

To overcome this problem, modern IGTs are fabricated by using P-base regions with two portions as illustrated in Fig. 7.39 [12]. In this structure, the P region is used to control the threshold voltage whereas the deeper, high-concentration P^+ region is used to reduce the lateral resistance R_P of the P-base region under the N^+ emitter. For a typical device, the sheet resistance of the P^+ region is two orders of magnitude lower than that of the P region. The depth and location of the P^+ region must be carefully controlled to minimize the P-base resistance R_P. To achieve this, the P^+ diffusion window must be perfectly centered with respect to the P-base and N^+-emitter regions, that is,

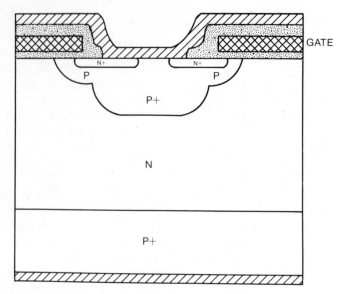

Fig. 7.39. Insulated-gate-transistor structure with deep P^+ region in DMOS cell.

the edge of the polysilicon window of the DMOS cell. The P^+ region should then be diffused as far as possible to the edges of the cell [23]. It should be noted that if the P^+ diffusion extends too far, it will raise the channel doping level and adversely impact the threshold voltage. With effective utilization of the P^+ region, the latching current density of IGTs can be raised to above 1000 A/cm^2.

7.2.7.2. Channel Length.
In general, the channel length for IGTs should be keep as small as possible to reduce the channel resistance. This is especially important for higher-frequency and higher-voltage devices in which a larger fraction of the collector current flows through the channel.

The minimum achievable channel length is governed by the same punch-through restrictions that were discussed for power MOSFETs with regard to the P-base region. In addition, the reduction of channel length is accompanied by a reduction of the depth of the P-region. This has the adverse effect of raising the lateral base resistance R_P for the portion of the N^+ emitter not over the deep P^+ region. Furthermore, the current gain of the upper $N-P-N$ transistor will increase with reduction in the P-base thickness because of both an increase in the injection efficiency resulting from a smaller integrated base charge and an increase in base transport factor due to the smaller base width. In state-of-the-art IGTs, the channel length is generally about 1 μm long. Devices with shorter channels have been fabricated and found to exhibit a higher transconductance at the expense of a reduction in the latching current density.

7.2.7.3. Gate Oxide Thickness. As CMOS technology evolves, there is a trend toward reduced gate oxide thickness while scaling lateral device dimensions. Power MOS devices have generally lagged behind this technology because the channel widths are very large compared with those encountered in devices used for integrated circuits. State-of-the-art devices are generally fabricated with a gate oxide thickness of 1000 Å. A reduction in oxide thickness has the obvious benefit of increasing the transconductance [see Eq. (7.32), in which $C_{ox} = (\epsilon_{ox}/t_{ox})$]. This is highly desirable for reducing the gate drive voltage.

In the case of the IGT, a decrease in the gate oxide thickness has the additional, and perhaps more important, advantage of raising the latching current density. To understand this effect, it is necessary to examine Eq. (6.28) for the threshold voltage:

$$V_T \simeq \frac{t_{ox}}{\epsilon_{ox}} \sqrt{4\epsilon_s kTN_A \ln(N_A/n_i)} + 2\frac{kT}{q}\ln\left(\frac{N_A}{n_i}\right) \tag{7.73}$$

Ignoring the relatively small contributions from the logarithmic terms, it can be concluded that the P-base doping level must be increased inversely as the square of the gate oxide thickness to maintain a constant threshold voltage:

$$N_A \propto \frac{1}{t_{ox}^2} \tag{7.74}$$

The lateral resistance R_P of the P-base varies inversely proportional to its doping level:

$$R_P \propto \frac{1}{N_A} \propto t_{ox}^2 \tag{7.75}$$

In combination with Eq. (7.43)

$$I_L \propto \frac{1}{t_{ox}^2} \tag{7.76}$$

Thus, for a constant threshold voltage, the latching current density will vary inversely as the square of the gate oxide thickness. This relationship provides a very powerful approach to raising the latching current density as experimentally confirmed for n-channel IGTs [24].

The smallest gate oxide thickness that is practical for a power device is decided by the reduction in device yield arising from pinholes in the gate oxide. In addition, as the gate oxide thickness decreases, its breakdown voltage is reduced. A typical gate oxide breakdown strength is about 10^7 V/cm. A 1000-Å gate oxide thickness will not be susceptible to damage unless the gate voltage exceeds 100 V. This provides a large margin over the gate drive voltage of nominally 15 V. As the gate oxide thickness is reduced to below 500 Å, its breakdown voltage falls to less than 50 V. Since noise spikes in the gate drive circuit may approach this value in high-speed switching circuits, extra care is required during device application to utilize the advantages of thin gate oxides.

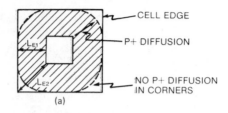

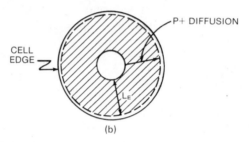

Fig. 7.40. Comparison of square (*a*) and circular (*b*) cell layouts for IGTs.

7.2.7.4. Cell Design. The layout of the IGT cell can have an important impact on its forward conduction characteristics and its latching current density. As an example, consider the case of a square cell versus a circular cell design. These two cases are illustrated in Fig. 7.40. In this figure, the shaded area indicates the location of N^+-emitter region. The dashed lines indicate the lateral extension of the deep P^+ diffusion for each of the cells from a perfectly aligned central location. Despite this assumption, there exists a portion at each corner of the square cell where the P^+ diffusion cannot penetrate below the N^+ emitter. The N^+-emitter length at these corners L_{E2} is also 40% greater than at the edges L_{E1}. These factors make the square cell vulnerable to latch-up at the corners. In comparison, the circular cell has a uniform N^+-emitter length L_E with the P^+ region extending all the way to the edge of the cell. On the basis of these arguments, it can be concluded that the circular cell should exhibit superior latch-up performance when compared to the square cell.

Another cell design that can be utilized to improve the latching current density is to use a stripe geometry with the N^+ emitter confined to only one edge of each cell [19]. A cross section of this type of cell is shown in Fig. 7.41. By confining the N^+ emitter to only one edge of each cell, the hole current can be collected at the opposite edge of each cell. This provides a path for the hole current that would otherwise flow under the N^+ emitter, leading to latch-up. Furthermore, in the linear cell topology, it is possible to provide another path for the hole current by breaking the N^+ emitter along the stripe length. Although these two modifications of the cell geometry have been found to provide

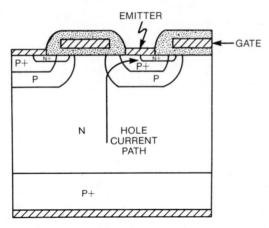

Fig. 7.41. Stripe cell layout for IGTs with N^+ emitter confined to one edge. (After Ref. 19, reprinted with permission from the IEEE © 1984 IEEE.)

a latching current density of over 375 A/cm² at 125°C [19], they degrade the device transconductance because of the much smaller channel length than that in the square cell.

7.3. MOS TURN-OFF THYRISTOR

The MOS turn-off thyristor differs from the MOS gated thyristor described earlier in the chapter in its ability to both turn on and turn off the current by using a signal applied to an MOS gate. These devices do not exhibit the fully gate controlled output characteristics that are exhibited by the bipolar transistor, power MOSFET, and IGT. Instead, they can be regeneratively turned on with the S-type negative resistance characteristic of thyristors and can be turned off like a gate turn-off thyristor (GTO) [5] without the need to reverse the anode voltage. In comparison to a GTO, these devices have the advantage of a high input impedance MOS gate. In comparison with the IGT, they can operate at an even higher forward conduction current density.

The basic structure of a MOS turn-off thyristor is shown in Fig. 7.42 and its equivalent circuit is provided in Fig. 7.43. The device consists of an FET structure integrated into the conventional $P-N-P-N$ thyristor structure in such a manner that the emitter-base junction of the upper $N-P-N$ transistor can be short-circuited by the application of a gate voltage to the FET. In the absence of the gate voltage, the device can be switched on either like a conventional thyristor or by utilizing the MOS gate as described before for the MOS gated thyristor. The current flow in the on-state occurs through the two coupled transistors as indicated by the arrow in Fig. 7.42. As a result of the strong injection of minority carriers into the N-drift region, its resistance becomes very

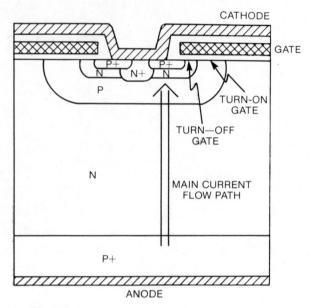

Fig. 7.42. Basic structure of the MOS turn-off thyristor.

small during current flow. In the on-state, the forward conduction characteristics approach those of a conventional thyristor, and these devices can operate at much (two to five times) greater current densities than an IGT.

To achieve forced gate turn-off, the *p*-channel FET within the device structure, which forms an active short circuit between the N^+-cathode and *P*-base regions, is turned on. Holes that enter the *P* base now have an alternate path

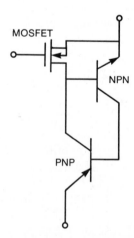

Fig. 7.43. Equivalent circuit for the MOS turn-off thyristor.

to the cathode contact; that is, they can bypass the N^+-P junction of the upper $N-P-N$ transistor. To accomplish forced gate turn-off, the resistance in the hole current path must be so low that, when all the hole current is diverted to the p-channel FET, the forward bias of the N^+-P junction must remain below 0.7 V. This condition can be stated as

$$I_{off} = \frac{0.7}{\alpha_{PNP}(R_{CH} + R_L + R_S)} \tag{7.77}$$

where R_{CH} is the channel resistance of the p-channel FET, R_L is the lateral resistance of the P-base region, and R_S is the FET spreading resistance. It should be noted that the high carrier density prevalent during forward conduction modulates the P-base (R_L) and spreading (R_S) resistances. Computer modeling indicates that turn-off can be accomplished at very high current densities (>2000 A/cm²) with proper cell design [21]. Since the mobility for electrons is higher than that for holes in silicon, the complementary structure with an n-channel FET is expected to exhibit even higher turn-off current density.

Experimental demonstration of the operation of the MOS turn-off thyristor concept has been obtained on rudimentary n-channel devices fabricated with only small active areas. It has been found that, for a single-cell device with active area of $20 \times 20\,\mu m$, a current density of 6000 A/cm² can be turned off at 25°C and at low anode voltages for a resistive load by using a gate drive of 60 V. The maximum turn-off current density decreases to 2000 A/cm² for a more practical gate drive of 15 V. This indicates that the channel resistance is an important limitation to the turn-off process. A bigger problem that has been encountered is the severe reduction in the turn-off current density with an increasing number of cells. This phenomenon is believed to be due to nonuniform distribution of current during device turn-off. This factor, coupled with the additional derating of the maximum turn-off current density at high anode voltages [25] under inductive loads and at elevated temperatures, makes the MOS turn-off thyristor current handling capability too poor for circuit applications at the present time. It is also worth pointing out that the regenerative turn-on of these devices does not make them suitable as a substitute for the bipolar transistor, power MOSFET, or the IGT. The MOS turn-off thyristor should be considered for use only in circuits where GTOs are presently being applied.

7.4. TRENDS

The devices discussed in this chapter represent an important new class of power devices that are expected to play an increasing role in power electronic systems. Their MOS gate provides an easy interface that can be addressed by use of low-cost integrated circuits. Their bipolar current conduction allows operation at very high current densities, which makes high-current, high-voltage devices available to circuit designers at a relatively low cost. The MOS gated thyristor and IGT are already available as commercial products. Their current and voltage ratings are expected to increase rapidly. The application of the IGT to power circuits used for motor drives, appliance control, lighting, and factory automation has already begun. More applications can be anticipated because of the very favorable cost reduction made possible with the utilization of these new power devices.

REFERENCES

1. N. Zommer, "The monolithic HV BIPMOS," *IEEE International Electron Devices Meeting Digest*, Abstract 11.5, pp. 263–266 (1981).

2. D. Y. Chen and S. Chin, "Design considerations for FET–gated power transistors," *IEEE Trans. Electron Devices*, **ED-31**, 1834–1837 (1984).

3. B. J. Baliga, "Enhancement and depletion mode vertical channel MOS gated thyristors," *Electron. Lett.* V. **15**, 645–647 (1979).

4. (a) B. J. Baliga, M. S. Adler, P. V. Gray, R. Love, and N. Zommer, "The insulated gate rectifier (IGR): A new power switching device," *IEEE International Electron Devices Meeting Digest*, Abstract 10.6, pp. 264–267 (1982); (b) J. P. Russell, A. M. Goodman, L. A. Goodman, and J. M. Nielson, "The COMFET: A new high conductance MOS gated device," *IEEE Electron Device Lett.*, **EDL-4**, 63–65 (1983)

5. S. K. Ghandhi, *Semiconductor Power Devices*, Wiley, New York, 1977.

6. J. D. Plummer and B. W. Scharf, "Insulated gate planar thyristors," *IEEE Trans. Electron Devices*, **ED-27**, 380–394 (1980).

7. M. F. Chang, G. C. Pifer, B. J. Baliga, M. S. Adler, and P. V. Gray, "25 amp, 500 volt insulated gate transistors," *IEEE International Electron Devices Meeting Digest*, Abstract 4.4, pp. 83–86 (1983).

8. M. F. Chang, G. C. Pifer, H. Yilmaz, R. F. Dyer, B. J. Baliga, T. P. Chow, and M. S. Adler, "Comparison of *N* and *P* channel IGTs," *IEEE International Electron Devices Meeting Digest*, Abstract 10.6, pp. 278–281 (1984).

9. B. J. Baliga, M. S. Adler, R. P. Love, P. V. Gray, and N. Zommer, "The insulated gate transistor: A new three terminal MOS controlled bipolar power device," *IEEE Trans. Electron Devices*, **ED-31**, 821–828 (1984).

10. B. J. Baliga, "Analysis of insulated gate transistor turn-off characteristics," *IEEE Electron Device Lett.*, **EDL-6**, 74–77 (1985).

11. B. J. Baliga, M. Chang, P. Shafer, and M. W. Smith, "The insulated gate transistor (IGT)—a new power switching device," *IEEE Industrial Applications Society Meeting Digest*, pp. 794–803 (1983).

12. B. J. Baliga, M. S. Adler, P. V. Gray, and R. P. Love, "Suppressing latch-up in insulated gate transistors," *IEEE Electron Device Lett.*, **EDL-5**, 323–325 (1984)

13. A. M. Goodman, J. P. Russell, L. A. Goodman, J. C. Neuse, and J. M. Neilson, "Improved COMFETs with fast switching speed and high current capability," *IEEE International Electron Devices Meeting Digest*, Abstract 4.3, pp. 79–82 (1983).

14. B. J. Baliga "Collector output resistance of insulated gate transistors," IEEE Electron Device Lett., **EDL-7**, December (1986).

15. J. W. Slotboom and H. C. deGraff, "Measurements of band-gap narrowing in Si bipolar transistors," *Solid State Electron.*, **19**, 857–862 (1976).

16. B. J. Baliga and T. P. Chow, "Dynamic latch-up in insulated gate transistors," to be published.

17. B. J. Baliga, "Switching speed enhancement in insulated gate transistors by electron irradiation," *IEEE Trans. Electron Devices*, **ED-31**, 1790–1795 (1984).

18. T. P. Chow and B. J. Baliga, "Comparison of 300, 600, 1200 Volt N-channel insulated gate transistors," *IEEE Electron Device Lett.*, **EDL-6**, 161–163 (1985).

19. A. Nakagawa, H. Ohashi, M. Kurata, H. Yamaguchi, and K. Watanabe, "Non-latch-up, 1200 volt bipolar mode MOSFET with large SOA," *IEEE International Electron Devices Meeting Digest*, Abstract 16.8, pp. 860–861 (1984).

20. B. J. Baliga, "Temperature behavior of insulated gate transistor characteristics," *Solid State Electron.*, **28**, 289–297 (1985).

21. V. A. K. Temple, "MOS controlled thyristors," *IEEE International Electron Devices Meeting Digest*, Abstract 10.7, pp. 282–285 (1984).

22. A. Nakagawa, Y. Yamaguchi, K. Watanabe, H. Ohashi, and M. Kurata, "Experimental and numerical study of non-latch-up bipolar-mode MOSFET characteristics," *IEEE International Electron Devices Meeting Digest*, Abstract 6.3, pp. 150–153 (1985).

23. T. P. Chow, B. J. Baliga, P. V. Gray, M. F. Chang, G. C. Pifer, and H. Yilmax, "A self-aligned short process for insulated gate transistors," *IEEE International Electron Devices Meeting Digest*, Abstract 6.2, pp. 146–149 (1985).

24. T. P. Chow and B. J. Baliga, "The effect of channel length and gate oxide thickness on the performance of insulated gate transistors," *IEEE Trans. Electron Devices*, **ED-32**, 2554 (1985).

25. M. Stoisek and H. Strack, "MOS GTO—a turn-off thyristor with MOS-controlled emitter shorts," *IEEE International Electron Devices Meeting Digest*, Abstract 6.5, pp. 158–161 (1985).

26. D. S. Kuo, J. Y. Choi, D. Giandomenico, C. Hu, S. P. Sapp, K. A. Sassaman, and R. Bregar, "Modelling the turn-off characteristics of the bipolar–MOS transistor," *IEEE Electron Device Lett.*, **EDL-6**, 211–214 (1985).

27. V. A. K. Temple and F. W. Holroyd, "Optimizing carrier lifetime profile for improving trade-off between turn-off time and forward drop," *IEEE Trans. Electron Devices*, **ED-30**, 782–790 (1983).

28. A. Mogro-Campero, R. P. Love, M. F. Chang, and R. F. Dyer, "Shorter turn-off times in insulated gate transistors by proton implantation," *IEEE Electron Device Lett.* **EDL-6**, 224–226 (1985).

PROBLEMS

7.1. Determine the increase in gate drive current for a thyristor with conventional gate structure when the (dV/dt) ratio is increased from 1000 to 10,000 V/μsec.

7.2. What are the gate drive currents in the preceding case for a junction capacitance of 1000 pF?

7.3. Consider an MOS gated thyristor fabricated with a P-base region that is homogeneously doped (epitaxial base) at a concentration of $10^{17}/cm^3$ and has a thickness of 5 μm. The N-type drift layer is doped at $10^{14}/cm^3$. Determine the (dV/dt) capability at an anode voltage of 1000 V assuming that the N^+-emitter length is 200 μm. Assume a linear cell topology. Note that the device threshold voltage was determined in Problem 6.2 for a 1000-Å gate oxide.

7.4. What is the N-base width for a symmetrical (reverse blocking) IGT designed to breakdown at 600 V if the minority carrier lifetime in the base region is 1 μ?

7.5. What is the width of the lightly doped portion of the N-base region for an asymmetric IGT designed to breakdown at 600 V if the lightly doped portion of the N-base has an impurity concentration of $10^{13}/cm^3$?

7.6. What is the total base width of the asymmetric IGT structure described in Problem 7.5 if the N-buffer layer has a doping of $10^{17}/cm^3$ and a minority carrier lifetime of 0.01 μsec?

7.7. Compare the forward voltage drop of the symmetrical IGT structure described in Problem 7.4 with a 600-V power MOSFET when operated at a current density of 200 A/cm^2. Ignore the channel resistance contribution for both devices. Assume that the diffusion length is equal to one-quarter of the width of the N-base region for the IGT.

7.8. For the same diffusion length, what is the forward voltage drop for the asymmetric IGT structure described in Problem 7.6?

7.9. Determine the increase in the transconductance at low voltages for the symmetrical IGT structure described in Problem 7.4 compared to a MOSFET with identical channel parameters when the diffusion length is 40 μm.

7.10. Repeat the calculation of the increase in transconductance for the symmetrical IGT structure discussed in Problem 7.6. Assume that the lightly doped portion of the N base is depleted at low voltages.

7.11. An IGT is fabricated with a homogeneously doped P-base region with a concentration of $10^{17}/cm^3$ and a thickness of 5 μm. The device has a symmetrical blocking voltage capability of 600 V. The N-drift layer has a width and lifetime as described in Problem 7.4. Calculate the steady-state latching current density, assuming that the length of the N^+-emitter is 200 μm as in the case of the MOS gated thyristor described in Problem 7.3.

7.12. What is the steady-state latching current density of the IGT described in Problem 7.11 if thee N^+-emitter length is reduced too 5 μm?

7.13. Determine the reduction in latching current density for the IGT described in Problem 7.12 when it is used in the gate-controlled turn-off mode for a 400-V DC supply with resistive load.

7.14. Repeat Problem 7.13 for the case of an inductive load.

8 NEW RECTIFIER CONCEPTS

The availability of the fast-switching three-terminal power devices described in earlier chapters for power control applications creates the option of raising the operating frequency of power circuits. The trend toward higher operating frequency is already evident in motor drives and switching power supplies. Operation at higher frequencies is attractive because of the reduction in size of passive components (inductors and capacitors), which leads to a more efficient, compact system design. To accomplish higher-frequency operation in power circuits, it is essential to use power rectifiers with improved switching performance.

In the past, only $P–i–N$ rectifiers were available for use in power circuits. The performance of these diodes has been continually improving as a result of optimization of device structure and the lifetime control process that is used to adjust the switching speed. The maximum operating frequency of these bipolar devices is ultimately limited by the large reverse recovery current that flows through the diodes and produces an undesirable stress on the power transistors operating in the circuits. Within the last few years, several new rectifier concepts based on unipolar current flow have been developed. This is expected to produce a revolutionary advancement in high-speed rectifier performance because these devices do not exhibit large reverse current flow during recovery.

In this chapter the progress in improving $P–i–N$ rectifier performance is first described. In doing this, it is assumed that the reader is familiar with the physics of operation of the $P–i–N$ diode under high injection levels [1]. The new device concepts for unipolar rectification are then analyzed, beginning with the Schottky rectifier and the MOSFET synchronous rectifier. The junction barrier-controlled Schottky (JBS) rectifier and the gallium arsenide rectifier described in Sections 8.4 and 8.5 allow the fabrication of unipolar rectifiers with improved performance, making them an attractive option for high-frequency circuits.

8.1. *P–i–N* RECTIFIERS

The $P–i–N$ rectifier was one of the very first semiconductor devices developed for power circuit applications. In this device the i region is flooded with minority carriers during forward conduction. The resistance of the i region becomes very

407

small during current flow, allowing these diodes to carry a high current density during forward conduction. The on-state characteristics of a P–i–N rectifier have been treated in detail elsewhere [1] and reviewed in Chapter 5 on the FCD. The on-state characteristics of an ideal diode can be described by the relationship

$$J_F = \frac{2qD_a n_i}{d} F\left(\frac{d}{L_a}\right) e^{qV_a/kT} \tag{8.1}$$

where the function $F(d/L_a)$ is as shown in Fig. 5.14. It is worth remembering that the current density is reduced when the diffusion length L_a decreases. Thus, as the speed of the rectifiers is increased by lifetime reduction, it is accompanied by a deterioration in the forward conduction characteristics.

A trade-off between the switching speed and the forward voltage drop is essential during power rectifier design. This trade-off is dependent on a number of factors such as the N-base width, the recombination center position in the energy gap, the distribution of the deep-level impurities, and the doping profile in the i region. In addition, an improved ohmic contact structure has been developed for rectifiers that can greatly improve the trade-off characteristics. These recent developments are discussed in the paragraphs that follow.

To start with, two important drawbacks of the P–i–N rectifier are worth discussing because the improvements in structure mentioned in the preceding paragraphs are designed to influence them. First, when a P–i–N rectifier is turned on with a high (di/dt), its forward voltage drop has been found to initially exceed its voltage drop during current conduction at the same level under steady-state conditions. This phenomenon is called *forward voltage overshoot during the turn-on transient*. This overshoot arises from the existence of the highly resistive i region. Under steady-state current conduction, the i-region resistance is drastically reduced by the injected minority carriers. During high-speed turn-on, however, the current rises at a faster rate than the diffusion of the minority carriers injected from the junction. A high voltage drop develops across the i region for a short time until the minority carriers can diffuse into the i region and reduces its resistance. The current and voltage waveforms observed during the forward recovery process are schematically illustrated in Fig. 8.1. In practical devices, the voltage overshoot V_{peak} can be an order of magnitude larger than the forward drop V_F. Under very high (di/dt) ratios, voltage overshoots in excess of 30 V have been observed.

The magnitude of the voltage overshoot is dependent on the resistivity and thickness of the i region. In general, the i region should be designed to minimize its resistance, as discussed in Chapter 4 for the power JFET, within the constraints of achieving the necessary reverse blocking capability. A high forward overshoot in the rectifier can be a serious problem in power circuits because this voltage may appear across the emitter–base junction of a bipolar transistor used in the circuits and exceed its breakdown voltage.

The second and more serious drawback of the P–i–N rectifier is its poor reverse recovery characteristics. Reverse recovery is the process whereby the

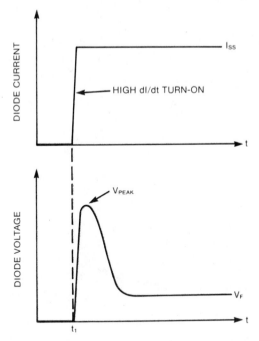

Fig. 8.1. Forward recovery waveforms for a *P–i–N* rectifier.

rectifier is switched from its on-state to its reverse blocking state. To undergo this transition, the minority carrier charge stored in the *i* region during forward conduction must be removed. The removal of the stored charge occurs by means of two phenomena—the flow of a large reverse current followed by recombination. The reverse recovery process in a *P–i–N* diode was discussed in Chapter 5.

The reverse recovery waveforms for a *P–i–N* rectifier are illustrated schematically in Fig. 8.2. Note the existence of a large reverse current pulse during the turn-off process. The peak reverse current I_{RP} is typically equal to the forward current I_F. This current flows through the transistors used in the circuit, adding to power dissipation and degrading their reliability. An additional concern is the large voltage overshoot represented by the peak V_{RP}. This voltage overshoot is caused by the reverse recovery (di/dt) current flow through inductances in the circuit.

When the switching frequency of a power circuit increases, the turn-off (di/dt) must be increased. It has been found that this causes an increase in both the peak reverse recovery current I_{RP} and the ensuing reverse recovery di/dt. If the reverse recovery (di/dt) is large, an increase in the breakdown voltage of all the circuit components becomes essential. As discussed in previous chapters, raising the breakdown voltage capability causes an increase in the forward voltage drop of power transistors, which degrades system efficiency. Much of the recent

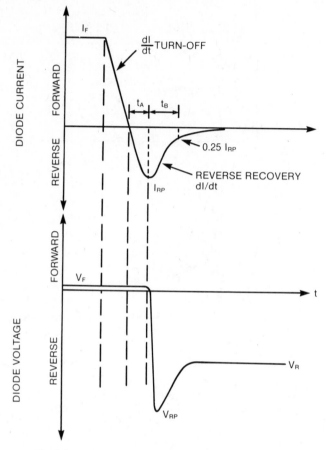

Fig. 8.2. Reverse recovery waveforms for a $P-i-N$ rectifier.

work on $P-i-N$ rectifiers has focused on improving the reverse recovery characteristics.

8.1.1. Lifetime Control

The most popular approach to controlling the switching speed of rectifiers has been by the introduction of recombination centers in the i region. In Chapter 5, it was shown that an approximate relationship between the injected carrier density in the i region and the forward conduction current density is

$$\bar{n} = \frac{\tau_{HL}}{2qd} J_F \qquad (8.2)$$

From this equation, it can be seen that as the high-level lifetime τ_{HL} decreases, the average carrier density in the i region will decrease proportionately, if the operating current density J_F is maintained constant. The net charge stored in the *i* region

$$Q_s = 2d\bar{n} \tag{8.3}$$

$$Q_s = \frac{\tau_{HL}}{q} J_F \tag{8.4}$$

will also decrease as the lifetime is reduced.

The time taken for the reverse recovery process is directly dependent on the stored charge. If the reverse recovery current waveform in Fig. 8.2 is approximated as a triangle of base $(t_A + t_B)$ and height I_{RP}, the reverse recovery charge is given by

$$Q_R = \tfrac{1}{2} I_{RP}(t_A + t_B) \tag{8.5}$$

By equating the reverse recovery charge to the stored charge, a relationship between the reverse recovery time and the lifetime can be obtained:

$$t_{rr} = (t_A + t_B) = 2 \frac{\tau_{HL}}{q} \left(\frac{J_F}{J_{RP}} \right) \tag{8.6}$$

where J_{RP} is the current density corresponding to the peak reverse current I_{RP}. From this equation, the reverse recovery time can be expected to be linearly proportional to the high-level lifetime and inversely proportional to the peak reverse current.

A reduction in the lifetime can be accomplished by the introduction of recombination centers into the *i* region. Among the many possible approaches, the diffusion of gold and platinum and the use of high-energy electron irradiation have been studied most carefully [2]. A discussion of the relative merits of these process techniques was provided in Chapter 2. A comparison of the trade-off curve between forward voltage drop and reverse recovery time for these three cases is shown in Fig. 8.3 [2]. The reasons for the differences between these methods of lifetime control are analyzed in Chapter 2. Note that the trade-off curve is the best for gold doping and the worst for electron irradiation performed at room temperature using 1.5-MeV electron energy. By using higher electron energies and performing the irradiation at elevated temperatures (typically 300°C), it has been found that the electron irradiation trade-off curve can be made to approach that for gold doping and superior to that for platinum doping [3].

The introduction of recombination centers into the *i* region also leads to an undesirable increase in leakage current as a result of enhanced space-charge generation at high operating temperatures. The magnitude of the increase in leakage current is dependent on the position of the recombination level in the energy gap. The leakage current for gold- and platinum-doped rectifiers has been compared to that in electron-irradiated devices [2]. The electron-irradiated devices exhibit an order of magnitude lower leakage current than do the gold-doped devices but have a leakage current higher than that for platinum doped

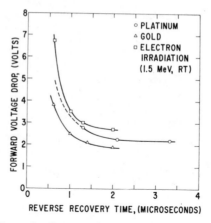

Fig. 8.3 Comparison of the trade-off curves between forward voltage drop and reverse recovery time for gold- and platinum-doped rectifiers with electron-irradiated devices. (After Ref. 2, reprinted with permission from the IEEE © 1977 IEEE.)

devices. In addition, it has been reported that electron-irradiated rectifiers exhibit a "snappy" recovery; that is, their reverse recovery (di/dt) is large compared with the "soft" recovery exhibited by platinum-doped devices [4]. This factor must be balanced against the significant processing convenience offered by electron irradiation.

An interesting approach to improving the trade-off curve between forward drop and reverse recovery speed is by using an inhomogeneous distribution of recombination centers. Computer simulation of various lifetime profiles indicates that the preferred location for the positioning of the recombination centers is in the middle of the N-base region and away from the $P–N$ junction [32]. This can be intuitively deduced by examining the carrier distribution within the N base during reverse recovery as shown in Fig. 5.21. From the time decay of the carriers, it can be seen that there is a high concentration of carriers in the central region of the N-base during reverse recovery. By locating a high density of recombination centers at the peak of the carrier distribution, the reverse recovery process can be accelerated with less penalty on forward voltage drop. Computer modeling indicates that a two fold improvement in switching speed can be obtained over a uniform recombination center distribution for the same forward drop. Unfortunately, it is difficult to achieve such a narrow distribution of recombination centers because the deep-level impurities have a very high diffusion coefficient in silicon and electron irradiation produces a uniform distribution of recombination centers. One promising approach is the use of high-energy proton implantation [5].

This discussion has been directed toward improving the reverse recovery characteristics. The introduction of recombination centers also affects the forward recovery performance. As discussed earlier, the forward voltage overshoot can be minimized by lowering the resistivity of the i region. Because of the higher mobility for electrons, an *N*-type drift region is preferable. The presence of a high concentration of recombination centers in the band gap leads to a compensation effect that raises the *N*-base resistance. For a given lifetime reduction, the compensation effect is dependent on the capture cross section of the recombination level as discussed in Chapter 2. In this regard, gold doping produces a larger change in *N*-base resistivity than does platinum doping for achieving the same reverse recovery speed.

8.1.2. Doping Profile

The analysis of the reverse recovery of a *P–i–N* rectifier indicates that a $P^+–\pi–N^+$ diode (π refers to a lightly doped *P*-type *i* region) will exhibit a faster recovery than a $P^+–v–N^+$ diode (v refers to a lightly doped *N*-type *i* region) [6]. The difference between these cases arises from the faster removal of charge in the $P^+–\pi–N^+$ diode. Although this may at first appear to favor the use of a π-type *i* region for power rectifiers, the opposite is generally done for two reasons. First, the reverse recovery of the π-base device is unacceptably abrupt, as shown in Fig. 8.4, leading to a snappy recovery [6]. The v-base diodes exhibit the desired soft recovery characteristic, which is favored despite the longer reverse recovery time. The second reason for preference of a v-base is that it is less prone to surface inversion so that it can be passivated much more easily than a π-base structure to achieve high, stable breakdown characteristics.

Apart from the semiconductor type used for the *i* region, the tailoring of its doping profile has been used to improve the reverse recovery characteristics. An important technique for improving the reverse recovery speed is by using very abrupt profiles for the P^+-anode and N^+-cathode regions. In the past, power rectifiers were made by using high-resistivity bulk, *N*-type material and diffusing the P^+ and N^+ regions from opposite sides of the wafer. This process

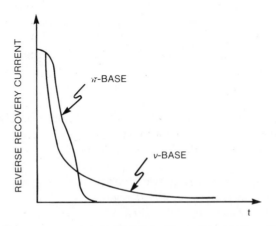

Fig. 8.4. Reverse recovery current waveforms for the $P^+–\pi–N^+$ and $P^+–v–N^+$ rectifiers. (After Ref. 6, reprinted with permission from the IEEE © 1967 IEEE.)

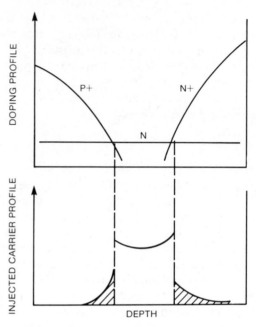

Fig. 8.5. Double-diffused $P-i-N$ diode doping profile and the corresponding injected charge distribution.

produces highly graded diffusion profiles and a relatively wide N-base region as shown in Fig. 8.5. The graded junction profiles are desirable for increasing breakdown voltage as discussed in Chapter 3. During forward conduction, however, there is a significant amount of minority carrier injection into the end regions, as shown in the lower portion of Fig. 8.5. The existence of the stored charge in the diffused region slows down the reverse recovery process. This effect is particularly important for low voltage rectifiers with narrow N-base regions.

With the improvement of silicon epitaxial growth technology, the development of power rectifiers with abrupt profiles is feasible [7]. The doping profile of a rectifier fabricated by diffusion of a P^+ region into an N-type epitaxial layer grown on an N^+ substrate with an abrupt interface is shown in Fig. 8.6 with the corresponding injected carrier profile. Note the much smaller injected carrier density in the end regions indicated by the shaded areas. Epitaxial diodes, with breakdown voltages of 400 V, have been found to exhibit a reduction in stored charge by over a factor 2 and a decrease in reverse recovery time by 50% [7]. The epitaxial process is particularly attractive for low-breakdown-voltage diodes that have narrow N-base regions.

A further improvement in the reverse recovery characteristics has been achieved by the use of an N base consisting of two portions [4]. A comparison between the conventional doping profile and the improved two-region profile

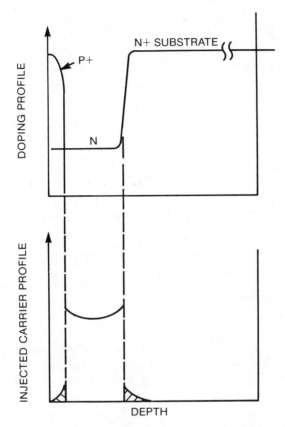

Fig. 8.6. Epitaxial *P–i–N* diode doping profile and the corresponding injected charge distribution.

is provided in Fig. 8.7. The improved N-base design consists of the conventional lightly doped region N_1 designed to support the reverse blocking voltage plus a more heavily doped region N_2. The doping of the second region (N_2) must be high enough to limit depletion layer spreading but low enough to still allow conductivity modulation. A typical doping level for this region is in the mid-10^{14}/cm^3 range. During reverse recovery, charge stored in region N_2 is not swept out rapidly, resulting in the desired soft recovery. Devices made with this profile by using epitaxial growth have employed the relationship $t_B > t_A$ (see Fig. 8.2 for the definition of t_A and t_B), which is a criterion used to identify a soft recovery rectifier.

8.1.3. Ideal Ohmic Contact

In the past, the contact to the drift region of the P^+–N rectifier has been made by using a high-concentration N^+ layer, which is formed by either diffusion in the case of the double-diffused rectifiers or by using a highly doped substrate in

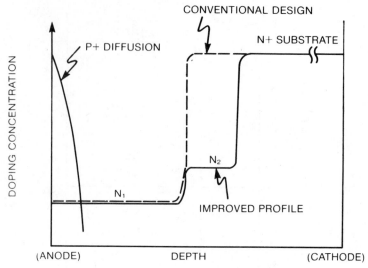

Fig. 8.7. Comparison of an improved two-region N-base profile (solid line) with conventional N-base profile (dashed line).

the case of epitaxial rectifiers. The N–N^+ interface produces an ohmic contact for majority carriers, that is, electrons in this case. This contact allows the transport of electrons across the interface but creates an electric field due to the concentration gradient:

$$\mathscr{E} = \frac{kT}{q}\frac{1}{N_\mathrm{D}}\frac{dN_\mathrm{D}}{dx} \tag{8.7}$$

The presence of the electric field reflects any minority carriers (holes in this case) that approach the interface as illustrated in the band diagram shown in Fig. 8.8. The minority carrier reflecting property of a highly doped contact layer has been utilized to improve the performance of solar cells [8]. In the case of power rectifiers, the reflection of minority carriers can be detrimental to the achievement of fast turn-off because they become trapped in the N-base region.

To obtain an "ideal" ohmic contact that can simultaneously allow the transport of holes and electrons across its interface, a contact consisting of a mosaic of P^+ and N^+ regions has been proposed [9]. In this contact, the P^+ regions provide an ohmic path for holes (minority carriers here) whereas the N^+ regions provide the conventional ohmic path for electrons (majority carriers here). The ideal ohmic contact formed with the mosaic of P^+ and N^+ regions and its composite band diagram are shown in Fig. 8.9. It should be noted that this contact differs from the well-known short-circuited emitter structure used for improving the performance of thyristors [1]. In the case of the short-circuited emitter structure, holes are injected from the P^+ region of the mosaic contact shown in Fig. 8.9 into the N layer during forward conduction. To achieve this, the trans-

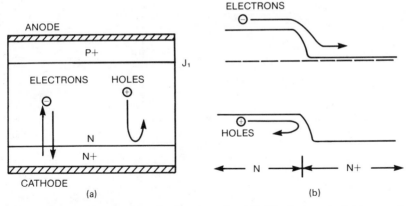

Fig. 8.8. Structure and band diagram for the conventional $N–N^+$ ohmic contact: (*a*) conventional ohmic contact; (*b*) band diagram.

verse ohmic voltage drop across the P^+ regions must be sufficient to forward-bias the $P^+–N$ junction, so that the width W_P must be made large. In contrast, for the ideal ohmic contact, the minority carrier injection used to modulate the resistance of the N-base region occurs through the forward-biased junction (J1) of the rectifier. The ideal ohmic contact is designed with small widths (W_P and W_N) for the P^+ and N^+ regions to suppress any transverse voltage drop across the P^+ region since no injection from the P^+ region is necessary.

A modification of the $P^+–N^+$ mosaic contact that achieves the same purpose is the substitution of the P^+ regions with a Schottky contact [9]. This ohmic contact design is illustrated in Fig. 8.10. It should be noted that the contact does not have to support any reverse voltage, which alleviates the stringent process requirements generally needed to suppress the soft, high-leakage, break-down characteristics of Schottky diodes.

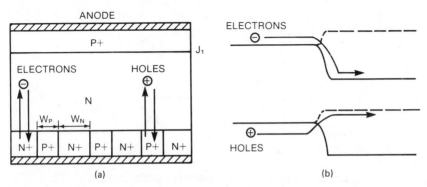

Fig. 8.9. Structure of the ideal ohmic contact containing a mosaic of P^+ and N^+ regions (*a*) and its composite band diagram (*b*).

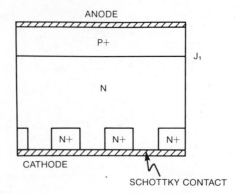

Fig. 8.10. Ideal ohmic contact structure utilizing a Schottky interface.

For rectifiers fabricated by the double diffusion process, the ideal contact can be achieved by masking portions of the N^+ contact diffusion and performing a P^+ diffusion in these portions. To create the ideal ohmic contact with an epitaxial process, the conventional process (with an N-type epitaxial layer grown on an N^+ substrate) must be altered to a N-type epitaxial layer grown on a P^+ substrate. This makes the surface of the N-type drift layer accessible to subsequent diffusion of the P^+ and N^+ regions to form the mosaic contact on the top surface. The drawback of the process is the need to form grooves around the edge of each device extending to the interface between the N-type epitaxial layer and the P^+ substrate to create individual diodes. A cross section of the completed epitaxial rectifier structure with the ideal ohmic contact is provided in Fig. 8.11.

This ideal ohmic contact concept has been experimentally tested by the fabrication of epitaxial diodes for which a P-type epitaxial layer was grown on an N^+ substrate [9]. Ideal ohmic contacts with the P^+–N^+ mosaic structure

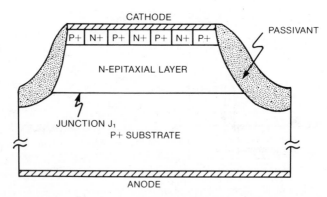

Fig. 8.11. Cross section of an epitaxial rectifier with ideal ohmic contact.

and the N^+–Schottky structure were both evaluated. The contact geometry used consisted of long thin stripes with a P^+ region width of 10 μm and a 30-μm width for the N^+ or Schottky regions. No lifetime control was used. The characteristics of these diodes were compared with those of diodes simultaneously fabricated with the conventional P^+ contact to the P-type drift region.

The reverse recovery time for the diodes with ideal ohmic contact was found to be 60 nsec, compared with about 500 nsec for the diodes with conventional contacts. A peak reverse recovery current I_{RP} equal to the forward conduction current I_F was used during these measurements. In addition to the improved reverse recovery characteristic, the forward voltage drop of the diodes with the ideal ohmic contact was found to be 0.1 V lower than that for diodes with conventional ohmic contacts that were processed with gold doping to achieve fast reverse recovery. Furthermore, in the absence of gold doping, diodes with ideal ohmic contacts exhibit a lower leakage current at elevated temperatures when compared with fast recovery diodes. These features allow their operation at higher ambient temperatures.

The maximum operating temperature of a rectifier is limited by thermal runaway. The temperature rise is determined by the power dissipation in the diode. This is comprised of three components—the power dissipation due to current conduction in the on-state, the power dissipation during reverse blocking, and the power dissipation during switching. As the switching speed improves, the steady-state power losses become dominant. The steady-state power dissipation in the on-state is determined by the forward voltage drop. As temperature increases, the forward drop of a $P–i–N$ rectifier decreases, leading to a reduction in the power dissipation. Concurrently, the leakage current grows exponentially with increasing temperature. The leakage current arises from two sources—space-charge generation in the depletion layer and the generation of carriers in the neutral base region within a diffusion length from the depletion layer edge:

$$I_L = I_{L,SC} + I_{LD} \tag{8.8}$$

$$= \frac{qW_D n_i A}{\tau_{SC}} + \frac{qAD_p n_i^2}{L_p N_D} \tag{8.9}$$

At high temperatures, the diffusion current term becomes predominant. Using the temperature dependence of the intrinsic carrier concentration from Eq. (2.28), it can be shown that

$$I_L \propto e^{-(E_g/kT)} \tag{8.10}$$

The leakage current grows exponentially with temperature with an activation energy equal to the energy gap E_g.

The power dissipation due to these two phenomena determines the total power dissipation:

$$P_D = I_F V_F \frac{t_{on}}{T} + I_L V_R \frac{(T - t_{on})}{T} \tag{8.11}$$

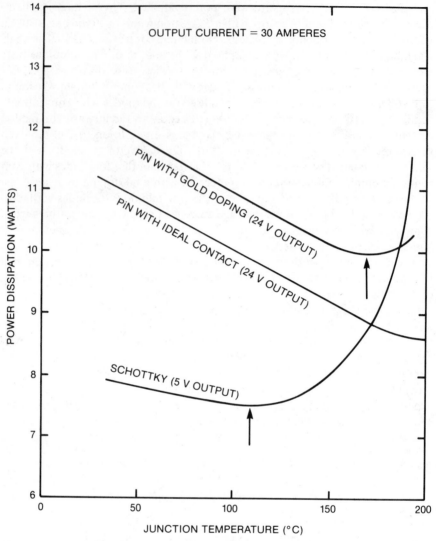

Fig. 8.12. Comparison of the power loss–temperature curves for the $P–i–N$ rectifier containing the ideal ohmic contact with a gold-doped $P–i–N$ rectifier and a Schottky rectifier. (After Ref. 9, reprinted with permission from the IEEE © 1982 IEEE.)

where t_{on} is the time the diode is in the on-state and T is the total period; I_F, V_F are the forward conduction current and forward voltage drop, respectively; and I_L, V_R are reverse leakage current and applied reverse voltage, respectively. At low temperatures, the leakage current I_L is small and the first term dominates, so that the power dissipation decreases with temperature because of a decrease in forward voltage drop. At high temperatures, the leakage current becomes large and the second term grows until it exceeds the first term in Eq. (8.11). Now, the power dissipation in the diode increases with increasing temperature. This can lead to a thermal runaway situation. The maximum operating temperature of the rectifier is determined by the point at which the curve relating power dissipation and temperature goes through a minimum.

The power dissipation–temperature curves for the $P–i–N$ rectifier containing the ideal ohmic contact, the gold-doped fast recovery diode, and the Schottky rectifier are compared in Fig. 8.12. It should be noted that these curves are calculated on the basis of a 30-A output current and a 24-V output voltage with 50% duty cycle. At low temperatures ($< 100°C$), the power dissipation in the Schottky rectifier is much superior to the $P–i–N$ rectifiers as a result of its low forward drop (typically 0.55 V at room temperature, compared with 0.8 V for the diode with an ideal contact and 0.9 V for the gold doped diode). As temperature increases, the leakage current in the Schottky rectifier increases at a faster rate than that for the $P–i–N$ rectifiers as the result of a smaller activation energy determined by the Schottky barrier height. Its power dissipation curve goes through a minimum at a much lower temperature ($\sim 100°C$) compared with the $P–i–N$ rectifiers. The $P–i–N$ rectifier with ideal ohmic contact exhibits a lower power dissipation and a higher maximum operating temperature than does the gold-doped fast recovery diode because of its slightly lower forward voltage drop and much lower leakage current. The higher leakage current of the conventional fast recovery diode is due to an enhanced space-charge generation produced by the gold doping. From these results, it can be concluded that the $P–i–N$ rectifiers with ideal ohmic contact are attractive for high-voltage applications (> 100 V) with high ambient temperatures in which Schottky rectifiers do not provide satisfactory performance.

8.2. SCHOTTKY BARRIER RECTIFIERS

The nonlinear current transport across a metal semiconductor contact has been known for a long time. The potential barrier responsible for this behavior was ascribed to presence of a stable space-charge layer by Schottky in 1938. The fundamental principles that describe current transport in the metal–semiconductor contacts has been treated in detail in several places [10, 11]. In this section, the application of the Schottky barrier to achieve high-power rectification is discussed. It is assumed that the reader is already familiar with the principles of the formation of the metal–semiconductor barrier and the various current transport mechanisms across the metal–semiconductor interface. Only those aspects of the Schottky barrier that are significant to power rectifiers are discussed here.

8.2.1. Forward Conduction

There are four basic processes for the transport of current across a metal N-type semiconductor contact: (1) the transport of electrons over the potential barrier into the metal, resulting in *thermionic emission* current; (2) quantum mechanical tunneling of carriers through the barrier, resulting in *field emission or tunneling current*, (3) recombination current flow in the space-charge region; and (4) current flow by the injection of minority carriers (holes) from the metal into the semiconductor. In the case of Schottky power rectifiers, current flow by means of the thermionic emission process is dominant. Because of the relatively low doping levels in the semiconductor required to support high reverse blocking voltages, the potential barrier is not sufficiently narrow to produce significant current flow by field emission. The space-charge recombination current is similar to that observed in a *P–N* junction diode and is significant only at very low current densities. The current transport due to minority carrier injection becomes increasingly significant at large forward-bias voltages. However, even at a current density of over 200 A/cm², the hole current contribution is less than a few percent and can be neglected.

In the case of silicon and gallium arsenide at relatively low doping levels, the thermionic emission theory [12] can be used to describe current flow across the Schottky barrier interface:

$$J = AT^2 \, e^{-(q\phi_{Bn}/kT)}(e^{qV/kT} - 1) \tag{8.12}$$

where A is the effective Richardson constant, T is the absolute temperature, q is the electron charge, k is Boltzmann's constant, ϕ_{Bn} is the barrier height between the metal and N-type semiconductor, and V is the applied voltage. For N-type silicon, an effective Richardson constant of 110 A/(cm²·K²) can be used [13]. For N-type gallium arsenide, an effective Richardson constant of 140 A/(cm²·K²) is more appropriate [14]. When a forward bias is applied, the first term in the square brackets becomes dominant and the current flow across the Schottky barrier under forward conduction is given by

$$J_F = AT^2 \, e^{-(q\phi_{Bn}/kT)} \, e^{qV_{FB}/kT} \tag{8.13}$$

where V_{FB} is the voltage drop across the Schottky barrier. In the case of Schottky power rectifiers, a thick, lightly doped drift layer must be used for the semiconductor region to support the reverse blocking voltage. The diode current flows through the drift layer, producing a resistive voltage drop as illustrated in Fig. 8.13. It should be noted that there is no modulation of the drift region resistance in these devices because of the negligible minority carrier injection.

The Schottky rectifier is made by growing a thin epitaxial layer on a thick, highly doped N^+ substrate. The resistance of the substrate can be significant in low-voltage rectifiers for which the drift layer resistance becomes small. The substrate resistance component is shown as R_S in Fig. 8.13.

By using Eq. (8.13) and including the resistive voltage drop, the total forward voltage drop in the Schottky rectifier can be shown to be given by

$$V_F = \frac{kT}{q} \ln\left(\frac{J_F}{J_s}\right) + R_{sp}J_F \tag{8.14}$$

where R_{sp} is the specific resistance of the rectifier. The specific resistance R_{sp} is the total series resistance for an area of 1 cm². It includes contributions from the drift region, the substrate, and any contact resistances.

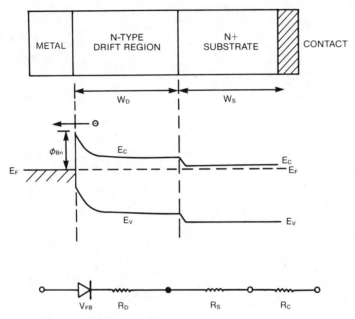

Fig. 8.13. Schottky rectifier structure and its band diagram.

In Eq. (8.14), the term J_s refers to the saturation current of the Schottky barrier:

$$J_s = AT^2 \, e^{-(q\phi_{Bn}/kT)} \tag{8.15}$$

The saturation current is a strong function of the barrier height and temperature. Its value determines the forward conduction and reverse leakage current (as discussed later) characteristics of the Schottky barrier diode. The calculated saturation current densities at four temperatures are provided in Fig. 8.14 as a function of barrier height ϕ_{Bn} for the range of 0.4–1.0 V of interest to power rectifiers. In performing this calculation, a Richardson constant of 120 A/(cm^2·K^2) was assumed.

To calculate the forward conduction characteristics of a Schottky barrier rectifier, the specific resistance of the diode must first be evaluated. For a N^+ substrate with a typical resistivity of 0.01 Ω·cm and thickness of 500 μm (20 mils), the contribution from the substrate is $R_s = 5 \times 10^{-4}$ Ω·cm^2. Although this may appear at first sight to be negligible, it can contribute to an increase in forward voltage drop by 50 mV at a typical forward conduction current density of 100 A/cm^2, which is not insignificant for rectifiers designed to operate at forward voltage drops of less than 500 mV. This contribution can be made negligible by increasing its doping level and reducing its thickness, but this requires the acquisition of specially prepared wafers at a higher cost.

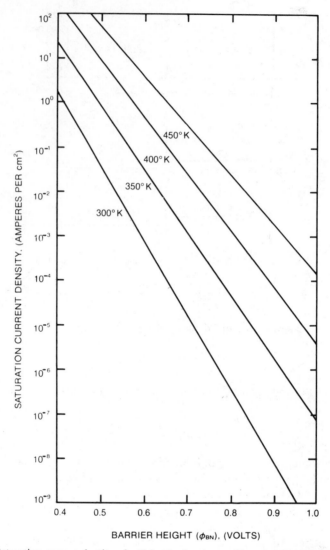

Fig. 8.14. Saturation current density of a Schottky barrier rectifier as a function of the barrier height and temperature.

The contribution to the specific resistance of the diode from the drift region R_D is dependent on the reverse blocking voltage. To achieve higher reverse blocking capability, it is essential to increase the resistivity and thickness of the drift region. For the ideal case, where the edge effects are assumed not to influence the breakdown voltage, the specific resistance of the drift region is equal to that derived for the ideal JFET in Chapter 4:

$$R_D = 5.93 \times 10^{-9} (BV_{PP})^{2.5} \tag{8.16}$$

for an N-type drift region. From this expression and Fig. 4.13, it is seen that the series resistance contribution from the dirft region will increase very rapidly as the reverse breakdown voltage increases. For the case of a typical Schottky power rectifier with a reverse breakdown voltage of 50 V, the drift region specific resistance is calculated to be $1.05 \times 10^{-4}\ \Omega \cdot cm^2$. In this case, the substrate contribution can be a major proportion of the device resistance.

To assess the impact of the breakdown voltage on the forward conduction characteristics of a Schottky rectifier, consider a typical case with a barrier height of 0.8 V. The calculated forward conduction characteristics at room temperature are provided in Fig. 8.15 for several breakdown voltages. In performing these calculations, it has been assumed that the contributions from the substrate and contact can be made negligible. These contributions can be factored into the characteristics by adding their voltage drop to it. From Fig. 8.15, it can be seen that the forward voltage drop of a Schottky power rectifier designed with reverse blocking capability of 50 V is between 0.5 and 0.6 V at a typical forward conduction current density of 100 A/cm². This low forward voltage drop compared to a P–i–N rectifier (whose forward drop is typically 0.9 V) makes the Schottky rectifier attractive as an output rectifier in switch-mode power supplies. As the reverse blocking capability is increased to 200 V, the forward drop of the Schottky rectifier approaches that of the P–i–N rectifier. In addition, their reverse blocking characteristics become soft, making them generally unacceptable for use in high-voltage power circuits.

The forward voltage drop of the Schottky rectifier is a function of temperature. As temperature increases, the saturation current density increases and the voltage drop across the metal–semiconductor barrier decreases. Concurrently, the series resistance of the drift region R_D increases because the mobility decreases. For low reverse blocking rectifiers (e.g., 50 V), the contribution from the series resistance is small. The calculated variation of the forward voltage drop of a Schottky rectifier between 300 and 450 K is provided in Fig. 8.16 for such low-voltage rectifiers. By using Eqs. (8.14) and (8.15), it can be shown that

$$V_F = \phi_{Bn} + \frac{kT}{q} \ln\left(\frac{J_F}{AT^2}\right) \tag{8.17}$$

if the series resistance is negligible. Since the logarithmic term is nearly constant and negative, the forward voltage drop decreases linearly with increasing temperature.

8.2.2. Reverse Blocking

When a reverse voltage is applied to the Schottky rectifier, a depletion layer extends into the semiconductor. The reverse blocking voltage of the Schottky rectifier is equal to that of an abrupt P–N junction rectifier. The design of the drift region can be performed by using the criteria derived in Chapter 3. The doping concentration and width of the drift region must be scaled with the

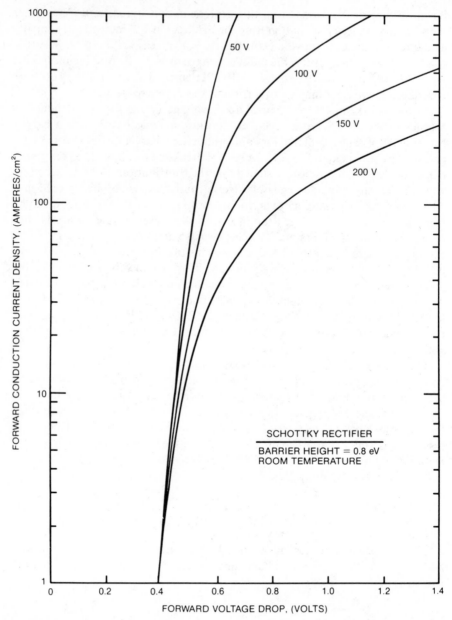

Fig. 8.15. Forward conduction characteristics of a silicon Schottky rectifier at room temperature calculated for an ideal device with various breakdown voltages.

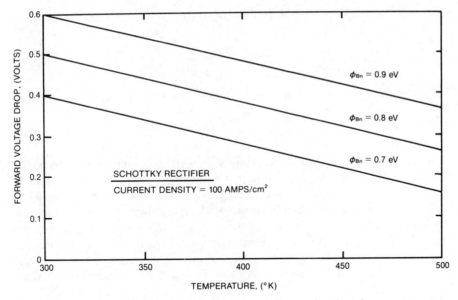

Fig. 8.16. Forward voltage drop of a Schottky rectifier as a function of temperature.

breakdown voltage:

$$N_D = 2 \times 10^{18} (BV_{PP})^{-4/3} \tag{8.18}$$

and

$$W_D = W_{C,PP} = 2.58 \times 10^{-6} (BV_{PP})^{7/6} \tag{8.19}$$

In general, the actual breakdown voltage of the Schottky rectifier is typically about one-third that for the abrupt parallel-plane junction. The doping level and width of the drift region must be adjusted to compensate for this.

In addition, the leakage current of Schottky rectifiers is substantially greater than that for P–i–N rectifiers. This has been an important limitation to their performance. It is caused by the large saturation current densities J_s. As discussed earlier with the aid of Fig. 8.14, the saturation current density increases with a reduction in the Schottky barrier height and increase in temperature. When a reverse bias voltage is applied to the Schottky barrier, Eq. (8.12) can be rewritten as

$$J_R = AT^2 e^{-(q\phi_{Bn}/kT)} \left[e^{-(qV_R/kT)} - 1 \right] \tag{8.20}$$

$$= -AT^2 e^{-(q\phi_{Bn}/kT)} \tag{8.21}$$

$$= -J_s \tag{8.22}$$

because the exponential term within the square brackets becomes negligible as the reverse-bias voltage increases. In Eq. (8.22), the barrier height ϕ_{Bn} is shown as a constant. Under reverse-bias voltage, it has been found that there is a reduction in the Schottky barrier height [10, 11]. The barrier height reduction is given by

$$\Delta\phi_{Bn} = \sqrt{\frac{q\mathscr{E}_m}{4\pi\epsilon_s}} \qquad (8.23)$$

where $\mathscr{E}_m$ is the maximum electric field at the metal–semiconductor interface:

$$\mathscr{E}_m = \sqrt{\frac{2qN_D}{\epsilon_s}(V_R + V_{bi})} \qquad (8.24)$$

The Schottky barrier height lowering is important in contributing to an increase in the reverse leakage current with increasing reverse-bias voltage. In addition, the leakage current of the Schottky rectifier contains the space-charge generation component (which increases with reverse bias voltage) and the diffusion component, as in the case of the $P-N$ junction rectifiers. These components of the leakage current represent only a small fraction of the total leakage current and can usually be neglected unless the barrier height is very large and the lifetime in the drift region is very low. In Fig. 8.17, the calculated leakage

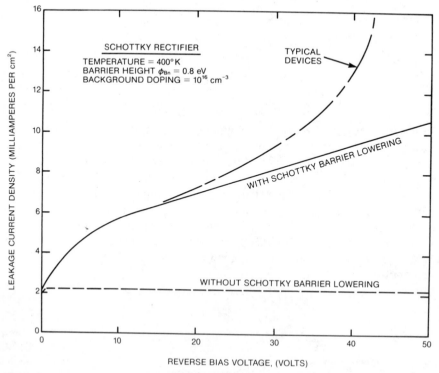

Fig. 8.17. Leakage current of a Schottky rectifier with and without the Schottky barrier lowering effect.

current of a Schottky rectifier with and without the Schottky barrier lowering effect are compared. Note the considerable increase in the reverse leakage current due to the Schottky barrier lowering at high reverse voltages. At high reverse voltages, the actual leakage current of Schottky rectifiers is even larger than that predicted by the Schottky barrier lowering.

The leakage current of the Schottky rectifier increases rapidly with temperature. To stress the importance of this phenomenon, a plot of the calculated leakage current density as a function of temperature is provided in Fig. 8.18, using a Richardson constant for silicon of 110 A/(cm$^2 \cdot$ K^2). It should be noted that, as the Schottky barrier height is reduced, a sharp increase in the leakage current occurs at lower temperatures. This reduces the maximum operating temperature of the Schottky rectifier as limited by thermal runaway.

8.2.3. Trade-off Curves

From the previous sections, it can be concluded that the optimization of the characteristics of the Schottky power rectifier requires a trade-off between forward voltage drop and reverse leakage current. As the Schottky barrier height ϕ_{Bn} is reduced, the forward voltage drop decreases but the leakage current increases and the maximum operating temperature decreases. Low barrier heights

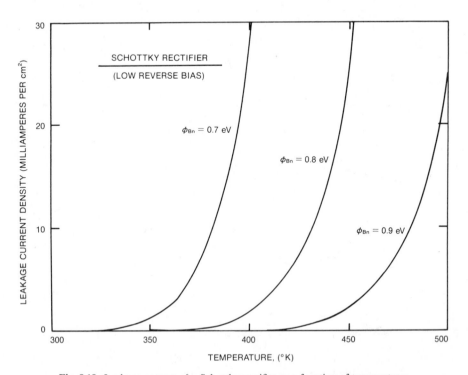

Fig. 8.18. Leakage current of a Schottky rectifier as a function of temperature.

should be used for Schottky rectifiers intended for high current operation with large duty cycles, where the power losses during forward conduction are dominant. Larger barrier heights should be used for Schottky rectifiers intended for applications with higher reverse-bias stress and higher ambient temperatures. A trade-off curve that can be useful during device design is the relationship between the forward voltage drop and the reverse leakage current. This relationship can be derived by using Eqs. (8.17) and (8.21) to eliminate ϕ_{Bn}:

$$J_R = J_F \, e^{-(qV_F/kT)} \tag{8.25}$$

The calculated trade-off curves for Schottky rectifiers at several ambient temperatures are provided in Fig. 8.19. It should be noted that these curves are applicable only to low-breakdown-voltage rectifiers because the series resistance of the diode was neglected in Eq. (8.17). Within this limitation, it is significant that the trade-off curves are not dependent on the semiconductor material used for device fabrication. The trade-off curves between forward voltage drop and reverse leakage current are determined solely by the Schottky barrier height, and no improvement can be expected by turning to other semiconductors, such as gallium arsenide, for these low-voltage rectifiers.

The ultimate limiting factor that determines the choice of the Schottky barrier height is the power dissipation in the rectifier. The power dissipation during forward conduction depends on the forward conduction current, the forward

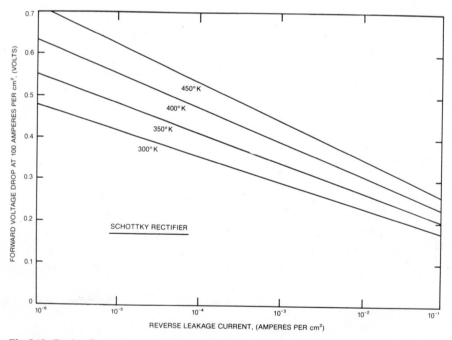

Fig. 8.19. Trade-off curve between forward voltage drop and reverse leakage current for Schottky rectifiers.

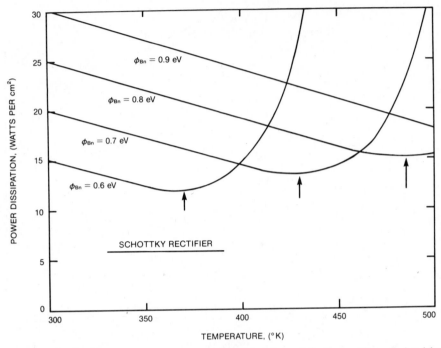

Fig. 8.20. Power dissipation as a function of temperature for a Schottky rectifier calculated by using a forward current density of 100 A/cm², reverse blocking voltage of 20 V, and a conduction duty cycle of 0.5.

voltage drop, and the duty cycle. The power dissipation during reverse blocking depends on the leakage current and the reverse-bias voltage. In choosing the optimum Schottky barrier height, it is important to calculate the total power dissipation as a function of temperature with the use of Eq. (8.11). The calculated power dissipation as a function of temperature for the case of four Schottky barrier heights is provided in Fig. 8.20. These curves were calculated by using a forward current density of 100 A/cm², a reverse blocking voltage of 20 V, and a duty cycle of 0.5. It can be seen that the total power dissipation can be reduced by lowering the Schottky barrier height, but this is accompanied by a reduction in the maximum operating temperature as indicated by the arrows. The effect of changing duty cycle is shown in Fig. 8.21. In this case, a fixed Schottky barrier height of 0.8 eV was used with other parameters remaining the same as those used for calculating the curves in Fig. 8.20. An important point to note in these curves is that although the total power dissipation decreases as the duty cycle (t_{on}/T) is reduced, the maximum operating temperature also decreases. To maintain a high operating temperature, it becomes necessary to raise the barrier height and incur an increase in total power dissipation. Calculations performed for a variety of switching power supply cases [15] indicate that the lowest Schottky barrier height for these rectifiers should be in excess of 0.7 eV.

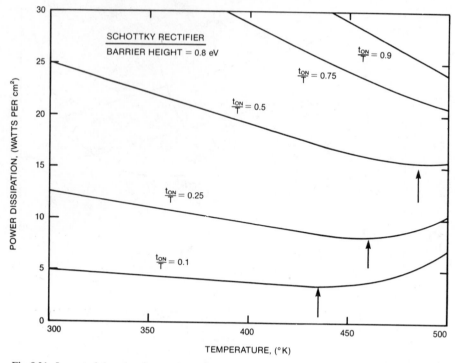

Fig. 8.21. Impact of changing duty cycle on the power dissipation–temperature curve for a Schottky rectifier.

8.2.4. Switching Behavior

Forward current transport in the Schottky rectifier occurs primarily through majority carriers with a very small minority carrier injection [11]. As a result, these devices exhibit an extremely fast reverse recovery behavior. Furthermore, there is no forward overvoltage transient as experienced with P–i–N rectifiers.

A problem that has been encountered during the application of these very-fast-recovery diodes is severe ringing in the current and voltage waveforms during the turn-off transient. This behavior is not unique to Schottky rectifiers and is even observed with the use of very-fast-recovery P–i–N rectifiers [16]. The ringing occurs due to the existence of a series resonant circuit formed by the capacitance of the rectifier and any series inductance in the circuit. This phenomenon can be reduced by the connection of an R–C snubber in parallel with the rectifier [16].

8.2.5. Device Technology

The fabrication of Schottky power rectifiers has been accomplished by using a variety of metals by different manufacturers. The Schottky barrier height ϕ_{Bn} is dependent on the metal by means of its work function [10, 11]. The forward conduction and reverse blocking characteristics obtained by using a variety of metals for the Schottky barrier are shown in Fig. 8.22 [17]. Among these metals, the platinum (Pt) and molybdenum (Mo) barrier devices exhibit the best-behaved forward conduction characteristics. From Fig. 8.22a, it can be seen that their forward voltage drop at 125°C is about 0.5–0.6 V—it will be larger at lower temperatures. The leakage current for these rectifiers is lower than

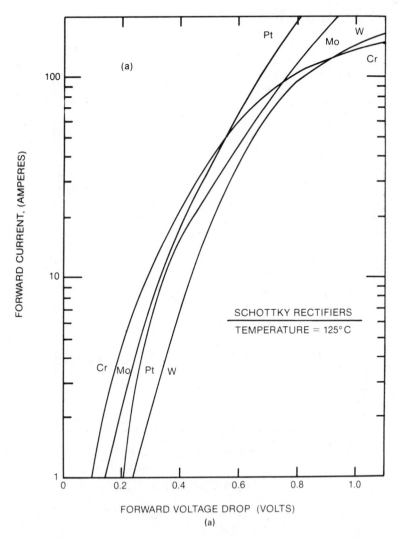

Fig. 8.22. Typical characteristics of Schottky power rectifiers fabricated by using a variety of metals: (a) forward conduction and (b) reverse blocking. (After Ref. 17, reprinted with permission from the IEEE © 1982 IEEE.)

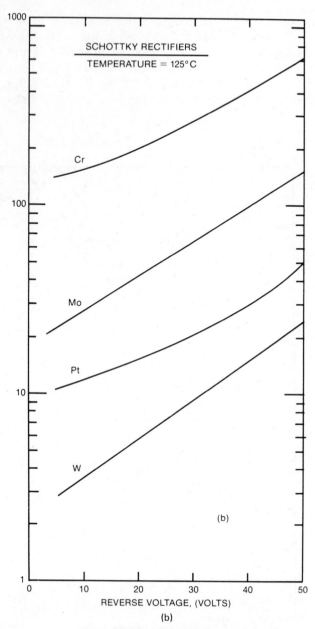

Fig. 8.22. (Continued)

that for the chrominum (Cr) case despite a nearly equal forward drop. This is because the series resistance of the chromium Schottky rectifier is much greater than that for the platinum and molybdenum rectifiers. The leakage current of tungsten (W) Schottky barrier rectifiers is lower, but their forward voltage drop is greater than for the platinum and molybdenum devices. An important point to note in the reverse blocking characteristics shown in Fig. 8.22b is that, in all cases, the leakage current grows nearly exponentially with reverse-bias voltage. This is consistent with the sharply rising leakage current indicated for typical Schottky rectifiers shown in Fig. 8.17. This soft breakdown characteristic causes excess power dissipation during reverse blocking, promoting instability.

The trade-off between forward voltage drop and reverse leakage current for Schottky power rectifiers is generally performed by changing the metal used to form the barrier. An alternative processing approach is the use of a very shallow ion implant at the surface of the semiconductor. The introduction of a thin layer (<100 Å) at the surface of the drift layer with a carefully controlled dose can be used to change the effective barrier height between the metal and the semiconductor [18]. For an N-type drift layer, an N-type layer at the surface will lower the barrier height whereas a P-type layer at the surface will raise it. This processing approach is attractive because it allows the selection of a metal based on the metallurgical properties of the interface, which will produce the most reliable operation while allowing the tailoring of the barrier height by controlling the ion-implant dose.

Since the optimization of the Schottky barrier can be best achieved by starting with a larger Schottky barrier height and reducing it, consider the case of an N-type semiconductor with a thin N^+ layer at the surface. The resulting electric field and band profiles are shown in Fig. 8.23. The electric field profile can be described by the equations

$$\mathscr{E}(x) = -\mathscr{E}_m + \frac{qN_sx}{\epsilon_s} \tag{8.26}$$

from $x = 0$ to $x = a$, and

$$\mathscr{E}(x) = -\frac{qN_D}{\epsilon_s}(W - x) \tag{8.27}$$

from $x = a$ to $x = W$, where $\mathscr{E}_m$ is the maximum electric field at the metal–semiconductor interface given by

$$\mathscr{E}_m = \frac{q}{\epsilon_s}[N_sa + N_D(W - a)] \tag{8.28}$$

The Schottky barrier lowering due to the presence of the surface layer can be obtained by substituting this maximum electric field expression into Eq. (8.23). For the case where the implanted charge N_sa is much greater than the charge

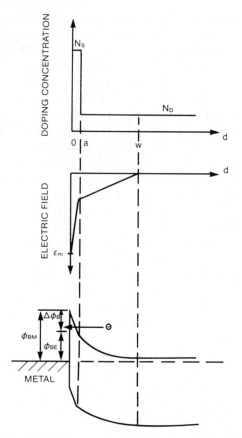

Fig. 8.23. Schottky barrier contact with highly doped, thin surface layer.

in the zero-bias depletion layer formed in the lightly doped drift region, it can be shown that

$$\Delta\phi_B \simeq \frac{q}{\epsilon_s}\sqrt{\frac{aN_s}{4\pi}} \tag{8.29}$$

The implantation of a shallow charge at the surface with doses ranging from 10^{12} to $10^{13}/\text{cm}^2$ can be used to decrease the barrier height by 0.05–0.2 eV.

It is important to maintain the implanted charge close to the surface. Antimony implantation at energies of 5–10 keV is very effective for accomplishing barrier height tailoring in N-type silicon because its large mass results in a shallow implantation depth, whereas its low diffusion coefficient prevents redistribution during subsequent implant activation or other high-temperature processing steps. It has been observed that any residual ion-implant damage can also produce a reduction in Schottky barrier height [19], indicating that care must be taken to ensure complete annealing of the damage if reproducible results are to be achieved.

The method of termination of the Schottky barrier rectifier has been found to be of importance in obtaining acceptable reverse blocking characteristics. A severe electric field crowding occurs at the edge of the metal in the absence of an edge termination structure, leading to very high leakage and soft breakdown well below the parallel-plane breakdown voltage. For this reason, special edge termination techniques have been developed compatible with the Schottky rectifier fabrication process. The more commonly used approaches are illustrated in Fig. 8.24. The metal field plate structure, illustrated in Fig. 8.24a, is based on

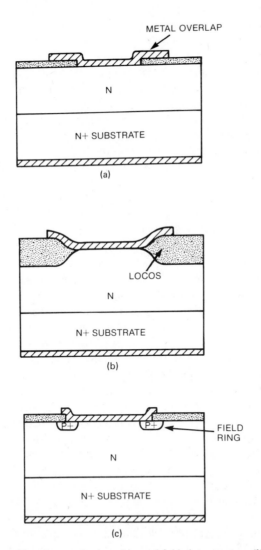

Fig. 8.24. Schottky rectifier edge terminations: (a) metal field plate structure; (b) LOCOS structure; (c) diffused field ring structure.

on extension of the Schottky barrier metal over an oxide layer at the edges. The design of this structure is discussed in Chapter 3. A tapered gate oxide achieved by using a high etch rate phosphosilicate glass has been found to provide improved breakdown characteristics [20]. An alternate approach uses the local oxidation of silicon (LOCOS) process to create a tapered oxide at the edges [21] as shown in Fig. 8.24b. The most commonly used approach is to place a P-type diffused field ring at the edges of the Schottky barrier metal [22]. The field ring greatly reduces the electric field crowding at the edges. It should be noted that a P–N junction is now formed in parallel with the Schottky rectifier. Since the forward voltage drop of the Schottky rectifier is in the range of 0.5–0.6 V, the P–N junction does not inject any significant amount of minority carriers that would adversely impact the high speed switching capability of the Schottky rectifier.

The low forward voltage drop and high switching speed of the Schottky barrier rectifier have made it attractive for high-frequency inverters. Until recently, the reverse breakdown voltage of these rectifiers has been limited to less than 50 V. Devices with breakdown voltages approaching 100 V are now emerging. The relatively low breakdown voltage of the Schottky rectifier has limited its application mainly as an output rectifier for high-frequency switching power supplies. For this application, devices with current ratings of over 50 A are available.

8.3. SYNCHRONOUS RECTIFIERS

As discussed in Chapters 4 and 6, the power JFET and power MOSFET exhibit a resistive on-state characteristics at low voltages. If the size of these devices is made sufficiently large, the forward voltage drop during high current flow can be made lower than that for a P–i–N or Schottky rectifier. It has been suggested that these devices be used as output rectifiers in switching power supplies to achieve a reduction in the power losses during current conduction [23]. To accomplish the operation of the power MOSFET as a rectifier, it is essential to provide a gate drive voltage whenever the drain voltage is negative and to turn off the gate voltage whenever the drain voltage is positive; in other words, these devices require a synchronous gate signal.

The use of a power MOSFET as a synchronous rectifier was discussed in Chapter 6. To achieve a sufficiently low on-resistance to maintain the forward voltage drop below 0.3 V, very large (300 × 300 mil) chips have been fabricated by the DMOS process to achieve an on-resistance of 12 mΩ at room temperature [24]. For the low-breakdown-voltage design, the channel resistance of the power MOSFET comprises a large fraction of the total on-resistance. To reduce this resistance, a power MOSFET structure with vertically walled grooves, as illustrated in Fig. 8.25, has been developed to achieve an increase in channel packing density [25]. An on-resistance of 50 mΩ at room temperature has been reported for a 140 × 140-mil chip designed with a breakdown voltage of 50 V. If

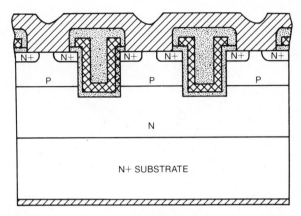

Fig. 8.25. Vertically walled channel structure for achieving low on-resistance in a power MOSFET.

these power MOSFETs are used as output rectifiers, a reduction of the forward conduction losses can be expected in power supplies. However, the relatively high cost of the MOSFET chip arising from a low-yield, complex technology and the need to provide a synchronous gate signal have restricted its application to high-performance systems where cost is not a primary consideration.

8.4. JBS RECTIFIERS

With the trend toward lower operating voltages for VLSI circuits, there is an increasing demand to reduce the forward voltage drop in rectifiers used for switching power supplies. The forward voltage drop of a Schottky rectifier can be reduced by decreasing the Schottky barrier height. Unfortunately, a low barrier height results in a severe increase in leakage current and reduction in maximum operating temperature. Furthermore, Schottky power rectifiers fabricated with barrier heights of less than 0.7 eV have been found to exhibit extremely soft breakdown characteristics, which makes them prone to failure.

The JBS rectifier is a Schottky rectifier structure with a $P-N$ junction grid integrated into its drift region. This device structure has also been called a *pinch rectifier* [26]. A cross section of the device structure is provided in Fig. 8.26. The junction grid is designed so that its depletion layers do not pinch off under zero- and forward-bias conditions. (The zero-bias depletion width can be obtained from Fig. 4.23.) When designed in this manner, the device contains multiple conductive channels under the Schottky barrier through which current can flow during forward-biased operation. When a positive bias is applied to the N^+ substrate, the $P-N$ junctions and the Schottky barrier become reverse-biased, and the depletion layers formed at the $P-N$ junctions spread into the channel. In the JBS rectifier, the junction grid is designed so that the depletion layers will intersect under the Schottky barrier when the reverse bias exceeds a few

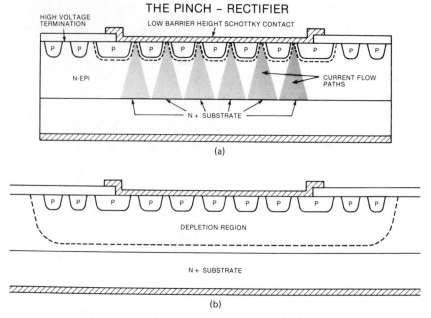

Fig. 8.26. Junction barrier-controlled Schottky rectifier structure: (*a*) forward conduction mode; (*b*) reverse blocking mode. (After Ref. 26, reprinted with permission from the IEEE © 1984 IEEE.)

volts. After depletion layer pinch-off, a potential barrier is formed in the channel as in the case of the power JFET described in Chapter 4. Once the potential barrier is formed, further increase in applied voltage is supported by it with the depletion layer extending toward the N^+ substrate.

The potential barrier shields the Schottky barrier from the applied voltage. This shielding prevents the Schottky barrier lowering phenomenon and eliminates the large increase in leakage current observed for conventional Schottky rectifiers. Once the pinch-off condition is established, the leakage current remains constant except for the small increasing contribution from the space-charge generation component. This allows the JBS rectifier to be operated right up to the avalanche breakdown point without the onset of the thermal runaway experienced in Schottky rectifiers because of their very soft breakdown characteristics.

Because of the suppressed leakage current, the Schottky barrier height used in the JBS rectifier can be significantly less than that for the conventional Schottky rectifiers. This has allowed a reduction in the forward voltage drop while maintaining an acceptable reverse blocking characteristic [26]. A comparison of the forward conduction and reverse blocking characteristics of the JBS rectifier with the conventional Schottky rectifiers fabricated with and without antimony implants is provided in Fig. 8.27. The JBS rectifier characteristics were obtained by reducing the Schottky barrier height by use of a shallow, anti-

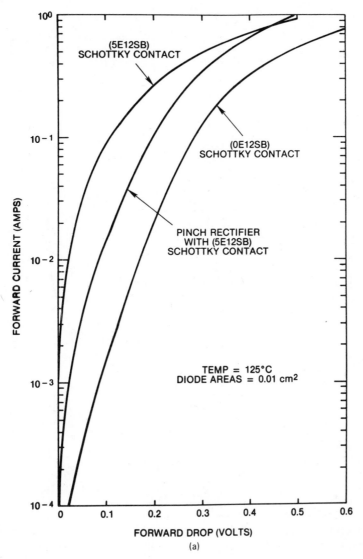

Fig. 8.27. Comparison of (*a*) the forward conduction and (*b*) reverse blocking characteristics of a JBS rectifier with the conventional Schottky barrier rectifier. (After Ref. 26, reprinted with permission form the IEEE © 1984 IEEE.)

mony implant. The lower barrier height causes a higher leakage current at low reverse-bias voltages when compared to the unimplanted Schottky rectifier. At reverse bias voltages typical of device operation, however, the leakage current of the Schottky rectifier becomes equal to that of the JBS rectifier. For the same leakage current, the JBS rectifier can be seen to provide a forward voltage drop of 0.2 V lower than that for the Schottky rectifier. It is worth pointing out that

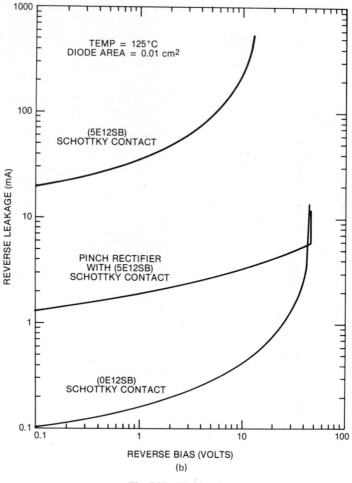

Fig. 8.27. (Continued)

the Schottky diode fabricated with antimony implant has an extremely soft breakdown characteristic as shown in Fig. 8.27*b* and will not block over 10 V.

8.4.1. Forward Conduction

An analysis of the forward conduction characteristics of the JBS rectifier can be performed along the same manner as used for the Schottky rectifier by allowing for the increase in the series resistance of the drift region as a result of current constriction in the channel structure [27]. Consider the case of a *P–N* junction grid with stripe geometry and cross-sectional dimensions defined in Fig. 8.28. In this structure, the junction grid is formed by planar diffusion through a diffusion window of width *s* with a masked region of width *m*. The lateral diffusion of the junction is assumed to be 85% of the vertical depth x_i.

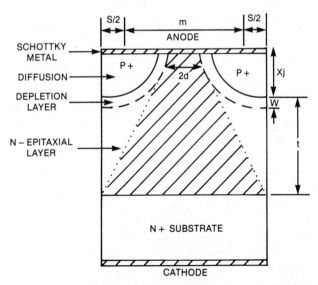

Fig. 8.28. Cross section of JBS rectifier with device dimensions defined for analysis. (After Ref. 27, reprinted with permission from *Solid State Electronics.*)

The relationship between the forward voltage drop and current density for a Schottky barrier [Eq. (8.17)] must be modified to allow for the area taken up by the P^+ regions in the JBS rectifier structure. The current density of the Schottky barrier J_{FS} is given by

$$J_{FS} = \frac{m + s}{2d} J_{FC} \tag{8.30}$$

where J_{FC} is the cell current density, that is, total JBS rectifier current divided by its active area. Using this expression in Eq. (8.17), it can be shown that

$$V_{FS} = \phi_B + \frac{kT}{q} \ln\left(\frac{m + s}{2d} \frac{J_{FC}}{AT^2}\right) \tag{8.31}$$

In this equation, the dimension $2d$, which defines the actual area of the Schottky barrier, is related to the structural dimensions and the junction depletion width:

$$2d \simeq m - 2W - 1.7x_j \tag{8.32}$$

where the junction depletion width is given by

$$W = \sqrt{\frac{2\epsilon_s}{qN_D}(V_{bi} - V_F)} \tag{8.33}$$

In addition to the voltage drop across the Schottky barrier, the voltage drop across the drift region must be accounted for. Using the spreading resistance analysis described in Chapter 4 for the power JFET, the drift region resistance is given by

$$R_D = \rho \frac{(x_j + t)(m + s)}{m + s - 2d} \ln\left(\frac{m + s}{2d}\right) \tag{8.34}$$

where ρ and t are, respectively, the resistivity and depletion layer width required to obtain the desired breakdown voltage. Combining the voltage drop across the Schottky barrier and the drift region, an expression for the forward voltage drop of the JBS rectifier can be derived:

$$V_F = \phi_B + \frac{kT}{q} \ln\left[\left(\frac{m + s}{2d} \frac{J_{FC}}{AT^2}\right)\right] + \rho \frac{(x_j + t)(m + s)}{m + s - 2d} \ln\left(\frac{m + s}{2d}\right) J_{FC} \tag{8.35}$$

With this equation, the forward conduction characteristics can be calculated. For an exact analysis, an iterative procedure is required because of the dependence of d on the forward voltage drop. In the case of JBS rectifiers intended for operation at very low forward voltage drops, the depletion layer width W can be assumed to be constant allowing a closed-form analytical solution of the forward conduction characteristics by using Equation (8.35). A very good agreement between the calculated curves and those measured in experimental devices has been observed [27].

8.4.2. Reverse Blocking

As in the case of the Schottky rectifier, the reverse leakage current of the JBS rectifier consists of two components. The first component arises from the injection of carriers across the Schottky barrier. Using Eq. (8.21), after accounting for the Schottky barrier lowering described by Eq. (8.23), and by including the effect of area taken up by the P^+ diffusion, it can be shown that:

$$J_L = \left(\frac{2d}{m + s}\right) AT^2 \exp\left[-\left(\frac{q\phi_B}{kT}\right)\right] \exp\left(\frac{q}{kT}\sqrt{\frac{q\mathcal{E}}{4\pi\epsilon_s}}\right) \tag{8.36}$$

where $\mathcal{E}$ is the electric field at the Schottky barrier given by Eq. (8.24). An important feature of the JBS rectifier is that the electric field $\mathcal{E}$ at the Schottky barrier remains constant (independent of the reverse bias) once the channel potential barrier forms. Its value corresponds to the voltage at which the channel pinch-off occurs:

$$\mathcal{E} = \sqrt{\frac{2qN_D}{\epsilon_s}(V_P + V_{bi})} \tag{8.37}$$

where V_P is the channel pinch-off voltage given by

$$V_P = \frac{qN_D}{8\epsilon_s}(m - 1.7x_j)^2 - V_{bi} \tag{8.38}$$

The second component of the leakage current arises from the space-charge generation and diffusion currents. This component is

$$J_{\mathrm{LD}} = q\sqrt{\frac{D}{\tau}\frac{n_i^2}{N_{\mathrm{D}}}} + \frac{qn_iW}{\tau} \tag{8.39}$$

where W is the depletion layer width given by

$$W = \sqrt{\frac{2\epsilon_s}{qN_{\mathrm{D}}}(V_{\mathrm{R}} + V_{\mathrm{bi}})} \tag{8.40}$$

This leakage current component is a small fraction of the total leakage current but must be included in the analysis of the reverse blocking characteristics to account for a slight increase in leakage current with reverse-bias voltage beyond the pinch-off point. An excellent agreement between the calculated and measured device leakage characteristics has been observed [27].

8.4.3. Device Ratings

By use of the preceding analysis of the forward conduction and reverse blocking characteristics of the JBS rectifier, an optimization of device characteristics can be performed. It has been found that the suppression of the reverse leakage current allows the reduction of the forward voltage drop to 0.3 V while retaining a maximum operating temperature of over 125°C. Large chips (0.5-cm² active area, 300 × 300-mil chip) have been fabricated that can conduct over 25 A at a forward drop of less than 0.3 V. If these rectifiers are used in switching power supplies, the power dissipation in the diodes can be reduced by over 50%. Since the power dissipation in the diodes is predominant, a reduction in the power loss not only allows improvement in power supply efficiency, but results in a reduction in its size because of the use of smaller heat sinks.

The concept of using the junction-induced barrier to reduce Schottky barrier leakage has also been used to achieve a higher reverse blocking capability [28, 29]. Devices with reverse blocking voltages of up to 200 V have been reported. A problem with applying the junction barrier-controlled pinch-off to high-voltage diodes is the increase in the forward voltage drop due to the increasing series resistance of the drift layer. When the reverse blocking voltage exceeds 150 V, the forward voltage drop exceeds 0.7 V before a reasonable forward conduction current density can flow. This results in the injection of minority carriers from the P–N junction during forward conduction, which destroys the high-speed switching capability. To overcome this problem, an alternative device structure can be used as shown in Fig. 8.29, where the P–N junction of the JBS rectifier has been replaced with an MOS structure [33]. The suppression of the reverse leakage current is achieved by the depletion layer spreading from the MOS interface as in the case of the JBS rectifier. However, the P–N junction is now eliminated, allowing operation of the MOS

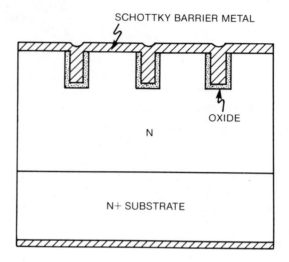

SCHOTTKY BARRIER METAL

OXIDE

N

N+ SUBSTRATE

Fig. 8.29. Metal-oxide–semiconductor barrier-controlled Schottky (MBS) rectifier structure.

barrier-controlled Schottky (MBS) rectifier at forward voltage drops of over 0.7 V while retaining unipolar operation and ensuring high switching speed. With the MBS structure, it is possible to fabricate ultra-high-speed rectifiers with high breakdown voltages. The forward voltage drop of these rectifiers is lower than in a *P–i–N* rectifier for breakdown voltages of up to 250 V. Above this voltage, their superior switching performance must be traded off against a higher forward voltage drop when compared with a *P–i–N* rectifier.

8.5. GALLIUM ARSENIDE RECTIFIERS

In the previous sections of this chapter, only silicon device structures were considered. It was pointed out that the silicon Schottky rectifier suffers from a high reverse leakage current and soft breakdown, which has limited its maximum reverse blocking voltage to less than 100 V. In addition, the high series resistance of the drift layer at higher voltages limits the maximum forward conduction current density. These two drawbacks of the silicon Schottky rectifiers can be addressed by fabricating the devices by using gallium arsenide as the semiconductor. As discussed in Chapter 4 for the power JFET, the resistance of the drift region can be reduced by a factor of 12.7 by replacing silicon with gallium arsenide. In addition, the pinning of the Fermi level at the gallium arsenide surface makes the fabrication of stable Schottky barriers relatively easy, resulting in the achievement of breakdown voltages close to the ideal parallel-plane case.

In Fig. 8.30, the calculated forward conduction characteristics of gallium arsenide Schottky rectifiers as provided for a variety of reverse blocking voltages [30]. In performing these calculations, a Schottky barrier height of 0.7 eV was

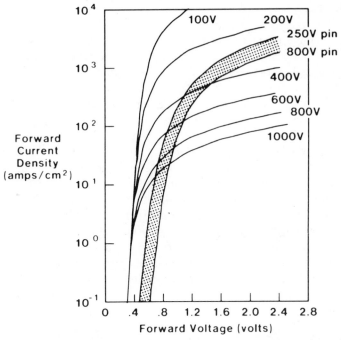

Fig. 8.30. Forward conduction characteristics of gallium arsenide Schottky rectifiers. (After Ref. 30, reprinted with permission from the IEEE © 1985 IEEE.)

used. For comparison, the forward conduction characteristics of $P–i–N$ rectifiers are also included. It can be seen that the gallium arsenide rectifiers should have a lower forward drop than silicon $P–i–N$ rectifiers for breakdown voltages of up to about 500 V at a typical operating current density of 100–200 A/cm^2. In this voltage range, the gallium arsenide Schottky rectifiers offer a clear advantage over $P–i–N$ rectifiers because of their inherently higher switching speed.

The fabrication of gallium arsenide Schottky power rectifiers has been accomplished by using a relatively simple fabrication process [30]. After the growth of the N-type epitaxial layer on an N^+ substrate, the ohmic contact is formed on the back of the wafer. This is followed by the evaporation of the Schottky barrier metal onto the surface of the drift layer through a shadow mask. To obtain diodes with breakdown voltages within 90% of the ideal parallel-plane case, it is important to etch the gallium arsenide surface with Caro's solution (a mixture of sulfuric acid, hydrogen peroxide, and water), followed by a dip in warm water and rinse in hydrofluoric acid, just prior to metal deposition. In addition, the best yield of high-voltage diodes has been obtained by using aluminum as the Schottky metal with the evaporation conducted while the gallium arsenide wafers are heated to 200°C. Gallium arsenide Schottky rectifiers with breakdown voltages of up to 200 V have been fabricated. These diodes exhibit excellent high-speed switching characteristics [30, 31].

8.6. TRENDS

As VLSI technology moves toward smaller device dimensions to achieve a higher packing density, there will be an increasing need to lower the power supply voltage. At present most circuits are operated from a 5-V power supply. In the future, the power supply voltage is expected to drop to about 3 V. The forward voltage drop of the diodes used as output rectifiers in switching power supplies then becomes increasingly important. The new devices developments discussed in this chapter are expected to make rectifiers with lower forward voltage drops available, resulting in major reduction in the power losses within the power supply. This will aid the power supply designer in achieving higher system efficiency, together with a reduction in size and weight. In addition, the availability of ultra-high-speed, high-voltage (100–600-V) rectifiers, which match the performance of the new transistors discussed in earlier chapters, will aid the trend toward increasing circuit operating frequency. Thus the slow evolutionary progress in improving the performance of the $P–i–N$ rectifiers in the past has been replaced by the introduction of several new device concepts that can be expected to have a strong impact on power systems in the future.

REFERENCES

1. S. K. Ghandhi, *Semiconductor Power Devices*, Wiley, New York, 1977.
2. B. J. Baliga and E. Sun, "Comparison of gold, platinum and electron irradiation for controlling lifetime in power rectifiers," *IEEE Trans. Electron Devices*, **ED-24**, 685–688 (1977).
3. R. O. Carlson, Y. S. Sun, and H. B. Assalit, "Lifetime control in silicon power devices by electron or gamma irradiation," *IEEE Trans. Electron Devices*, **ED-24**, 1103–1108 (1977).
4. E. D. Wolley and S. F. Bevacqua, "High speed, soft recovery, epitaxial diodes for power inverter circuits," *IEEE Industrial Application Society Meeting Digest*, pp. 797–800 (1981).
5. D. Silber, D. W. Novak, W. Wondrak, B. Thomas, and H. Berg, "Improved dynamic properties of GTO–thyristors and diodes by proton implantation," *IEEE International Electron Devices Meeting Digest*, Abstract 6.6, pp. 162–165 (1985).
6. H. Benda and E. Spenke, "Reverse recovery processes in silicon power rectifiers," *Proc. IEEE*, **55**, 1331–1354 (1967).
7. R. J. Grover, "Epi and Schottky diodes," in R. Sittig and P. Roggwitter, Eds., *Semiconductor Devices for Power Conditioning*, Plenum, New York, 1982, pp. 331–356.
8. J. R. Hauser and P. M. Dunbar, "Minority carrier reflecting properties of semiconductor high–low junctions," *Solid State Electron.*, **18**, 715–716 (1975).
9. Y. Amemiya, T. Sugeta, and Y. Mizushima, "Novel low-loss and high speed diode utilizing an ideal ohmic contact," *IEEE Trans. Electron Devices*, **ED-29**, 236–243 (1982).
10. E. H. Rhoderick, *Metal–Semiconductor Contacts*, Clarendon, Oxford, 1978.

11. S. M. Sze, *Physics of Semiconductor Devices*, Wiley, New York, 1981, Chapter 5.

12. C. R. Crowell and S. M. Sze, "Current transport in metal–semiconductor barriers," *Solid State Electron.*, **9**, 1035–1048 (1966).

13. J. M. Andrews and M. P. Lepselter, "Reverse current–voltage characteristics of metal–silicide Schottky diodes," *Solid State Electron.*, **13**, 1011–1023 (1970).

14. C. R. Crowell, "The Richardson constant for thermionic emission in Schottky barrier diodes, "*Solid State Electron.*, **8**, 395–399 (1965).

15. D. J. Page, "Theoretical performance of the Schottky barrier power rectifier," *Solid State Electron.*, **15**, 505–515 (1972).

16. B. Bixby, B. Hikin, and V. Rodov, "Application considerations for very high speed fast recovery power diodes," *IEEE Industrial Applications Society Meeting*, pp. 1023–1027 (1977).

17. B. R. Pelly, "Power semiconductor devices—a status review," *IEEE Industrial Semiconductor Power Conversion Conference Digest*, pp. 1–19 (1982).

18. J. M. Shannon, "Reducing the effective height of a Schottky barrier using low energy ion implantation," *Appl. Phys. Lett.*, **24**, 369–371 (1974).

19. S. Ashok and B. J. Baliga, "Effect of antimony ion implantation on Al–silicon Schottky diode characteristics," *J. Appl. Phys.*, **56**, 1237–1239 (1984).

20. Y. I. Choi, "Enhancement of breakdown voltages of Schottky diodes with a tapered window," *IEEE Trans. Electron Devices*, **ED-28**, 601–602 (1981).

21. N. G. Anantha and K. G. Ashar, "Planar mesa Schottky barrier diode," *IBM J. Res. Devel.*, **15**, 442–445 (1971).

22. M. P. Lepselter and S. M. Sze, "Silicon Schottky barrier diode with near-ideal $I–V$ characteristics," *Bell Syst. Tech. J.*, **47**, 195–208 (1968).

23. R. Severns, "The power MOSFET as a rectifier," *Power Conversion International*, 49–50 (March–April 1980).

24. R. P. Love, P. V. Gray, and M. S. Adler, "A large area power MOSFET designed for low conduction losses," *IEEE International Electron Devices Meeting*, Abstract 17.4, pp. 418–421 (1981).

25. D. Ueda, H. Takagi, and G. Kano, "A new vertical power MOSFET structure with extremely reduced on-resistance," *IEEE Trans. Electron Devices*, **ED-32**, 2–6 (1985).

26. B. J. Baliga, "The pinch rectifier: A low forward drop, high speed power diode," *IEEE Electron Device Lett.*, **EDL-5**, 194–196 (1984).

27. B. J. Baliga, "Analysis of junction barrier controlled Schottky rectifier characteristics," *Solid State Electron.*, **28**, 1089–1093 (1985).

28. B. M. Wilamowski, "Schottky diodes with high breakdown voltages," *Solid State Electron.*, **26**, 491–493 (1983).

29. Y. Shimuzu, M. Naito, S. Murakami, and Y. Terasawa, "High speed, low loss, PN diode having a channel structure," *IEEE Trans. Electron Devices*, **ED-31**, 1314–1319 (1984).

30. B. J. Baliga, A. R. Sears, M. M. Barnicle, P. M. Campbell, W. Garwacki, and J. P. Walden, "Gallium arsenide Schottky power rectifiers," *IEEE Trans. Electron Devices*, **32**, 1130–1134 (1985).

31. A. R. Sears, B. J. Baliga, M. M. Barnicle, P. M. Campbell, and W. Garwacki, "High voltage, high speed, GaAs Schottky power rectifiers," *IEEE International Electron Devices Meeting Digest*, Abstract 9.7, pp. 229–232 (1983).

32. V. A. K. Temple and F. W. Holroyd, "Optimizing carrier lifetime profiles for improving trade-off between turn-off time and forward drop," *IEEE Trans. Electron Devices*, **ED-30**, 782–790 (1983).

33. B. J. Baliga, "New power rectifier concepts," *Proceedings of Third International Workshop on the Physics of Semiconductor Devices*, November 24–28, World Scientific Publ., Singapore, 1985.

PROBLEMS

8.1. Determine the barrier height for a Schottky rectifier designed to operate at a forward voltage drop of 0.3 V at 400 K for a current density of 100 A/cm². The reverse blocking voltage is 50 V.

8.2. Determine the reverse leakage current density at 400 K for the Schottky rectifier described in Problem 8.1 at a reverse bias voltage of 40 V.

8.3. What is the increase in leakage current density if the forward voltage drop of the device described in Problem 8.1 is reduced to 0.25 V?

8.4. Calculate the shallow *n*-type ion implant dose required to achieve the decrease in forward voltage drop from 0.3 to 0.25 V for the Schottky rectifier described in Problem 8.1.

8.5. A power MOSFET synchronous rectifier is designed to block 50 V. Determine its forward conduction current density for a forward drop of 0.25 V. Assume that its specific on-resistance is 20 times the ideal case. (This is typical for state-of-the-art devices.)

8.6. A JBS rectifier designed to block 50 V is fabricated by using a linear cell geometry with diffusion window of 3 μm, repeat spacing of 6 μm, and junction depth of 1 μm. Determine the barrier height for achieving a forward voltage drop of 0.3 V at 400 K for a current density of 100 A/cm².

8.7. Determine the reverse leakage current of the JBS rectifier described in Problem 8.6 at 40 V assuming negligible contribution from space-charge generation.

8.8. Compare the reverse leakage current of the JBS rectifier described in Problem 8.6 to that of the Schottky rectifier calculated in Problem 8.2.

9 SYNOPSIS

This book has focused on providing a discussion of the physics and technology of recently developed power semiconductor devices. With the development of these new devices, it becomes important to consider their relative merit for circuit applications. In this chapter, a comparison of the characteristics of three-terminal (active) devices is provided. In performing the comparison, the gate turn-off thyristor (GTO) and the bipolar transistor have been included because the new devices must compete with or displace these devices in circuit applications. This chapter also includes a summary of the power ratings that have been achieved in the three-terminal devices discussed in this book and provides a projection of their growth in power handling capability in the future.

9.1. COMPARISON OF GATE-CONTROLLED DEVICES

With the development of many alternative devices to the power bipolar transistors, the device designer is faced with the task of making a judicious choice between these devices. As an aid to device selection, the gate-controlled devices discussed in the previous chapters are compared here on the basis of several criteria [1]. To begin with, normally-on devices have been found to be undesirable for power switching because of the need to ensure that a negative gate drive is available during circuit power-up. Since the power JFET and the FCD exhibit this characteristic, their use is limited to those applications in which some of their unique characteristics [such as the very-high-frequency response of the JFET or the high (dV/dt) and radiation tolerance of the FCD] are necessary. Since these requirements do not exist in most applications, the circuit designer is left with a choice between the remaining normally-off devices—namely, the bipolar transistor, the GTO, the power MOSFET, and the IGT. For very-high-power levels, such as AC traction, only GTOs with sufficient current and voltage ratings are available, making the choice quite limited. However, at lower power levels, where gate-controlled operation of the devices is highly desirable, the circuit designer must choose between the bipolar transistor, the power MOSFET, and the IGT. The relative merits of these three devices are discussed here.

One criterion in selecting a power device is its gate drive power requirement. Since the power bipolar transistor is a current-controlled device with a typical current gain of 10, it requires a relatively high gate drive power during steady-state current conduction as well as during turn-off. The gate drive circuitry for the power transistor becomes complex and expensive. In contrast, both the power MOSFET and the power MOS–IGT are voltage-controlled devices with very high input impedance. The gate drive power required to control these devices is relatively small. This eliminates complexity in the gate drive and often allows control of these devices directly from an integrated circuit since the gate circuit must merely provide enough current to charge and discharge the input capacitance of these devices. In this regard, the IGT is even superior to an equivalent power MOSFET because its input capacitance is an order of magnitude smaller for the same power rating. Further, since the technology for the fabrication of the power MOSFET and IGT is similar, the IGT offers a lower cost to the circuit designer because the chip area is an order of magnitude smaller than that for the power MOSFET.

The performance of a power device is ultimately limited by the power dissipation, which determines the temperature rise:

$$\Delta T = T_J - T_A = P_D R_\theta \tag{9.1}$$

where P_D is the power dissipation and R_θ is the thermal resistance. From reliability considerations, the maximum junction temperature T_J of a power device is generally maintained below 125°C. For an ambient temperature T_A of 25°C and a typical thermal resistance of 1°C/W, the maximum power dissipation must be maintained below 100 W/cm^2.

The power dissipation in a power device can arise during steady-state and switching conditions. The leaking currents of the power devices are generally so low that the power losses incurred in their steady-state forward blocking modes can be neglected. Device analysis can be performed by considering the sum of the power loss during the steady-state on-state and during switching:

$$P_D = V_F I_F \frac{t_{on}}{T} + \frac{1}{2} I_F V_S \frac{\tau}{T} \tag{9.2}$$

where V_F is the forward voltage drop at a current I_F and V_S is the blocking voltage. In this equation, the device is assumed to be operated with a constant current I_F in the on-state for a fraction t_{on} of the total period T, and τ is the switching time from the on-state to the off-state.

The forward voltage drop of the power device depends on its blocking voltage capability. To perform a comparison between the devices, it is necessary to specify the breakdown voltage. The device comparison must be performed separately for each blocking voltage rating dictated by the application. Here the comparison will be done for three cases—blocking voltage of 100, 600, and 1200 V. To facilitate the selection between the devices, a plot of power dissipation versus frequency, as shown in Fig. 9.1, is useful. In calculating these curves, it was assumed that all the devices are operating at a current density

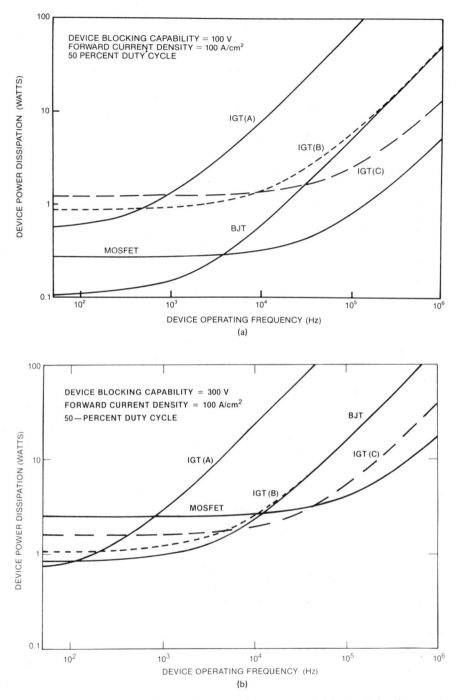

Fig. 9.1. Power dissipation as a function of frequency for gate-controlled devices designed to operate at (*a*) 100 V, (*b*) 300 V, (*c*) 600 V, and (*d*) 1200 V.

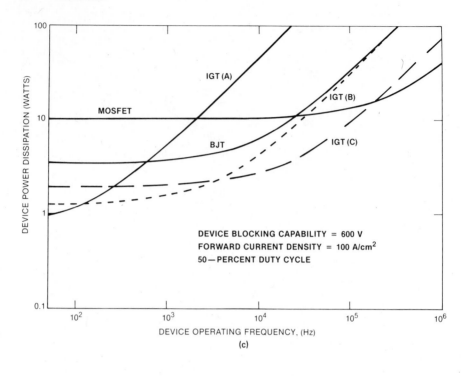

(c)

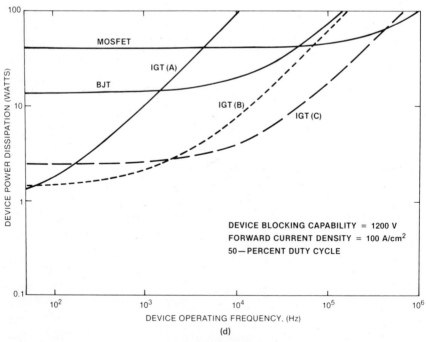

(d)

Fig. 9.1. (Continued)

TABLE 9.1 Transistor Parameters Used for Calculation of Power Dissipation

Device	Forward Voltage Drop at 100 A/cm²(V)			Forced Gate Turn-off Time (μsec)
	100 V	600 V	1200 V	
Bipolar transistor	0.20	6.65	26.70	1.00
MOSFET	0.55	20.00	80.00	0.10
IGT (A)	1.10	1.45	1.70	15.00
IGT (B)	1.75	2.45	2.85	1.00
IGT (C)	2.50	3.80	4.80	0.25

of 100 A/cm² with a duty cycle of 50%. Other cases can be analyzed by the same approach. The turn-off time of the devices used for the calculations is provided in Table 9.1.

In terms of both device and system considerations, the best power device is the one that provides the lowest power dissipation. For the 600- and 1200-V applications, the power dissipation in the IGT is clearly much lower than that for the power MOSFET and bipolar transistor, even at frequencies as high as 100 kHz. As the blocking voltage is lowered to 300 V, the power losses in the bipolar transistor and IGT become comparable but lower than those for the power MOSFET. In this case, the higher input impedance of the MOS gate in the IGT makes it a better choice than the bipolar transistor despite the lower processing cost and higher yield of bipolar transistors at the present time. This is due to the very large cost reduction in the gate drive circuitry, which can be integrated on a single chip for the IGT but must consist of many discrete parts for bipolar transistors. When the blocking voltage capability is lowered to 100 V, the forward drop of the power MOSFET and bipolar transistor become very low whereas the existence the diode drop in the IGT maintains its power dissipation higher than for these devices. For these low voltage circuits, the power MOSFET becomes the most attractive device because of its high input impedance. On the basis of these calculations, it can be concluded that the IGT is preferable for circuits operating at over 200 V and the power MOSFET will be preferable for circuits operating at lower voltages. Exceptions to this statement are very-high-frequency circuits for which only power MOSFETs are suitable and AC circuits (in which reverse blocking capability is required) for which only the IGT is suitable.

9.2. DEVICE POWER HANDLING CAPABILITY

Gate-controlled devices are required for a very wide variety of power conditioning applications that range in power ratings from 100 W to over 1 MW, as well as over a broad range of frequencies extending from 60 Hz to over 1 MHz.

The present ratings of commercially available power transistors are given in Fig. 9.2. The lines for each device define the boundary within which devices are available. For high-voltage applications (> 1500 V) at high power levels (> 100 kW), only GTOs with adequate ratings are available today. Power bipolar transistors have the next highest power ratings. Devices that can handle power levels of up to about 50 kW have been developed by use of the Darlington configuration. The power MOSFET ratings have been increasing rapidly, but the highest power handling capability today is limited to about 5 kW. Although the power MOS–IGT was commercially introduced very recently (1983), its power handling capability already exceeds that of the power MOSFETs. Devices are now available that can switch between 10 and 15 kW of power.

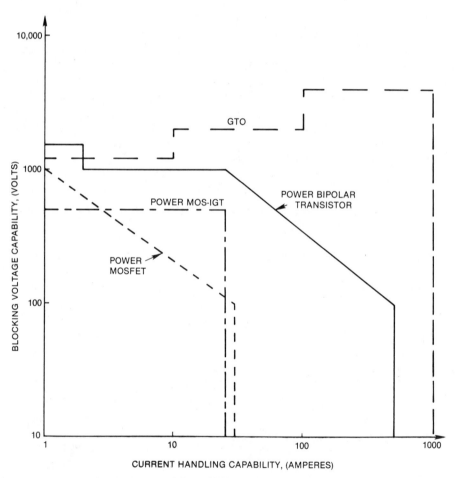

Fig. 9.2. Commercially available power device ratings in 1983.

It can be anticipated that the power handling capability of all these devices will grow as improvements are made in device processing technology. The development of the new high-input-impedance power MOS devices will have a strong impact on GTOs and bipolar transistors. It can be projected that power MOS–IGTs will be available with voltage blocking capabilities of up to 2500 V that rival that of the GTO. Since these high-voltage devices are expected to have a current handling capability ranging up to 100 A as indicated in Fig. 9.3, they will replace low-current GTO because of their simpler gate drive requirements. The GTO device ratings can, therefore, be expected to be confined to over 100 A and 2500 V. A similar conclusion can be made regarding power bipolar transistors with current ratings of less than 100 A. These devices are not expected to compete with the power MOS–IGT because of their relatively

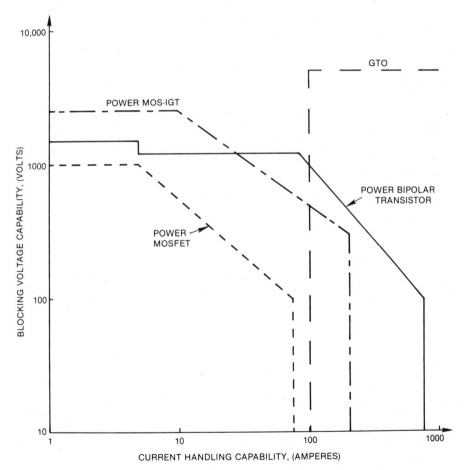

Fig. 9.3. Projected power device ratings in 1990.

high gate drive power requirements. They will continue to have a role only at higher power levels that cannot be served by the power MOS–IGT or power MOSFET. The ratings of the power MOSFET are also expected to increase to serve the low-voltage (<100-V) and high-frequency applications, but this increase will occur at a more modest pace since it is restrained by the ability to fabricate larger area devices with adequate yields.

9.3. POWER INTEGRATED CIRCUITS

In this book, the physics and technology of recently developed discrete power devices have been treated. The advent of the MOS-controlled power device technology has led to a tremendous reduction in the power level required to actuate power devices in control circuits. Gate drive circuits needed for power devices have undergone a significant simplification when compared with circuits used in conjuction with the bipolar transistor. This has led to a large reduction in the number of components in the gate circuitry, and their power requirements have decreased by several orders of magnitude. The integration of the entire gate drive circuit on a single chip has now become feasible.

An essential technological breakthrough, which was required to create power integrated circuits, was the ability to fabricate high-voltage and logic elements on the same chip with isolation voltages of over 600 V. This was accomplished by using the RESURF, or lateral charge-control technique [2, 3]. With this technology, high-voltage devices that can be referenced with respect to each other can be fabricated by using junction isolation techniques. It is now possible to integrate level-shifting elements for use in circuits that drive power devices connected in a totem-pole configuration, such as the half bridge used for motor drives.

Processes have been developed that can simultaneously create logic elements and analog devices on the same chip with the high-voltage devices [3–9]. These power integrated circuits can replace the multicomponent circuits that were used in the past, resulting in a large cost reduction. They also offer the opportunity to perform on-chip current and temperature sensing with the objective of providing protection against destructive failure under adverse circuit operating conditions.

The combination of the new MOS–bipolar discrete device innovations and the power integrated circuit developments has resulted in a system cost reduction by over two orders of magnitude for many applications [4]. This new technology offers much greater reliability because of a large reduction in the number of components and interconnections. In addition, it allows greatly simplified procedures for system diagnosis and a much less expensive field test, maintenance, and replacement option. These features are expected to fuel a rapid growth in the application and extension of power electronics in both industrial and consumer sectors.

REFERENCES

1. B. J. Baliga and D. Y. Chen, *Power Transistors: Device Design and Applications*, IEEE Press, 1984.
2. J. A. Appels and H. M. J. Vaes, "High voltage thin layer devices (RESURF DEVICES)," *IEEE International Electron Devices Meeting Digest*, Abstract 10.1, pp. 238–241 (1979).
3. E. J. Wildi, T. P. Chow, M. S. Adler, M. E. Cornell, and G. C. Pifer, "New high voltage IC technology," *IEEE International Electron Devices Meeting Digest*, Abstract. 10.2, pp. 262–265 (1984).
4. M. S. Adler, K. W. Owyang, B. J. Baliga, and R. A. Kokosa, "The evolution of power device technology," *IEEE Trans. Electron Devices*, **ED-31**, 1570–1591 (1984).
5. W. Wakaumi T. Suzuki, M. Saito, and H. Sakuma, "Highly reliable 16 output high voltage NMOS/CMOS logic with shielded source structure," *IEEE International Electron Devices Meeting Digest*, Abstract. 16.3, pp. 416–420 (1983).
6. W. T. Weston, H. W. Becke, J. E. Berthold, J. C. Gammel, A. R. Hartman, J. E. Kohl, M. A. Shibib, R. K. Smith, and Y. H. Wong, "Monolithic high voltage gated diode crosspoint array IC," *IEEE International Electron Devices Meeting Digest*, Abstract 4.5, pp. 85–88 (1982).
7. T. Kamei, "High voltage integrated circuits for telecommunications," *IEEE International Electron Devices Meeting Digest*, Abstract 11.2, p. 254 (1981).
8. Y. Sugawara, K. Miyata, and M. Okamura, "350V analog–digital compatible power IC technologies," *IEEE International Electron Devices Meeting Digest*, Abstract 30.2, pp. 728–731 (1985).
9. W. G. Meyer, G. W. Dick, K. H. Olson, K. H. Lee, and J. A. Shimer, "Integrable high voltage CMOS," *IEEE International Electron Devices Meeting Digest*, Abstract 30.3, pp. 732–735 (1985).

INDEX